EXPERIMENTATION WITH ANIMAL MODELS IN SPACE

ADVANCES IN SPACE BIOLOGY AND MEDICINE

Volume 10

Series Editor

Augusto Cogoli

Space Biology Group
ETH Technopark
Zurich, Switzerland

Experimentation with Animal Models in Space

Editor

Gerald Sonnenfeld

Vice President for Research
Binghamton University
State University of New York
Binghamton, New York, U.S.A.

ELSEVIER

2005

Amsterdam – Boston – Heidelberg – London – New York – Oxford
Paris – San Diego – San Francisco – Singapore – Sydney – Tokyo

ELSEVIER B.V.
Radarweg 29
P.O. Box 211, 1000 AE
Amsterdam, The Netherlands

ELSEVIER Inc.
525 B Street, Suite 1900
San Diego, CA 92101-4495
USA

ELSEVIER Ltd.
The Boulevard, Langford Lane
Kidlington, Oxford OX5 1GB
UK

ELSEVIER Ltd.
84 Theobalds Road
London WC1X 8RR
UK

First edition 2005

Library of Congress Cataloging in Publication Data

Experimentation with animal models in space / editor, Gerald Sonnenfeld.
p. cm. – (Advances in space biology and medicine)
Includes bibliographical references.
ISBN 0-444-51907-6
1. Space biology–Research–Methodology. 2. Animal experimentation. I. Sonnenfeld, Gerald. II. Series.

QH327.E956 2005
571'.0919–dc22 2005040631

British Library Cataloguing in Publication Data
A catalogue record is available from the British Library.

ISBN: 0-444-51907-6
ISSN: 1569-2574

Transferred to digital printing 2007

Contents

Foreword of the Series Editor to Volume 10

I am proud to have the pleasure to introduce Volume 10, the third after I took over the series editorship of the ASBM series. I was very happy when Gerald Sonnenfeld accepted my invitation to be the editor of a volume dedicated to animal experimentation in space not only because he is a good friend of mine but also because he is one of the most prominent scientists who contributed to the advancement of biological and medical research in space and is also an internationally renowned immunologist.

Gerry succeeded in collecting ten review articles written by scientists with direct experience in space experimentation and covering all disciplines and physiological functions affected by the conditions of space flight.

Animal experimentation is and has been necessary to prepare for human space exploration. The veterans among space scientists well remember the historical flight of the dog Laika in the 1950s that paved the way to the flight of Gagarin, the first human in space. Many other missions with animals on board followed in the last forty years. Animal research facilities will be installed soon on the International Space Station.

I am convinced that volume 10, as the previous volumes of this series, will contribute to the advancement of space biology and medicine in space and to disseminate important information and data in a broad scientific community.

Augusto Cogoli
Zurich, October 2004

Experimentation with Animal Models in Space
G. Sonnenfeld (editor)

DOI: 10.1016/S1569-2574(05)10001-X

Overview

Gerald Sonnenfeld

Department of Biological Sciences and Vice President for Research, Binghamton University, State University of New York, Binghamton, New York, USA

Introduction

Animal models have been utilized to study the effects of space flight on physiological systems since early on in the exploration of space by humans (Sonnenfeld, 2003). They were utilized before it was possible for humans to travel in space (Sonnenfeld, 2003).

The rationale for the early use of animals was to allow for assessment of the ability of humans to explore space. Before humans could be placed at risk in the space flight environment, it was important to establish that the space flight environment would not cause irreparable harm to the potential human travelers. Often, in the case of clinical medicine drug and device development, animal models are utilized to ensure efficacy and safety prior to allowing human trials and use (Polk *et al.*, 1991). This same principle was applied to space exploration, and animal studies paved the way for human exploration of space.

Additionally, animal models have been used to study the basic biology and mechanisms of the effects of space flight on physiological systems (Sonnenfeld, 2003). The limited availability of human subjects in space has made the use of animal models one of the best ways to discover and understand the mechanisms by which space flight conditions affect biological systems.

As we enter the era of long-term, and possibly interplanetary, exploration class space flight missions, the need for continued and expanded animal research in and about space again must move to the forefront of our consideration. In order to meet the mandates of the new exploration initiatives (Lawler, 2004), animal experiments are required. First, the limited number of potential human subjects available for study makes meeting the mandates extraordinarily difficult without the use of animal models (Committee on Space Biology and Medicine, 1998). Secondly, exploration class mission astronauts and cosmonauts may meet hazards that have never before been met, including novel types of radiation exposure (Committee on Space Biology and Medicine, 1998). It would not be ethical to expose humans on earth to some of these hazards; therefore, the only way to study them effectively is by use of animal models. Rapid and effective establishment of the risk of interplanetary space travel will require the use of animals. Additionally, the development of counter measures to eliminate or ameliorate the risks will also require animal models to establish the safety and efficacy of the agents, devices, and techniques to be utilized as counter measures. Therefore, it is likely that we are entering an age where there will be even greater need for utilization of animal models to facilitate exploration of space.

The goal of this volume is to review the different types of animal models available. The data that have been generated utilizing animal models for studying the effects of space flight on physiological systems are reviewed, along with the positive benefits obtained as well as the limitations of the models. In this way, we hope to reinforce the need for and benefit of the use of animal models to determine the effects of space flight on physiological systems, as well as to develop counter measures against any deleterious effects of the space flight environment.

Animal models

It is clear that a wide variety of animal models have been utilized in space. The animals that have been studied include: canines, non-human primates, mice, rats, fish, amphibians, invertebrates and insects (Committee on Space Biology and Medicine, 1998). The vertebrate animal that has been utilized most often for space flight studies is the rat (Committee on Space Biology and Medicine, 1998).

Additionally, most physiological systems have been studied in space utilizing animal models (Committee on Space Biology and Medicine, 1998). These include, the musculoskeletal system, the haematological system, the neurological and neurovestibular system, the cardiovascular system, the immune system, and the circadian/bioregulation system (Committee on Space Biology and Medicine, 1998). Additionally, the effects of space flight on developmental biology has also been extensively studied using animal models (Committee on Space Biology and Medicine, 1998). Developmental biology could not be studied in space without the use of animal models.

This volume has attempted to cover all these areas. Cardiovascular and muscle studies are not covered in individual chapters in this volume due to the lack of availability of qualified authors within the time limitations for publication of this volume. However, the general chapters on system (Chapters 2, 3, 9, 10 and 11) should provide sufficient information and references to allow for a basic understanding of results with those two systems. The other areas mentioned are covered with specific chapters in this volume.

Focus of the chapters

At the present time, 2004–2005, the opportunity to carry out space flight studies with animals is severely limited. This is because the construction phase of the International Space Station is ongoing, there is no animal facility for vertebrate animals yet available on the International Space Station, and resources are very limited so that very little space or time can be allocated for animal experiments (Committee on Space Biology and Medicine, 1998). Even if more resources for space flight experiments were available, there would still not be enough flight opportunities to allow for completion of the necessary experiments to determine the effects of space flight on physiological systems and to allow for development of counter measures. For this reason, ground-based models of space flight conditions have been developed (Committee on Space Biology and Medicine, 1998).

It is not possible to completely model or simulate space flight conditions on earth, simply because it is not possible to isolate and eliminate the force of gravity. Therefore, several models have been developed for animals that replicate some of the conditions that occur during space flight and microgravity. For small invertebrate animals as well as for animal cell culture, a clinostat or bioreactor or HARV model has been developed (Committee on Space Biology and Medicine, 1998). This model allow for a repeated change in the direction of the vector of gravity, and results in a system that mocks some of the changes in sheer forces and other factors that occur during microgravity (Committee on Space Biology and Medicine, 1998).

However, the most frequently used and widely accepted ground-based animal model for space flight conditions is hindlimb unloading of rodents (the name recommended by the US National Academy of Sciences). This model, also known as tail suspension or antiorthostatic, hypokinetic, hypodynamic suspension, allows for no load-bearing on the hindlimbs of rodents, and a head-down shift of fluids to the head. This causes many changes similar to those observed during space flight (Committee on Space Biology and Medicine, 1998). The use of hindlimb unloading is fully explored in Chapter 2 of this volume, authored by Morey-Holton, Globus, Kaplansky and Durnova. They are the pioneers in the development and exploitation of this model. The model will also be referred to routinely in many of the other chapters of this volume.

The Cosmos or Bion space flights were a series of biosatellites flown by the Soviet Union and, later, Russia to allow for regular space flight involving animals. Using this biosatellite, flight conditions that were not acceptable for humans could be used for study of animal models. The Soviets and Russians invited world-wide collaboration for these flights, and a large volume of very useful data was generated. The results of these flights are summarized in Chapter 3, authored by Grindeland, Ilyin, Holley and Skidmore. These authors played a major role in the success of these flights.

The immune system has been shown to be challenged by space flight conditions, and generally suppressed (Sonnenfeld, 2002; Sonnenfeld and Shearer, 2002; Sonnenfeld *et al.*, 2003). In Chapter 4, Chapes and Ganta review the use of murine models that have provided substantial data in this area. They have played a major role in generating those data. The mouse is the animal model of choice for study of the immune system on earth, and it is likely that future space flight studies will involve mouse models (Sonnenfeld, 2002; Sonnenfeld and Shearer, 2002; Sonnenfeld *et al.*, 2003). It should be noted, however, that most of the space flight studies have been carried out utilizing rats. Those flight studies have recently been reviewed elsewhere and the reader is referred to those reviews for more information (Sonnenfeld, 2002; Sonnenfeld and Shearer, 2002; Sonnenfeld *et al.*, 2003).

Chapter 5 covers a review of animal model studies to determine the effects of space flight on the neurovestibular system. There have been extensive studies in this area to determine the causes of and counter measures for space motion sickness and other potential disorders. The chapter is authored by pioneer researchers in that area, Cohen, Yakushin, Holstein, Dai and Kozlovskaya.

Tosini and Aguzzi review the effects of space flight on circadian rhythms in Chapter 6. Loss of sleep due to alterations in circadian rhythms during space flight travels is a potential serious problem for performance and survival of crews, and has been under extensive study utilizing animal models.

Developmental biology can only be studied in space using animal models. The gestation time and life span for humans is too long and the risk of teratogenic events too great to allow for meaningful human studies. In Chapter 7, Alberts and Ronca, two of the pioneers in the field, thoroughly review the results in that area.

Potential loss of bone is a great concern as humans contemplate exploration class missions. Breakage of weakened bones could severely impair both function and safety of crews, if they land on a new body in space for exploration. Doty, Vico, Wronski and Holton review the animal studies in this area in Chapter 8.

Wade, in Chapter 9, explores the effects of the continuum of gravity exposure, He reviews the animal studies that have used hypergravity to explore potential mechanisms and effects of gravity on animal physiologic systems.

There are several advantages in using aquatic animals to study the effects of space flight on physiological systems. They are generally smaller than the land vertebrates and the life support systems required for their maintenance in space can be less complex and require less maintenance than would the system to maintain land animals. Aquatic animals have been extensively studied in space, and Horn extensively reviews the results obtained with them in Chapter 10.

The final chapter of the book, Chapter 11, is dedicated to a review of the use of the animal model closest to humans, non-human primates, in space. Because of their closeness to humans, their use has allowed for precision of the relevance to humans of the effects of space flight on animal models. The results using primates have been expertly reviewed by Hoban-Higgins, Robinson and Fuller.

Perspectives

It is hoped that the reader, after completing this volume, will fully understand the great value of animal models in facilitating space exploration. It is hoped that this volume will prove to be a definitive reference for the design of future animal studies for space exploration.

Acknowledgments

The US National Aeronautics and Space Administration (NASA) partially supported the author's work through the NASA Cooperative Agreement NCC 9-58 with the National Space Biomedical Research Institute. The Amino-Up Chemical Company of Sapporo, Japan also partially supported this work.

References

Committee on Space Biology and Medicine (1998) *A Strategy for Research in Space Biology and Medicine in the New Century*. National Research Council, Space Studies Board, National Academy Press, Washington, DC.

Lawler, A. (2004) Remaking NASA: How much space for science? Science 303, 610–612.

Polk, H.C. Jr., Galandiuk, S., Hershman, M.J. and Sonnenfeld, G. (1991) Immune restoration and stimulation, vaccines and biological modifiers. *In* Pollock, A.V. (ed.). *Surgical Immunology*, pp. 254–264. Edward Arnold Publishers, Sevenoaks, Kent, UK.

Sonnenfeld, G. (2002) The immune system in space and microgravity. Med. Sci. Sports Exerc. 34, 2021–2027.

Sonnenfeld, G. (2003) Animal models for the study of the effects of spaceflight on the immune system. Adv. Space Res. 32, 1473–1476.

Sonnenfeld, G. and Shearer, W.T. (2002) Immune function during space flight. Nutrition 18, 899–903.

Sonnenfeld, G., Butel, J.S. and Shearer, W.T. (2003) Effects of the space flight environment on the immune system. Rev. Environ. Health 18, 1–17.

Experimentation with Animal Models in Space
G. Sonnenfeld (editor)
2005. Published by Elsevier B.V.
DOI: 10.1016/S1569-2574(05)10002-1

The Hindlimb Unloading Rat Model: Literature Overview, Technique Update and Comparison with Space Flight Data

Emily Morey-Holton[1], Ruth K. Globus[1], Alexander Kaplansky[2] and Galina Durnova[2]

[1]*NASA Ames Research Center, Moffett Field, CA 94035-1000, USA*
[2]*Institute of Biomedical Problems, 123007, Moscow, Russia*

Abstract

The hindlimb unloading rodent model is used extensively to study the response of many physiological systems to certain aspects of space flight, as well as to disuse and recovery from disuse for Earth benefits. This chapter describes the evolution of hindlimb unloading, and is divided into three sections. The first section examines the characteristics of 1064 articles using or reviewing the hindlimb unloading model, published between 1976 and April 1, 2004. The characteristics include number of publications, journals, countries, major physiological systems, method modifications, species, gender, genetic strains and ages of rodents, experiment duration, and countermeasures. The second section provides a comparison of results between space flown and hindlimb unloading animals from the 14-day Cosmos 2044 mission. The final section describes modifications to hindlimb unloading required by different experimental paradigms and a method to protect the tail harness for long duration studies. Hindlimb unloading in rodents has enabled improved understanding of the responses of the musculoskeletal, cardiovascular, immune, renal, neural, metabolic, and reproductive systems to unloading and/or to reloading on Earth with implications for both long-duration human space flight and disuse on Earth.

Introduction

Models for simulating certain aspects of space flight have been developed because decreased gravity, less than 1 unit gravity (G), is presently impossible to achieve on the surface of Earth for an appreciable length of time, and access to space is limited. The primary physiological changes resulting from space flight include fluid shifts, repositioning of certain organs or organelles, unloading of all systems, and lack of stimulus to gravity sensors. These changes may trigger additional metabolic and endocrine responses. The hindlimb unloading rodent model, first described in the 1970s, is used routinely to study changes that occur in the whole animal, both during and following a period of disuse. In the hindlimb unloading model, the hindquarters are elevated to prevent weight-bearing, while the forelimbs remain weight-bearing. This position results in a cephalic fluid shift and musculoskeletal unloading, which also occur during space flight. The intent of developing a hindlimb unloading model is to provide a ground-based system that simulates certain aspects of space flight in animals, similar to bedrest in humans, and enables study of disuse on Earth in a controlled manner.

Characteristics of publications

An Excel spreadsheet was populated from a database of 1064 references using or reviewing hindlimb unloading. The following search terms were entered into the search box on the web site http://www.ncbi.nlm.nih.gov/entrez/query.fcgi: (rats OR mice) AND (simulat* weightless* OR hindlimb* unload* OR hindlimb suspen* OR hindquarter* unload* OR tail suspen* OR altered gravity) OR disuse NOT antidepressant NOT human NOT (tail suspension test). References were downloaded into an EndNote library from this web site. Chapters and publications not cited by PubMed by April 1, 2004, may be missing from this database. All abstracts and many articles were read to assure that hindlimb unloading was used or reviewed; articles using solely centrifugation, space flight, confinement, nerve section, or limb casting were deleted from the database. Information in the spreadsheet included: first author, year of publication, journal, country of first author, rodent type, sex, genetic strain, age of animal at the beginning of the experiment, countermeasure (drug or exercise), and scientific focus of the article. Not all papers included the gender, age of the animals at the beginning of the experiment, or genetic strain, so these categories were incomplete and conclusions were drawn from the information available. The spreadsheet, list of references, and other information on the hindlimb unloading model can be found at the web site: http://lifesci.arc.nasa.gov/holton/index.html.

Papers published

Figure 1 depicts the numbers of papers using the hindlimb unloading model, published since 1976. The first rodent unloading publication described horizontal unloading of the entire animal (Saiki *et al.*, 1976). Three years later, the first peer-reviewed paper on head-down hindlimb unloading was published (Morey, 1979), and in 1980 five peer-reviewed publications appeared (Caren *et al.*, 1980; Deavers *et al.*, 1980; Il'in and Novikov, 1980; Jordan *et al.*, 1980; Musacchia *et al.*, 1980). From 1976 through 1983, 35 papers were published. In each succeeding five-year period, approximately 90 more papers

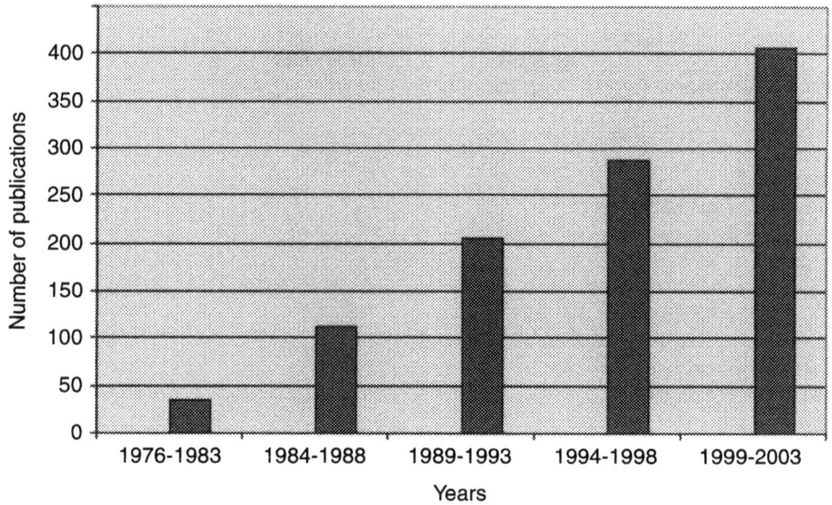

Fig. 1. The number of papers published on the hindlimb unloading rodent model from 1979–2003. Each bar represents a specific time period and the data are not cumulative. The first bar represents an eight-year period while the remaining bars represent a five-year period. The number of hindlimb unloading publications have increased from less than 50 to over 400 in a single five-year period.

were published compared to the preceding five-year period. In the 1999–2003 period, over 400 papers were published. In the first 3 months of 2004, 20 papers appeared. Thus, from a slow start in the 1970s, the dramatic increase in publications suggests that the hindlimb unloading model is useful for studying disuse either on Earth or associated with space flight.

Journals publishing hindlimb unlaoding articles

Journals publishing five or more papers are listed in Table 1. Hindlimb unloading articles have been published in at least 185 different journals. Approximately one-third of the 1064 hindlimb unloading publications are found in three physiology journals including Journal of Applied Physiology, American Journal of Physiology, and Journal of Gravitational Physiology. Space Med Med Eng (Beijing), a Chinese journal, has published 61 hindlimb unloading articles. Two Russian journals, Kosm Biol Aviakosm Med and Aviakosm Ekolog Med, have published 65 hindlimb unloading papers.

Hindlimb unloading publications from different countries

A consortium of NASA investigators developed the hindlimb unloading model in the early 1970s (Morey, 1979). Since then, the technique has spread to many laboratories across the globe. Table 2 lists the countries of first authors and the year of the first hindlimb unloading publication from that country. Seven

Table 1

Journals publishing articles using or reviewing hindlimb unloading

JOURNALS with 5 or more hindlimb unloading papers	Number of papers
Acta Astronautica	12
Acta Physiologica Scandinavia	10
Advances in Space Research	9
American Journal of Physiology	79
Aviakosm Ekolog Med	30
Aviation, Space, and Environmental Medicine	57
Biochemica Biophysica Research Communications	5
Biological Research in Nursing	5
Biological Sciences in Space	6
Bone	12
Brain Research	
Calcified Tissue International	7
Endocrinology	8
Environmental Medicine	17
European Journal of Applied Physiology and Occupational Physiology	16
Experimental Neurology	10
Faseb Journal	10
Journal of Applied Physiology	213
Journal of Biological Chemistry	5
Journal of Biomechanics	6
Journal of Bone and Mineral Research	19
Journal of Gravitational Physiology	68
Journal of Leukocyte Biology	6
Japanese Journal of Physiology	10
Kosm Biol Aviakosm Med	16
Life Sciences	8
Medical Sciences in Sports and Exercise	10
Metabolism	10
Muscle Nerve	9
Pflugers Archives	10
Physiologist	71
Research in Experimental Medicine (Berlin)	10
Space Medicine and Medical Engineering (Beijing)	66

Journals with 4 publications

Arch Phys Med Rehabil, Clin Exp Hypertens, Gravit Space Biol Bull, Int J Sports Med, J Bone Miner Metab, J Pharmacol Exp Ther, J Physiol, Mech Ageing Dev, Proc West Pharmacol Soc

Journals with 3 publications

Anat Rec, Biull Eksp Biol Med, Can J Physiol Pharmacol, Eur J Appl Physiol, FEBS Lett, Fiziol Zh, Indian J Physiol Pharmacol, J Muscle Res Cell Motil, J Nutr Sci Vitaminol (Tokyo), J Physiol Anthropol Appl Human Sci, Neurosci Lett, Nippon Seirigaku Zasshi, Prostaglandins

(*Continued*)

Table 1

Continued

Journals with 2 publications
Adv Myochem, Adv Space Biol Med, Am J Pathol, Ann Clin Lab Sci, Bioelectromagnetics, Biomed Sci Instrum, Brain, Comp Biochem Physiol A, Dokl Biol Sci, Exp Cell Res, Exp Physiol, J Androl, J Exp Biol, J Interferon Res, J Neurosci Res, J Nutr, J Orthop Res, J Orthop Sci, NASA TM, Neuroscience, Pharmacology, Sheng Li Xue Bao, West J Nurs Res
Journals with 1 publication
Acta Anat (Basel), Acta Physiol Hung, Adv Exp Med Biol, Aging (Milano), Alcohol Clin Exp Res, Am J Clin Nutr, Am J Phys Med Rehabil, Ann N Y Acad Sci, Arch Physiol Biochem, Basic Appl Myol, Behav Brain Res, Biochem Cell Biol, Biochem J, Biol Chem, Biol Pharm Bull, Biomed Biochim Acta, Biomed Environ Sci, Biomimetics, Biosci Rep, BioScience, Bone Miner, C R Seances Soc Biol Fil, Cells Tissues Organs, Chem Pharm Bull (Tokyo), Clin Chem Lab Med, Clin Chim Acta, Clin Nutr, Clin Orthop, Comp Biochem Physiol C, Comput Med Imaging Graph, Connect Tissue Res, Crit Rev Oral Biol Med, Curr Opin Clin Nutr Metab Care, Di Yi Jun Yi Da Xue Xue Bao, Dokl Akad Nauk, Drug Discov Today, Eksp Klin Farmakol, Endocrine, Endocrinologist, Exerc Sport Sci Rev, Exp Biol Med (Maywood), Exp Brain Res, Exp Gerontol, Exp Mol Pathol, Exp Toxicol Pathol, Free Radic Biol Med, Growth Horm IGF Res, Histochem J, Histol Histopathol, Horm Metab Res, Hum Gene Ther, Inflamm Res, Int J Biochem Cell Biol, Int J Sport Nutr Exerc Metab, Invest Radiol, Iowa Orthop J, J Allergy Clin Immunol, J Anat, J Biochem (Tokyo), J Biol Regul Homeost Agents, J Cell Biochem, J Clin Invest, J Clin Pharmacol, J Exp Med, J Exp Zool, J Gerontol A Biol Sci Med Sci, J Histochem Cytochem, J Korean Med Sci, J Lipid Res, J Mater Sci Mater Med, J Med Invest, J Mol Cell Cardiol, J Neurobiol, J Neurocytol, J Neurol Sci, J Nutr Biochem, Joint Bone Spine, Kaibogaku Zasshi, Lab Anim Sci, Magn Reson Med, Med Biol Eng Comput, Med Phys, Mol Biol Rep, Nat Cell Biol, Neurochem Res, Neuroimmunomodulation, Neurosci Res, Nichidai Igaku Zasshi, Nutr Res, Nutrition, Physiol Behav, Physiol Bohemoslov, Physiol Genomics, Plast Reconstr Surg, Proc Soc Exp Biol Med, Proteomics, Regul Pept, Res Commun Mol Pathol Pharmacol, Res Nurs Health, Ross Fiziol Zh Im I M Sechenova, Scand J Med Sci Sports, Spine, Sports Med, Stal, Trav Sci Cherch Serv Sante Armees, Uchu Koku Kankyo Igaku, Usp Fiziol Nauk, Zhongguo Ying Yong Sheng Li Xue Za Zhi

countries have published 97% of hindlimb unloading articles with most authors being from the US, followed by France, Japan, China, Russia, Canada, and Italy. Languages in which hindlimb unloading articles have been published, and cited by PubMed, include Chinese, English, French, Japanese, and Russian.

Major physiological category or technique modifications

Table 3 lists the major physiological categories or techniques providing hindlimb unloading data and the numbers of publications. Approximately half of the 1064 publications focus on some aspect of skeletal muscle mass or metabolism during hindlimb unloading and/or recovery following reambulation. About 20% of the papers deal with bone or calcium metabolism. Cardiovascular studies, including changes in vessels, comprise about 15% of hindlimb

Table 2

Countries of first authors on hindlimb unloading publications

Country of first author	Number	First pub year
Algeria	1	1999
Belguim	2	1999
Canada	19	1984
China	93	1991
France	137	1987
Greece	1	2004
Germany	5	1989
India	3	1997
Italy	13	1993
Japan	125	1987 (except Saiki)
Korea	1	1994
Netherlands	1	2001
New Zealand	1	2003
Poland	1	1997
Russia	80	1980
Slovakia	1	1990
Slovenia	2	2000
Sweden	1	1998
Switzerland	3	2000
Tunisia	2	1996
Turkey	1	2003
Ukraine	1	2000
US	571	1979

unloading papers. The remaining 15% is distributed amongst the areas of immunology, neurology including reflexes related to posture, renal/fluid and electrolyte physiology, metabolism, reproduction, connective tissue (including cartilage, ligaments, tendons, and spinal disks) function, methods, technique modifications, and other miscellaneous subjects. Approximately 50 papers report data on more than one physiological system. The impact of changes in one system on another is seldom discussed, and few authors study multiple physiological systems within the same experiment.

Reviews

Table 4 lists the number of reviews by scientific category. Of the 44 reviews in the database, 43% describe various aspects of skeletal muscle adaptation to hindlimb unloading. Excellent reviews of skeletal muscle adaptation to unloading include those by Adams *et al.* (2003), Baldwin and Haddad (2002), Desplanches (1997), Fitts *et al.* (2000), and Riley (1999), while Machida and Booth (2004) focus on recovery from disuse atrophy. Recent reviews on skeletal changes with hindlimb unloading include Morey-Holton and Globus (1998),

Table 3

Categories and numbers of papers using mice or rats

Category	Publications	Mice	Rats
Skeletal muscle	**518**	**26**	**467**
Bone/calcium metabolism	**211**	**23**	**174**
(Bone and skeletal muscle)	(17)	(1)	(10)
Cardiovascular	**165**	**1**	**158**
(CV/bone/muscle)	(5)		(3)
(CV/bone)	(1)		(1)
(CV/muscle)	(2)		(1)
Immunology/hematology	**52**	**23**	**29**
(Immune/muscle)	(1)		(1)
(Immune/bone)	(1)		(1)
Neural/posture	**32**	**0**	**31**
(Neural/muscle)	(2)		(2)
Renal/fluid/electrolyte	**32**	**1**	**24**
(Renal/CV/bone/muscle)	(2)		
(Renal/bone/muscle)	(1)		(1)
(Renal/muscle)	(4)	(1)	(3)
(Renal/CV)	(4)		(4)
Energy/metabolism	**23**		**22**
(Metabolism/muscle)	(4)		(4)
(Metabolism/CV)	(1)		(1)
Reproduction	**18**	**1**	**14**
(Repro/muscle)	(1)		(1)
(Neural/repro)	(1)		(1)
Connective tissues	**12**	**0**	**12**
Methods	**11**	**0**	**6**
(Methods/skeletal muscle)	(1)		(1)
Blood values	**6**	**0**	**6**
Other*	**33**	**1**	**29**

Numbers in parentheses indicate papers on multiple systems.
*Includes brain, brown adipose tissue, drug metabolism proteins, gut, learning, nutrition, organ weight/density, pituitary, serum values, zinc distribution, etc.

Vico *et al.* (1998) Giangregorio and Blimkie (2002), and Bikle *et al.* (2003). Mueller *et al.* (2003), and Zhang (1994) review cardiovascular responses to hindlimb unloading, while Musacchia and Fagette (1997) review both cardiovascular and skeletal muscle systems. Tipton (1996) published an excellent critique of hindlimb unloading and cardiopulmonary adaptation. Da Silva *et al.* (2002) reviewed nutrition and metabolism. Krasnov (1994) authored a review on gravitational neuromorphology. Sonnenfeld (2003) compared hindlimb unloading with space flight effects on the immune system. Two brief reviews discussed multiple systems (Musacchia, 1985 and 1992) with the latter involving brief reports from multiple investigators. Booth and Grindeland (1994) reviewed similarities and differences between the hindlimb unloaded control

Table 4

Cosmos 2044 mission parameters; Launch: September 15, 1989; Recovery: September 29, 1989; Duration: 14 days; Orbital Period (min): 89.3; Apogee (km): 294; Perigee (km): 216; Inclination (deg): 82.3

Group	Flight	Synchronous control	Hindlimb-unloaded	Vivarium control
Number of rats	10	10	10	10
Body mass, g	332±4.4	342±8.6	326±7.3	361±2.2*
Thymus, mg	171±8.1	227±15.5*	188±7.9	208±15.4*
Adrenals, mg	43.8±1.4	45±2.2	48.7±1.4*	40.7±0.8
Spleen, mg	489±37.8	553±15.6	572±16.7*	646±14.1*
Liver, g	9.8±0.4	10.4±0.3	8.1±0.2*	10.1±0.4
Kidneys, g	2.3±0.04	2.3±0.05	2.2±0.05	2.2±0.05
Testes, g	2.3±0.2	2.6±0.11	0.96±0.05*	2.5±0.23

* = Significantly different from flight.
Data are expressed as mean±SD.

group and rats flown on Cosmos 2044. The reviews provide excellent insights into the response of rodents to disuse and often evaluate the effectiveness of the model for predicting responses to space flight.

Species

Rats are routinely used both for hindlimb unloading and space flight. Mice are appealing for space flight because of their size and genetic characteristics. On the other hand, mice produce odors that challenge habitat designs destined for human space missions. Mice (female C57BL/6) have flown on a single Space Shuttle mission (Pecaut *et al.*, 2003), although pocket mice were carried onboard Apollo 17. Hamsters have been used in hindlimb unloading studies and might be of interest for space flight due to their minimal requirement for water and minimal excretion products; however, this rodent strain has not been tested in space flight experiments.

Table 3 shows the numbers of papers on studies using mice and rats arranged by topical category. Rats are used in studies described in 934 hindlimb unloading papers, mice in 71 papers, and 3 report the use of both species (Rivera *et al.*, 2003; Sudoh *et al.*, 1994; Tidball and Spencer, 2002). Two hindlimb unloading papers describe the use of hamsters (Corley *et al.*, 1984; Elder, 1988). The remaining papers either do not report the rodent used or are reviews.

Gender and Reproductive Function

The influence of hindlimb unloading has been studied in both male and female rodents, although only rarely are both genders evaluated in the same study. In mice, 30 papers report the use of females, 22 papers report males, and 5 additional papers report the use of both genders. In rats, 185 papers report the use of females, 606 report on males, and 7 additional papers report the use of both genders.

Testes size and testosterone levels in both serum and testes are reduced following 7 days of hindlimb unloading (Hadley *et al.*, 1992). However, ligation to keep the testes from descending into the abdomen during hindlimb unloading prevents a decline in serum testosterone levels, but not the adverse effects of hindlimb unloading on spermatogenesis (Hadley *et al.*, 1992; Tash *et al.*, 2002). Spermatogenesis appears normal following a 2-week space flight even though the size of the testis and plasma testosterone levels are reduced (Amann *et al.*, 1992; Deaver *et al.*, 1992). Furthermore, Tou *et al.*, 2004 demonstrate a disruption of estrus in females. Whether the hindlimb unloading-induced changes in testosterone levels, sperm production, and estrus also occur in adult rodents during space flight is not known. Reproductive status and steroid levels during hindlimb unloading should be considered as a potential factor contributing to hindlimb unloading-induced changes in other physiological systems such as bone and muscle.

Genetic strain

Humans have diverse genetic backgrounds yielding large variability to data from crew members. Specific rodent strains can be selected to minimize variability for hindlimb unloading studies. For example, different mouse strains vary greatly in their skeletal responses to unloading; mouse strains with the highest bone mass show the least change while those with the lowest density show the greatest change (Judex *et al.*, 2004; Simske *et al.*, 1994). Several hindlimb unloading articles described the use of C56Bl/6 mice that have features of skeletal ageing and unloading responses similar to those of humans (Halloran *et al.*, 2002; Judex *et al.*, 2004). The Fisher 344 rat is a smaller inbred genetic strain that is commonly used for adult hindlimb unloading studies.

Mouse strains used for hindlimb unloading studies include various inbred and hybrid laboratory strains, naturally occurring mutants, and strains generated by homologous recombination and transgenic methods. Strains studied include β293, Balb/C, Balb/C AnNHsd, Balb/CJ, Balb/C crossed with C57/BL10, BDF1, B6.129P2-Nos2tm1lau, C3H, C3HeB/FeJ, C3H/HeJ, C57BL, C57BL/6J, C57BL/6J±6-Cybbtrn1 (gp91phox), C57BL/6J into 129/SvJ strain (Mstn), 129/S3XC57BL/6F2±OPNDBA, CBA/J, DBA-2J, ICR or CD1, hIGF-I gene driven by regulatory regions from the a-actin gene, βMHC/CAT on FBV/n, MyoD WT/null, p53null or +, selectin deficient (C57BL/6 background), Swiss Webster, human TnIs gene linked to the bacterial CAT-coding region, WF/Hsd BR, and line 129B6F1 dystrophic. Several papers compare differences due to hindlimb unloading between at least two genetic mouse strains in skeletal tissue (Judex *et al.*, 2004; Simske *et al.*, 1994; Amblard *et al.*, 2003), the immune system (Kulkarni *et al.*, 2002; Miller *et al.*, 1999; Rivera *et al.*, 2003; Yamauchi *et al.*, 2002) and skeletal muscle (Stelzer and Widrick, 2003).

Rat strains used for hindlimb unloading experiments include dw4 (GH-), Fisher 344 (F344), Fisher 344 Brown Norway, Long Evans, Munich-Wistar, Sprague-Dawley (S/D), Wistar (W) and Wistar Hanover. The most commonly used rat strains are S/D and W; F344 are used for most adult studies.

Age

If one assumes a direct, linear relationship between species in life spans, then a one-year-old rat or mouse would be approximately equivalent to a 30 year old human, given that a mouse or rat lives about 2.5 years while a human lives about 75 years. This reasoning suggests a 1 : 30 age ratio of rodents : humans and suggests that rats or mice approximately one year old may provide the best models for adult crew members, many of whom are over 40 years of age. Some investigators assume that a sexually mature rat at 6 weeks of age is an adult. However, many other physiological systems are in a growth, rather than maintenance, phase at this age. Recent publications suggest that albino rats are not skeletally mature until the epiphyses close and bone elongation ceases, which occurs after the rat is 8 (male) or 10 (female) months old (Martin *et al.*, 2003). A Fischer 344/Brown Norway rat is not considered an adult until it is 18 months old (Machida and Booth, 2004).

Mice may reach peak bone density by 4 months, but their femora continue to lengthen for an additional 4 to 8 months (Beamer *et al.*, 1996). These data suggest that growth plates do not close until the animal is between 8 and 12 months of age when the animals can be considered adults. Few investigators use mice older than 4 months of age probably due to maintenance costs.

The vast majority of hindlimb unloading articles used rapidly growing rats between 1.5 and 4 months of age. The youngest rats were 2 days of age and were used to study critical developmental periods (Walton, 1998) while experiments with 4-day-old rats were used to investigate muscle and bone growth (Ohira *et al.*, 2001, 2003). The oldest rats studied were between 1 and 3 years of age. Of these 21 investigations on adult rats, 81% of the papers studied skeletal muscle (e.g., Brown *et al.*, 1999; Chen and Alway, 2001; Deschenes and Wilson, 2003; Herrera *et al.*, 2001; Ojala *et al.*, 2001; Sandmann *et al.*, 1998, and Stump *et al.*, 1997), while two articles reported on sympathetic responses (Kawano *et al.*, 1994, 1995), one studied bone (Simmons *et al.*, 1983), and one reported on reproductive capacity (Tash *et al.*, 2002). Rats between 6 and 11 months of age were used in an additional 29 studies investigating muscle (10 papers), bone (16 papers), metabolism (1), reproduction (1), and the anterior pituitary (1). Only 36 of 74 hindlimb unloading mouse papers reported ages of the animals. One paper used mice between 6 and 9 months of age to study muscle (Criswell *et al.*, 1998), while another author used mice from 1.2 to 12 months of age to investigate bone responses to hindlimb unloading (Simske *et al.*, 1990). All remaining articles reported the use of mice between 1 and 4.5 months of age.

The age of animal selected for hindlimb unloading studies is often a factor in the results obtained. For example, bone responses to hindlimb unloading are minimal in weanling rats compared to 6-week or 6-month-old animals and recovery is much more rapid (Dehority *et al.*, 1999; Morey-Holton *et al.*, 2003).

Duration of experiments

The duration of hindlimb unloading required for attaining a new steady state depends on the turnover rate of the specific system. Over 70 papers use multiple durations of hindlimb unloading ranging from 10 min to 14 weeks while most papers report durations between 1 and 4 weeks. Over 125 papers describe a reambulation period lasting from 10 min to 3 months following hindlimb unloading. The longest hindlimb unloading period lasted for 206 days beginning when rats were 18 days of age; the authors studied changes in skeletal muscle (Elder and McComas, 1987).

Countermeasures

A countermeasure is a therapy (drug and/or physical) devised to minimize or prevent adverse changes caused by hindlimb unloading or space flight. Over 220 hindlimb unloading articles describe the testing of countermeasures during hindlimb unloading: 139 papers describe the use of drugs, 68 articles describe exercise, and 14 articles describe both drugs and exercise as countermeasures. Physical countermeasures tested include centrifugation, electrical stimulation, far infrared irradiation, magnetic fields, venous ligation, vibration, or exercise regimes (before, during or after hindlimb unloading, e.g., normal weight bearing, treadmill, wheel running, ladder climbing $\pm$ weights, dorsiflexion). Drug therapy included anabolic steroids, angiotensin antagonists, bisphosphonates, calcineurin inhibitors, synthetic catecholamines, dietary supplements, gallium nitrate, factors that deplete high energy forms of phosphate, hormones or local factors, non-steroidal anti-inflammatory drugs, and inhibitors of nitric oxide synthase, serotonin, or prostacyclin synthase. The studies typically focused on a single system and did not attempt to determine if the countermeasure affected other physiological systems.

Summary

Over 1000 papers have been published on the response of diverse physiological systems to hindlimb unloading in rodents with approximately 400 publications appearing in the last five years. Most papers focus on muscle, bone, or cardiovascular responses to hindlimb unloading, yet other physiological systems have been studied. Rodent species include rats, mice, and hamsters of various strains and ages. More studies use male than female animals. Experimental duration varies from hours to months depending upon the physiological system studied. Over 200 publications have reported on potential physiological and/or pharmacological countermeasures during hindlimb unloading. Publications primarily in physiology journals are found not only in English, but also in Chinese, French, Japanese, and Russian demonstrating the utility of the hindlimb unloading model throughout the globe.

Cosmos 2044

Background

A detailed study of the effects of space flight factors, particularly microgravity, on mammalian systems is necessary to develop countermeasures and support health and performance of space crew members. Animal experiments in space are expensive and limited, so that ground simulation techniques allowing studies of various space flight effects are important. Comparison of space flight data with hindlimb unloading or other ground-based models is critical for understanding those aspects of space flight that are accurately mimicked by such models. Cosmos 2044 (Tables 4 and 5) was the first space flight experiment to offer hindlimb unloading as a separate control group to all investigators participating in the mission. Comparative analysis of hindlimb unloading and flight groups from this experiment is based on our own data (Table 4) and the results of other researchers who participated in the Cosmos 2044 flight (Table 5). The broad objectives of the Cosmos program include: (a) characterize the biological responses to space flight, (b) clarify the mechanisms mediating the responses, and (c) use the space environment as a tool to better understand adaptive and disease processes of disuse in Earth organisms. Specific objectives and results of many individual experiments conducted on this mission are found in a supplement to the August 1992 issue of Journal of Applied Physiology (volume 72, number 2).

Materials and methods

Cosmos 2044 consisted of five groups of male Wistar (Specific Pathogen Free) rats, with an average initial body mass of 320 g. The groups included (1) basal controls (i.e., animals were killed at the beginning of the experiment for baseline data), (2) flight, (3) synchronous controls (i.e., ground controls which were housed identically to flight animals but lagged 5 days behind the flight group, so that the inflight environmental parameters listed below and vibration and G-loads similar to those experienced by flight rats during launch and recovery could be simulated with high fidelity), (4) vivarium control (i.e., animals that were housed under standard vivarium conditions), and (5) hindlimb unloading.

Table 5

Plasma, liver, and testes values expressed as a percentage of synchronous control animals for Cosmos 2044

Group	Basal	F	S	HU	V	FWH	SWH	HUWH	VWH
Number of rats	10	5	5	5	5	5	5	5	5
Surgery date						12-Sep	17-Sep	21-Sep	19-Sep
Date of euthanasia	15-Sep	29-Sep	4-Oct	8-Oct	5-Oct	29-Sep	4-Oct	8-Oct	6-Oct
Age at euthanasia, days	109	123	127	131	129	123	127	131	129
Corticosterone, %S					92	142	100	276	155
(nmol/ml plasma)					(0.49±.11)	(0.75±0.08)	(0.53±0.16)	(1.10±0.2)	(0.82±0.2)
Corticosterone, %S	12	167	100	59	63				
(μg/dl plasma ±SE)	(2.97±1.5)	(41±7.45)	(24.9±5.05)	(14±6.13)	(15.7±3.54)				
Liver glycogen									
(concentration),		224	100	12	67				
%S (mg/g±SE)		(51.6±4.4)	(23±2.1)	(2.7±1)	(15.3±2)				
Liver glycogen (%),		255	100	9	36				
%S (%±SE)		(~28±6)	(~11±3)	(~1±0.5)	(~4±2)				
Kupffer cells, %S		88	100	134	102				
(#/reference area±SE)		(5.7±0.31)	(6.5±0.56)	(8.7±1.31)	(6.6±0.37)				
Liver cholesterol,		87	100	110	97				
%S (mg/g±SE)		(1.50±0.04)	(1.72±0.05)	(1.90±0.08)	(1.66±0.1)				
Insulin, %S					49	110	100	27	44
(μU/ml plasma)					(20±3)	(45±11)	(41±14)	(11±3)	(18±2)
Glucose, %S					93	138	100	79	88
(mmol/l plasma)					(6.3±0.3)	(9.4±0.9)	(6.8±0.3)	(5.4±0.3)	(6.0±0.3)
Glucose, %S	117	154	100	91	110				
(mg/dl plasma ±SE)	(143±5.4)	(188±19)	(122±1)	(111±4)	(134±5)				
Cholesterol, %S	77	117	100	79	68				
(mg/dl±SE)	(60±1.7)	(91.6±2.6)	(78±5.2)	(61.6±5.1)	(53.2±2.7)				
Spermatids, %S		76	100	1	88				
(g/testis×1,000,000)		(28.9)	(37.8)	(<0.5)	(33.2)				
Testosterone, %S	158	14	100	27	183				
(ng/ml plasma ±SE)	(3.02±0.99)	(0.27±0.05)	(1.91±0.4)	(0.51±0.27)	(3.5±0.66)				

F = Flight, S = Synchronous control, HU = Hindlimb unloading control, V = Vivarium control, WH = Wound healing.
Numbers in parentheses are data from Amann *et al.*,[36]; Macho *et al.*,[75]; Merrill *et al.*,[77,81]; Pedrini-Mille *et al.*,[89]; Racine and Cornier,[108].
The units for these numbers are in the parenthesis under the parameter measured.

Each group had 10 animals. The groups were further subdivided into intact or wound healing groups. Five of the 10 rats in each group underwent surgery for wound healing studies two days before the experiment was initiated. Prior to wounding, rats were anesthetized. Then the soleus and lateral head of the gastrocnemius muscle were crushed and fibulas (within the mid-third) were cut. All operated animals were able move freely within a day after surgery. The duration of the experiment was 14 days. The groups of 10 rats were housed together, except hindlimb-unloaded rats, which were housed individually. Basal, vivarium and hindlimb-unloaded animals received 55 g/rat/day of a paste-like diet, while flight and synchronous controls were given ~ 14 g aliquots of the same diet every 6 h. All groups had access to water *ad libitum*. The light–dark cycle was 16 on and 8 off. The inflight environmental parameters in the spacecrafts were as follows: pO_2 = 140–180 mm Hg; pCO_2 no more than 2 mm Hg; ambient temperature = 21–25 °C for the first three days, increased to 24–26 °C for nine days and was 26–29 °C for the last two days. The wounded flight rats were killed 4–7 h after landing, while the intact flight animals were euthanized 8–11 h after landing. Hindlimb-unloaded rats were weight-bearing for 15–60 min before euthanasia. The flight, synchronous, vivarium, and hindlimb-unloaded animals were decapitated and tissues processed at the same time of the day by the same team, so the groups were euthanized on landing day, or 5 days, 7 days, 9 days postflight, respectively.

Limitations

Cosmos 2044 was the first dedicated mission to include a hindlimb unloading control group. Although animal cohorts and experiment duration were identical, several confounding factors need to be considered when comparing the flight and hindlimb-unloaded groups. The flight group was housed 10 rats/cage while the hindlimb-unloaded animals were individually housed. We have previously shown that housing can affect the skeletal response to space flight (Morey-Holton *et al.*, 2000). The flight animals were fed aliquots every 6 h to total the same amount of food given to the hindlimb-unloaded group once a day. The flight group was killed 4–11 h after landing while hindlimb-unloaded rats were weight-bearing a maximum of 1 h prior to euthanasia. The flight group experienced reentry and landing stresses and the hindlimb-unloaded group did not. The entire animal was unloaded during space flight while only the hindquarters of the hindlimb-unloaded rats were unloaded. The forelimbs and upper spine of hindlimb-unloaded rats were weight-bearing (Globus *et al.*, 1986; Hargens *et al.*, 1984), serving as useful internal controls for possible systemic changes during hindlimb unloading. Also, during hindlimb unloading the jaw still opened with gravity whereas in space, the jaw did not have gravity forcing it open and, thus, mandibular changes seen in flight rats were not observed in hindlimb-unloaded animals (Simmons *et al.*, 1983). Foot position was different in space flight compared to hindlimb unloading (Huckstorf *et al.*, 2000). The head-down position of the hindlimb-unloaded animals probably induced a greater fluid shift toward the head than that experienced by the flight group (Tipton *et al.*, 1987). Hindlimb unloading also increased cerebral vascular pressure (Maurel *et al.*, 1996; Wilkerson *et al.*, 2002), likely related to the fluid shift. Even with these differences, the flight and hindlimb-unloaded groups showed many similar physiological responses at the end of the experiment.

Body mass and organ weights

Body mass and organ weights of rats in each group, whether wounded or not, were not significantly different at the end of the experiment so the subgroups were combined for a total of 10 animals per group (Table 4). The major differences between flight and hindlimb unloading were adrenal, spleen, liver, and testes weights (see discussions further in the chapter).

Endocrine stress measurements

No differences in body weight were found among the groups (Table 4). Flight rats had a smaller thymus than either control group, while hindlimb unloading thymus mass was only different from the vivarium controls. Hindlimb-unloaded rats did exhibit a small (11%), but statistically significant increase in adrenal weight compared to flight and synchronous control groups. The increase in adrenal weight in hindlimb-unloaded rats compared with flight suggests that hindlimb unloading caused a chronic stress that was not experienced by the flight group. The hindlimb-unloaded wound-healing group compared with the corresponding flight group exhibited a 47% increase in plasma corticosterone (Macho *et al.*, 1991) (Table 5), as well as neutrophilosis and lymphopenia. The intact hindlimb-unloaded animals compared to intact flight rats showed a 66% decrease in plasma corticosterone (Merrill *et al.*, 1992a) (Table 5). The differences in corticosterone levels between the wound-healing and intact flight and hindlimb unloading groups are difficult to explain. However, the corticosterone levels in the flight groups were higher than their corresponding synchronous controls indicating an "acute gravitational stress (Durnova *et al.*, 1977)" that may be caused by reentry and the sudden increase of weight-bearing upon return to Earth (Kaplanskii and Savina, 1981; Il'in *et al.*, 1989). This explanation is supported by the finding that tyrosine aminotransferase levels are elevated in livers of both flight and hindlimb-unloaded groups, relative to synchronous and vivarium groups ($T^{1/2}$ of tyrosine aminotransferase is 2 h) (Merrill *et al.*, 1992b). The data suggest that return from a 14-day space flight induced an acute stress, while chronic stress occurred during the 14-day hindlimb unloading experiment.

Skeleton

Decreased trabecular bone volume was noted in the intact tibia of flight and hindlimb-unloaded wound-healing groups compared to controls. Detailed histomorphometric analysis showed that osteopenia resulted from suppression of longitudinal bone growth and bone formation as well as increased bone resorption. Inhibition of longitudinal growth was indicated by the decreased height of the epiphyseal cartilage growth plate and primary spongiosa in the wound-healing animals euthanized 4–7 hours after flight (Durnova *et al.*, 1991; Kaplansky *et al.*, 1991). At the same time, no significant differences were noted in the total growth plate height in intact flight or hindlimb-unloaded animals compared to the other control groups, suggesting that the growth plates of the older rats in Cosmos 2044 were less responsive to unloading than the younger animals studied previously. However, an increased height of the proliferation zone and decreased height of the hypertrophic/calcification zone was found in flight rats compared to the hindlimb-unloaded rats (Montufar-Solis *et al.*, 1992) and may be due to reloading following flight (Montufar-Solis *et al.*, 2001).

The volume of primary spongiosa, as well as the number and thickness of trabeculae, showed a greater decrease in hindlimb-unloaded than in flight animals, while the distance between trabeculae increased. In the secondary spongiosa, these parameters decreased similarly in flight and hindlimb-unloaded rats compared to controls. Resorption surfaces of the secondary spongiosa were slightly greater in flight than in hindlimb-unloaded rats and slightly greater in hindlimb unloading than controls. Osteoid volume increased slightly only in the flight rats (Durnova *et al.*, 1991).

Additional results from the fibular fracture healing experiments showed that bone formation, and particularly cartilage formation, was reduced and the number of osteoclasts was increased in the fracture calluses of flight and hindlimb-unloaded animals relative to controls. Both flight and hindlimb-unloaded rats had minimal callus development, which resulted in a weaker bone at the fracture site (Durnova *et al.*, 1991). Delayed fracture repair occurred in parallel with metabolic changes. For example, hindlimb-unloaded and flight animals showed increases in bone aspartate aminotransferase, and isocitrate dehydrogenase compared to controls. Hindlimb-unloaded rats also showed increased bone lactate dehydrogenase and decreased alkaline phosphatase, while flight rats exhibited lower creatine phosphokinase, lactate dehydrogenase, alkaline phosphatase and acid phosphatase (Fedotova, 1991).

No difference in humeral cortical bone composition or strength was found between groups (Vailas *et al.*, 1992). Only collagen cross-link concentration in the fifth lumbar vertebra decreased in flight and hindlimb unloading, while all other parameters were unchanged from controls (Vailas *et al.*, 1994). L4-6 intervertebral disks from flight rats were significantly smaller (both wet and dry weights) compared to all other groups with a significantly greater collagen-to-proteoglycan ratio and no detectable changes in the relative proportions of type I or type II collagen or the number of collagen cross-links (Pedrini-Mille *et al.*, 1992) suggesting that flight, but not hindlimb unloading, may adversely affect the biomechanical function of spinal disks.

Muscle

Numerous muscles were analyzed including the adductor longus (AL, slow-twitch antigravity), extensor digitorum longus (EDL, fast-twitch), lateral and medial gastrocnemius (LG and MG, fast-twitch extensor), plantaris (PLAN, fast-twitch), soleus (SOL, mixed fiber types antigravity), tibialis anterior (TA, fast-twitch flexor), triceps brachii (TB, primarily fast-twitch), and vastus intermedius and medialis (VI and VM, primarily fast with mixed fibers closer to bone).

Significant reduction of myofibrillar cross section of the AL occurred in rats exposed to flight or hindlimb unloading (Riley *et al.*, 1992). Only hindlimb-unloaded rats also lost wet muscle mass. However, space flight compromised microcirculation causing thrombi formation post-flight. The thrombi apparently blocked capillaries and venules and led to edema, thereby increasing the weight of the flight AL and masking the loss of muscle mass.

Atrophy increased the susceptibility to form eccentric contraction-like lesions, tearing of connective tissue and microcirculation thrombosis. Such lesions were greater in the flight rats compared to hindlimb-unloaded rats, likely due to reentry forces and longer reloading time. Investigators who dissected tissues 48 h after flight (Ilyina-Kakueva *et al.*, 1976; Ilyina-Kakueva and Portugalov, 1977; Il'ina-Kakueva and Portugalov, 1981) previously reported microcirculatory changes in muscles. In both hindlimb unloading and flight, slow muscle fibers characterized by oxidative metabolism showed greater changes than either intermediate fibers with oxidative-glycolytic metabolism or fast fibers with glycolytic metabolism. In addition, light chains of slow myosin decreased and those of fast myosin increased indicating transformation of some slow fibers into fast or intermediate fibers in the AL (Riley *et al.*, 1992).

Ohira *et al.* (1992) investigated the soleus muscle, a slow antigravity muscle, in both flight and hindlimb-unloaded rats. The muscle mass decreased by 25% and 34%, respectively. Also, the cross-section area of slow and, to some extent, fast myofibers, as well as total protein and succinate dehydrogenase (SDH) activity decreased in these groups compared to controls. In myofibers containing both fast and slow myosins, SDH and α-glycerophosphate dehydrogenase (α-GPDH) slightly increased, indicating slow-to-fast myofiber transformation. This finding was supported by biochemical examinations of soleus myofibers, which showed atrophic changes in parallel with increases in pyruvate kinase, glycerol-3-phosphate dehydrogenase, 1-phosphofructokinase and hexokinase in both flight and hindlimb-unloaded rats (Chi *et al.*, 1992). These authors concluded that hindlimb unloading can adequately simulate microgravity-induced changes in slow-twitch muscles.

Fast-twitch muscles were less responsive than slow-twitch muscles to either real or simulated microgravity. The average fiber size in the medial head of MG and TA muscle remained unchanged in flight and hindlimb-unloaded rats (Jiang *et al.*, 1992). The same finding was true for the slow/fast ratio in these muscles. Only the intermediate fibers were different. They were smaller in the flight and hindlimb-unloaded rats compared to controls, with myosin ATPase in the flight animals being higher than in hindlimb-unloaded or control rats. α-GPDH in the TA muscle did not change in the flight rats, but increased in hindlimb-unloaded animals. VM muscle mass did not differ between flight, hindlimb-unloaded, or synchronous groups (Musacchia *et al.*, 1992). In the mixed fiber populations of VM, type I fibers of flight and hindlimb-unloaded rats were comparatively reduced compared to the other groups. Type II fibers in this area were smaller in flight rats than all other groups. Capillary density and fiber density were greatest in the mixed portion of the VM in flight rats compared to the other groups. In the homogenous portion of VM with predominately fast-twitch fibers, the cross-sectional area of fast fibers decreased in flight compared to hindlimb-unloaded but not in synchronous controls, while their number per area unit increased. Protein, RNA, DNA, and oxidative capacity in the VM did not differ between flight, hindlimb-unloaded, and synchronous controls while triglycerides were elevated only in the flight group (Musacchia *et al.*, 1992). In addition, α-actin mRNA expression in the VI and LG muscles decreased in flight and hindlimb unloading compared to controls while no change was noted in TB, suggesting differential responses to unloading between contractile muscles of the arms and legs (Booth *et al.*, 1994; Thomason *et al.*, 1992).

The post-traumatic repair of the soleus muscle and the lateral head of the gastrocnemius muscle (Il'ina-Kakueva and Burkovskaia, 1991; Stauber *et al.*, 1992) was similar in hindlimb-unloaded and flight animals, although it appeared slower in flight rats. Both investigations identified similarities in the repair patterns of extracellular matrix structure and myofibers.

In summary, Cosmos-2044 and hindlimb-unloaded rats showed similar atrophic and metabolic muscle changes, particularly of slow muscles, where a transformation from slow-to-fast fibers developed. Atrophic changes were accompanied by decreases in both strength and speed of contractions, as well as tension and performance (Cordonier *et al.*, 1992; Oganov *et al.*, 1991).

Heart

Cardiac muscle atrophy was more apparent in flight rats than in hindlimb-unloaded rats (Goldstein *et al.*, 1992). CSA of papillary but not ventricular muscle decreased significantly in flight compared to synchronous groups, with values for hindlimb-unloaded rats falling between the two groups. Whether the changes resulted from unloading, fluid shifts, or both is unclear.

Neural

Motoneurons were studied to learn if they adapted similarly to muscle. The ventral horn cells of the lumbosacral enlargement of the spinal cord showed no significant differences in CSA or SDH activity, but the population distributions of both parameters shifted significantly. Flight animals showed a shift toward smaller cells with higher SDH activity, decreased RNA and acetyl cholinesterase, diminished nucleoli, and greater number of capillaries and alkaline phosphatase compared to hindlimb-unloaded rats and synchronous controls. In contrast, hindlimb-unloaded rats showed an increase in CSA with lower SDH, and lower cytochrome oxidase compared with the other two groups. The lack of similarity in results between hindlimb-unloaded and flight animals led Jiang *et al.* (1992) and Krasnov *et al.* (1993) to hypothesize that the changes in flight rat motoneurons may be caused by factors other than solely musculoskeletal unloading.

Kidney

Kidney weight at the end of flight was similar among groups (Table 4). Renal measurements in flight rats were not different from controls, whereas hindlimb-unloaded rats showed increases in water, sodium, potassium, and magnesium compared to controls (Lavrova *et al.*, 1993).

Liver

Hindlimb unloading did not mimic flight rats in liver function. Liver weight was lowest in hindlimb unloading compared to all other groups suggesting glycogen loss, but liver weight was not different between flight, vivarium, and synchronous controls (Table 4) (Racine and Cormier, 1992). Compared to synchronous controls, flight increased liver glycogen by 124% and decreased numbers of Kupffer cells by 12%, while hindlimb unloading decreased glycogen by 88% and increased Kupffer cells by 34% (Table 5). Liver cholesterol was highest in hindlimb-unloaded and lowest in flight animals while plasma insulin, glucose, and cholesterol were highest in flight and lowest in hindlimb-unloaded animals compared with synchronous controls (Table 5) (Macho *et al.*, 1991; Merrill *et al.*, 1992a and b). At the same time, tyrosine aminotransferase and tryptophane pyrolase increased and aspartate aminotransferase remained unchanged in the liver of flight and hindlimb-unloaded rats compared with controls (Macho *et al.*, 1991). These data suggest that the metabolic activity of the liver in flight and hindlimb-unloaded rats is very different and may be influenced by circulating levels of insulin and other hormones.

Intestine

Jejunal cell proliferation was unaffected by space flight or hindlimb unloading (Sawyer *et al.*, 1992). Jejunal villi were longer in vivarium rats than in any other group. These data suggest that neither flight nor hindlimb unloading caused a change in proliferation and migration of jejunal mucosal cells.

Endocrine and reproductive function

Dramatic differences in testicular mass were found following hindlimb unloading compared with flight (Table 4). The change in testes weight in flight rats was minimal vs. a 63% decrease in hindlimb-unloaded animals compared to synchronous controls. In flight rats, the small weight change was associated with a reduction of testicular canaliculi and spermatogenic epithelium, while spermatogenesis was normal (Amann *et al.*, 1992). In hindlimb-unloaded rats, the loss in mass was caused by destruction of testicular epithelium with atrophy and depletion of testicular canaliculi, accompanied by loss of spermatogenic cells (Amann *et al.*, 1992; Serova, 1996, 1998). In Cosmos 2044, plasma testosterone decreased in both the flight and hindlimb-unloaded rats compared to synchronous controls (Table 5), while testis testosterone decreased only in flight rats (Amann *et al.*, 1992) even though spermatogenesis was normal. Interestingly, testosterone levels in testes and plasma were highest in the vivarium control group (Amann *et al.*, 1992). Changes in testosterone synthesis by Leydig cells due to the displacement of testicles into the abdominal cavity may also contribute to the decreased plasma testosterone levels. In other hindlimb unloading experiments, changes were found in the testicular epithelium with a marked proliferation of Leydig cells and decreased testosterone production leading to a reduction of serum testosterone (Kaplanskii *et al.*, 2003). Deaver *et al.* (1992) recognized that hindlimb unloading is an adequate model for studying the reproductive capacity of male rats provided that the testicles are prevented from shifting to the abdominal cavity by partial constriction of the inguinal canal with sutures. Inhibition of spermatogenesis has been reported in hindlimb-unloaded rats even when cryptorchism is prevented by testes ligation (Tash *et al.*, 2002); in these experiments, Sertoli and Leydig cell appearance, testosterone, luteinizing hormone, and follicle-stimulating hormone levels, and epididymal and seminal vesicle weight were unchanged from controls.

Changes in hormonal regulation and/or metabolism may contribute to musculoskeletal alterations during unloading (Grigoriev *et al.*, 1993). Growth hormone and prolactin accumulation in the pituitary changed in both flight and hindlimb-unloaded animals. In flight and hindlimb-unloaded rats, plasma immunoreactive growth hormone decreased relative to controls (Grindeland *et al.*, 1994; Hymer *et al.*, 1992; Merrill *et al.*, 1992a). In flight rats, decreased growth hormone secretion correlated with depletion of growth hormone releasing factor (GRF) and a significant reduction in the number and signal intensity of cells expressing mRNA for GRF; indices of somatostatin did not change (Sawchenko *et al.*, 1992). Unlike the flight rats, hindlimb-unloaded rats did not differ from controls in these parameters, suggesting that mechanisms of growth hormone regulation may differ between space flight and hindlimb unloading. Growth hormone and prolactin plasma levels showed inconsistencies between groups suggesting differences not due to actual or simulated flight (Merrill *et al.*, 1992a). Thyroxin levels differed only between basal and synchronous control groups. Thyroid C cells were reduced in numbers and nuclear volume (Loginov, 1993). Plasma calcitonin levels were reduced in both flight and hindlimb-unloaded rats indicating alterations in calcitonin production and/or secretion or a failure to show the normal age dependent increase in circulating calcitonin (Arnaud *et al.*, 1992). In hindlimb-unloaded rats, plasma PTH did not change significantly (Arnaud *et al.*, 1992), although parathyroid cell activity increased (Loginov, 1997). Vasopressin and oxytocin synthesis in the hypothalamic nuclei was decreased and their accumulation in the posterior pituitary was diminished only in flight rats (Keil *et al.*, 1992). Melatonin in the pineal gland of the flight and hindlimb-unloaded rats remained unchanged while serotonin increased in the flight rats (Holley *et al.*, 1994).

Blood cells

After space flight or hindlimb unloading, the reticulocyte count decreased relative to controls; in hindlimb-unloaded rats, erythropoietin also decreased (Chelnaya, 1991; Lange *et al.*, 1994). These changes were indirect evidence of an inhibition of erythropoiesis, which may result from lowered oxygen requirements due to musculoskeletal unloading. Flight rats showed a significant increase in numbers of neutrophils and a decrease in numbers of lymphocytes, possibly due to the acute stress of reentry and recovery. Such changes were not detected in hindlimb-unloaded rats. In bone marrow of both flight and hindlimb-unloaded rats, changes were less distinct and limited to an insignificant decline in numbers of immature red blood cells (Chelnaya, 1991; Lange *et al.*, 1994). Both flight and hindlimb-unloaded rats showed a significant decrease of spleen mass compared with vivarium controls (Table 4), suggestive of an inhibition of erythropoiesis and lymphopoiesis (Durnova *et al.*, 1977). Interestingly, flight spleen weights were significantly lower than hindlimb unloaded spleen weights. Hindlimb unloading did not predict the hematological changes observed during flight.

Plasma biochemistry

Biochemical measurements differed between hindlimb-unloaded and flight animals. For instance, the flight rats showed an increase in glucose, insulin, cholesterol, phosphate, creatinine, blood urea nitrogen, lactate dehydrogenase, and the aspartate aminotransferase/glutamine oxalacetate transaminase/LDH ratio relative to controls (Macho *et al.*, 1991; Merrill *et al.*, 1992a). Hindlimb-unloaded rats did not demonstrate any of these differences. The increase in both insulin and glucose in the flight animals is an interesting paradox.

Immunology

The activity of bone marrow and spleen killer cells from flight rats towards YAC-1 target cells was significantly suppressed compared to controls, whereas the cytotoxicity of splenocytes from hindlimb-unloaded rats remained normal. However, the cytotoxicity towards K-562 target cells did not change in the flight rats and increased in hindlimb-unloaded rats (Rykova *et al.*, 1992). Bone marrow and spleen cells of both flight and hindlimb-unloaded rats showed weaker responses to colony-stimulating factor compared to controls. The numbers of T-helpers and T-suppressors in the spleen of the flight animals increased. Leukocyte distribution in the bone marrow of hindlimb-unloaded rats differed from that in the flight rats (Sonnenfeld *et al.*, 1992). Flight rats did not show any changes in inguinal lymph node lymphocyte proliferation and interleukin-2 production compared to synchronous, vivarium, or hindlimb-unloaded rats, while lymphocyte proliferation in hindlimb-unloaded animals was greater than all other groups (Nash *et al.*, 1992). These data suggest that hindlimb unloading produces some (particularly dynamic immune responses), but not all, of the changes in immune function caused by space flight.

Summary

Cosmos 2044 was the first study to directly compare hindlimb-unloaded and flight animals. Body, thymus, and kidney weight did not differ between hindlimb unloading and flight, while adrenal and spleen weight increased in hindlimb unloading and liver and testes weights decreased in hindlimb unloading compared to flight. Many bone responses were similar in hindlimb-unloaded and flight rats, while changes in spinal disks and motoneurons were only found in the flight group. The changes slow-twitch muscles were very similar between flight and hindlimb unloading. Musculoskeletal repair was similar between hindlimb unloading and flight, but differed from group controls. Atrophy of heart muscle was more apparent in flight than hindlimb unloading. Liver function was altered in flight animals as indicated by an increase in liver glycogen and plasma insulin, glucose, and cholesterol, while the same parameters decreased in hindlimb unloading. The testes in hindlimb unloading were much smaller than in flight rats, although both groups exhibited decreased plasma testosterone; spermatogenesis appeared normal in flight and significantly suppressed in hindlimb unloading. In immunology, hindlimb unloading modeled the effects of space flight on functional immune responses, but not on leukocyte subset distribution. Data from this mission showed that many, but not all, hinidlimb unloading responses were similar to those observed following space flight in rodents or humans.

Hindlimb unloading model modifications

Model modifications

The first unloaded rat model suspended the entire animal (Saiki *et al.*, 1976). This model system did not allow for ambulatory activity and provoked a stress response in the animals. The head-down hindlimb unloading model was first described in 1979 (Morey, 1979) and the device used for lifting the hindquarters progressed from elastic straps surrounding the torso, to a plastic back harness, to a tail-traction system (Wronski and Morey-Holton, 1987), which was found to minimize stress to the animal. Il'in and Novikov (1980) first suggested tail casting and Musacchia *et al.* (1980) proposed a modification to hindlimb unloading that allowed either horizontal or head-down unloading to define responses to fluid shifts. Jaspers *et al.* (1985) cast one limb of the animal in dorsiflexion to test whether chronic shortening of posterior leg muscles affected the metabolic response to hindlimb unloading. Stump *et al.* (1990) introduced a modification that included support of one hindlimb during hindlimb unloading to differentiate between metabolism in freely hanging versus supported muscles. Modifications and technique updates continually occur as investigators refine hindlimb unloading for their specific applications.

Technique update

We recently published a detailed description of the hindlimb unloading technique (Morey-Holton and Globus, 2002). A lingering problem for long duration studies has been the integrity of the tail harness. The traction tape may slip over time, and both rats and mice have a tendency to chew on the traction tape, which can lead to release from hindlimb unloading. Dr. Marjolein van der Meulen of Cornell University found a simple solution to minimize this problem. The modification involves using a plastic syringe barrel to cover and protect the base of the tail where traction tape is applied. A plastic syringe barrel is selected to fit over the traction tape without

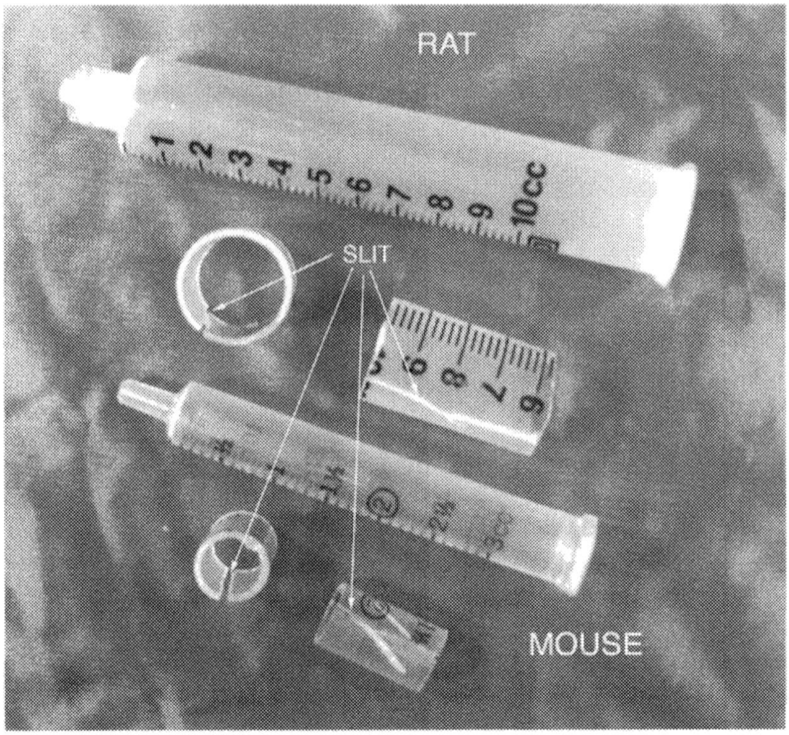

Fig. 2. A simple modification for hindlimb unloading using a tail guard for long term studies.

constricting the tail. Then, both ends of the syringe barrel are removed and the barrel is cut horizontally into 1-1.5″ cylinders (Fig. 2). The cylinders are slit longitudinally so that the barrel can be opened and slipped over the tail. The barrel is positioned over the traction tape at the proximal end of the tail distal to the anus and attached to the traction tape with small strips of medical paper tape to hold it in place. This minor modification minimizes the number of animals that slip out of the harness device and prevents removal of the tape by chewing. The modification was originally designed for mice and also works well for rats.

Summary

Hindlimb unloading continues to be modified as required for specific experimental paradigms. Hindlimb unloading in rodents has enabled increased understanding of the responses of the musculoskeletal, cardiovascular, immune, renal, neural, metabolic, and reproductive systems to unloading and/or to reloading on Earth, producing data with implications for long-duration human space flight.

Acknowledgments

The authors would like to thank Galina Tverskaya for her excellent translations and for her assistance in editing the text.

References

Adams, G.R., Caiozzo, V.J. and Baldwin, K.M. (2003) Skeletal muscle unweighting: Spaceflight and ground-based models. J. Appl. Physiol. 95, 2185–2201.

Amann, R.P., Deaver, D.R., Zirkin, B.R., Grills, G.S., Sapp, W.J., Veeramachaneni, D.N., Clemens, J.W., Banerjee, S.D., Folmer, J., Gruppi, C.M., Wolgemuth, D.J. and Williams, C.S. (1992) Effects of microgravity or simulated launch on testicular function in rats. J. Appl. Physiol. 73, 174S–185S.

Amblard, D., Lafage-Proust, M.H., Laib, A., Thomas, T., Ruegsegger, P., Alexandre, C. and Vico, L. (2003) Tail suspension induces bone loss in skeletally mature mice in the C57bl/6j strain but not in the C3h/Hej strain. J. Bone. Miner. Res. 18, 561–569.

Arnaud, S.B., Fung, P., Popova, I.A., Morey-Holton, E.R. and Grindeland, R.E. (1992) Circulating parathyroid hormone and calcitonin in rats after spaceflight. J. Appl. Physiol. 73, 169S–173S.

Baldwin, K.M. and Haddad, F. (2002) Skeletal muscle plasticity: Cellular and molecular responses to altered physical activity paradigms. Am. J. Phys. Med. Rehabil. 81, S40–S51.

Beamer, W.G., Donahue, L.R., Rosen, C.J. and Baylink, D.J. (1996) Genetic variability in adult bone density among inbred strains of mice. Bone. 18, 397–403.

Bikle, D.D., Sakata, T. and Halloran, B.P. (2003) The impact of skeletal unloading on bone formation. Gravit. Space. Biol. Bull. 16, 45–54.

Booth, F.W. and Grindeland, R.E. (1994) Comparison of the physiology of the spaceflight and hindlimb suspended rat. *In* Connolly, J.P., Grindeland, R.E. and Ballard, R.W. (eds.). *Final Reports of the U.S. Experiments Flown on the Soviet Biosatellite Cosmos 2044, Volume II: Experiments K-7-16-K-7-41*, pp. 275–281. NASA TM 108802, NASA Ames Research Center, Moffett Field.

Booth, F.W., Thomason, D.B., Morrison, P.R., Oganov, V.S., Ilyina-Kakueva, E.I., Smirnov, K.V., Wieisbrodt, N.W. and Lai, M. (1994) Part I: Some mRNA's decrease in skeletal muscle during spaceflight, Part II: mRNA levels in smooth muscle. *In* Connolly, J.P., Grindeland, R.E. and Ballard, R.W. (eds.). *Final Reports of the U.S. Experiments Flown on the Soviet Biosatellite Cosmos 2044, Volume I: Mission Description, Experiments K-7-01-K-7-15*, pp. 359–371. NASA TM 108802, NASA Ames Research Center, Moffett Field.

Brown, M., Fisher, J.S. and Salsich, G. (1999) Stiffness and muscle function with age and reduced muscle use. J. Orthop. Res. 17, 409–414.

Caren, L.D., Mandel, A.D. and Nunes, J.A. (1980) Effect of simulated weightlessness on the immune system in rats. Aviat. Space Environ. Med. 51, 251–255.

Chelnaya, N.A. (1991) Effect of space flight of varying duration upon rat blood and bone marrow. *Biosatellites Cosmos*, Abstracts, Leningrad, 12–15 August, pp. 126–128. Moscow.

Chen, K.D. and Alway, S.E. (2001) Clenbuterol reduces soleus muscle fatigue during disuse in aged rats. Muscle Nerve 24, 211–222.

Chi, M.M., Choksi, R., Nemeth, P., Krasnov, I., Ilyina-Kakueva, E., Manchester, J.K. and Lowry, O.H. (1992) Effects of microgravity and tail suspension on enzymes of individual soleus and tibialis anterior fibers. J. Appl. Physiol. 73, 66S–73S.

Cordonier, C.L., Stevens, L., Mounier, J. (1992) Compared contractile properties of soleus muscle fibers in real and simulated microgravity conditions. *In Book of Abstracts. World Space Congress: 43rd Congr. Inter. Astronaut. Fed. And 29th Plen. Meet. Comm. Space Res*, p. 537. Washington, D.C.

Corley, K., Kowalchuk, N. and McComas, A.J. (1984) Contrasting effects of suspension on hind limb muscles in the hamster. Exp. Neurol. 85, 30–40.

Criswell, D.S., Booth, F.W., DeMayo, F., Schwartz, R.J., Gordon, S.E. and Fiorotto, M.L. (1998) Overexpression of Igf-I in skeletal muscle of transgenic mice does not prevent unloading-induced atrophy. Am. J. Physiol. 275, E373–E379.

Da Silva, M.S., Zimmerman, P.M., Meguid, M.M., Nandi, J., Ohinata, K., Xu, Y., Chen, C., Tada, T. and Inui, A. (2002) Anorexia in space and possible etiologies: An overview. Nutrition 18, 805–813.

Deaver, D.R., Amann, R.P., Hammerstedt, R.H., Ball, R., Veeramachaneni, D.N. and Musacchia, X.J. (1992) Effects of caudal elevation on testicular function in rats. separation of effects on spermatogenesis and steroidogenesis. J. Androl. 13, 224–231.

Deavers, D.R., Musacchia, X.J. and Meininger, G.A. (1980) Model for antiorthostatic hypokinesia: head-down tilt effects on water and salt excretion. J. Appl. Physiol. 49, 576–582.

Dehority, W., Halloran, B.P., Bikle, D.D., Curren, T., Kostenuik, P.J., Wronski, T.J., Shen, Y., Rabkin, B., Bouraoui, A. and Morey-Holton, E. (1999) Bone and hormonal changes induced by skeletal unloading in the mature male rat. Am. J. Physiol. 276, E62–E69.

Deschenes, M.R. and Wilson, M.H. (2003) Age-related differences in synaptic plasticity following muscle unloading. J. Neurobiol. 57, 246–256.

Desplanches, D. (1997) Structural and Functional Adaptations of Skeletal Muscle to Weightlessness. Int. J. Sports Med. 18, S259–S264.

Durnova, G.N., Burkovskaia, T.E., Vorotnikova, E.V., Kaplanskii, A.S. and Arustamov, O.V. (1991) The effect of weightlessness on fracture healing of rats flown on the biosatellite Cosmos-2044. Kosm. Biol. Aviakosm. Med. 25, 29–33.

Durnova, G.N., Kaplanskii, A.S. and Portugalov, V.V. (1977) Changes in the lymphoid organs of rats during space flight. Arkh. Anat. Gistol. Embriol. 72, 14–20.

Elder, G.C. (1988) Effects of hind limb suspension on the development of dystrophic hamster muscle. Exp. Neurol. 99, 187–200.

Elder, G.C. and McComas, A.J. (1987) Development of rat muscle during short- and long-term hindlimb suspension. J. Appl. Physiol. 62, 1917–1923.

Fedotova, H.Y. (1991) Microgravity and hindlimb suspension effects on bone anzyme activites in normal rats and in healing bone fractures. *In Biosatellites Cosmos*, Abstracts, Leningrad, 12–15 August, p. 126. Moscow.

Fitts, R.H., Riley, D.R. and Widrick, J.J. (2000) Physiology of a microgravity environment invited review: microgravity and skeletal muscle. J. Appl. Physiol. 89, 823–839.

Giangregorio, L. and Blimkie, C.J. (2002) Skeletal adaptations to alterations in weight-bearing activity: a comparison of models of disuse osteoporosis. Sports Med. 32, 459–476.

Globus, R.K., Bikle, D.D. and Morey-Holton, E. (1986) The temporal response of bone to unloading. Endocrinology 118, 733–742.

Goldstein, M.A., Edwards, R.J. and Schroeter, J.P. (1992) Cardiac morphology after conditions of microgravity during Cosmos 2044. J. Appl. Physiol. 73, 94S–100S.

Grigoriev, A.I., Kaplansky, A.S. and Popova, I.A. (1993) Metabolic changes in weightlessness and mechanisms of their hormonal regulation. Aviakosm. Ekolog. Med. 27, 4.

Grindeland, R.E., Popova, I.A., Grossman, E. and Rudolph. (1994) Part I. Growth hormone regulation synthesis and secretion in microgravity. Part II: Plasma analysis hormone measurements. *In* Connolly, J.P., Grindeland, R.E., Ballard, R.W. (eds.). *Final Reports of the U.S. Experiments Flown on the Soviet Biosatellite Cosmos 2044, Volume II: Experiments K-7-16-K-7-41*, pp. 183–192, NASA TM 108802, NASA Ames Research Center, Moffett Field.

Hadley, J.A., Hall, J.C., O'Brien, A. and Ball, R. (1992) Effects of a simulated microgravity model on cell structure and function in rat testis and epididymis. J. Appl. Physiol. 72, 748–759.

Halloran, B.P., Ferguson, V.L., Simske, S.J., Burghardt, A., Venton, L.L. and Majumdar, S. (2002) Changes in bone structure and mass with advancing age in the male C57bl/6j mouse. J. Bone Miner. Res. 17, 1044–1050.

Hargens, A.R., Steskal, J., Johansson, C. and Tipton, C.M. (1984) Tissue fluid shift, forelimb loading, and tail tension in tail-suspended rats. Physiologist. 27, S37–S38.

Herrera, N.M. Jr., Zimmerman, A.N., Dykstra, D.D. and Thompson, L.V. (2001) Clenbuterol in the prevention of muscle atrophy: a study of hindlimb-unweighted rats. Arch. Phys. Med. Rehabil. 82, 930–934.

Holley, D.C., Asadi, H., Krasnov, I., Soliman, M.R.I. (1994) Pineal physiology in microgravity: Relation to rat gonadal function. *In* Connolly, J.P.,

Grindeland, R.E., Ballard, R.W., (eds.). *Final Reports of the U.S. Experiments Flown on the Soviet Biosatellite Cosmos 2044, Volume II: Experiments K-7-16-K-7-41*. pp. 183–192. NASA TM 108802, NASA Ames Research Center, Moffett Field.

Huckstorf, B.L., Slocum, G.R., Bain, J.L., Reiser, P.M., Sedlak, F.R., Wong-Riley, M.T. and Riley, D.A. (2000) Effects of HU on neuromuscular development of neonatal rats. Brain Res. Dev. Brain Res. 119, 169–178.

Hymer, W.C., Grindeland, R., Krasnov, I., Victorov, I., Motter, K., Mukherjee, P., Shellenberger, K. and Vasques, M. (1992) Effects of spaceflight on rat pituitary cell function. J. Appl. Physiol. 73, 151S–157S.

Il'in, E.A., Kaplanskii, A.S. and Savina, E.A. (1989) Experiments using rats on kosmos biosatellites: Morphologic and biochemical studies. Kosm. Biol. Aviakosm. Med. 23, 4–9.

Il'in, E.A. and Novikov, V.E. (1980) Stand for modelling the physiological effects of weightlessness in laboratory experiments with rats. Kosm. Biol. Aviakosm. Med. 14, 79–80.

Il'ina-Kakueva, E.I. and Burkovskaia, T.E. (1991) Effect of weightlessness on the course of the reparative process in the muscles of the biosatellite Kosmos-2044 rats. Kosm. Biol. Aviakosm. Med. 25, 24–29.

Il'ina-Kakueva, E.I. and Portugalov, V.V. (1981) Structural changes in the soleus muscle of rats on the kosmos-series biosatellites and in hypokinesia. Kosm. Biol. Aviakosm. Med. 15, 37–40.

Ilyina-Kakueva, E.I. and Portugalov, V.V. (1977) Combined effect of space flight and radiation on skeletal muscles of rats. Aviat. Space Environ. Med. 48, 115–119.

Ilyina-Kakueva, E.I., Portugalov, V.V. and Krivenkova, N.P. (1976) Space flight effects on the skeletal muscles of rats. Aviat. Space Environ. Med. 47, 700–703.

Jaspers, S.R., Fagan, J.M. and Tischler, M.E. (1985) Biochemical response to chronic shortening in unloaded soleus muscles. J. Appl. Physiol. 59, 1159–1163.

Jiang, B., Ohira, Y., Roy, R.R., Nguyen, Q., Ilyina-Kakueva, E.I., Oganov, V. and Edgerton, V.R. (1992a) Adaptation of fibers in fast-twitch muscles of rats to spaceflight and hindlimb suspension. J. Appl. Physiol. 73, 58S–65S.

Jiang, B., Roy, R.R., Polyakov, I.V., Krasnov, I.B. and Edgerton, V.R. (1992b) Ventral horn cell responses to spaceflight and hindlimb suspension. J. Appl. Physiol. 73, 107S–111S.

Jordan, J.P., Sykes, H.A., Crownover, J.C., Schatte, C.L., Simmons, J.B.2nd and Jordan, D.P. (1980) Simulated weightlessness: Effects of bioenergetic balance. Aviat. Space. Environ. Med. 51, 132–136.

Judex, S., Garman, R., Squire, M., Busa, B., Donahue, L.R. and Rubin, C. (2004) Genetically linked site-specificity of disuse osteoporosis. J. Bone. Miner. Res. 19, 607–613.

Kaplanskii, A.S., Alekseev, E.I. and Loginov, V.I. (2003) Morphofunctional changes in the endocrine system of male rats during microgravity and suspension. Aviakosm. Ekolog. Med. 37, 24–28.

Kaplanskii, A.S. and Savina, E.A. (1981) Morphological effects of weightlessness and their pathogenesis. Kosm. Biol. Aviakosm. Med. 15, 66–72.

Kaplansky, A.S., Durnova, G.N., Burkovskaya, T.E. and Vorotnikova, E.V. (1991) The effect of microgravity on bone fracture healing in rats flown on Cosmos-2044. Physiologist. 34, S196–S199.

Kawano, S., Ohmori, S., Kanda, K., Ito, T., Murata, Y. and Seo, H. (1994) Adrenocortical response to tail-suspension in young and old rats. Environ. Med. 38, 7–12.

Kawano, S., Ohmori, S., Kambe, F., Kanda, K., Murata, Y. and Seo, H. (1995) Catecholamine response to stress: Age related modifications in tail-suspended rats. Environ. Med. 39, 107–111.

Keil, L., Evans, J., Grindeland, R. and Krasnov, I. (1992) Pituitary oxytocin and vasopressin content of rats flown on Cosmos 2044. J. Appl. Physiol. 73, 166S–168S.

Krasnov, I.B. (1994) Gravitational neuromorphology. Adv. Space Biol. Med. 4, 85–110.

Krasnov, I.B., Poliakov, I.V. and Drobyshev, V.I. (1993) Effect of space flight and head-down position on motoneuron-glia-capillary system of the rat bone marrow. Aviakosm. Ekolog. Med. 27, 38–42.

Kulkarni, A.D., Yamauchi, K., Hales, N.W., Ramesh, V., Ramesh, G.T., Sundaresan, A., Andrassy, R.J. and Pellis, N.R. (2002) Nutrition beyond nutrition: plausibility of immunotrophic nutrition for space travel. Clin. Nutr. 21, 231–238.

Lange, R.D., Gibson, L.A., Driscoll, T.B., Servona, T., Mishurova, T., Vacek, A., Ichiki, A.T. and Khrushchov, N.G. (1994) Hemotology studies in rats flown on the soviet biosatellite Cosmos 2044. *In* Connolly, J.P., Grindeland, R.E., Ballard, R.W. (eds.). *Final Reports of the U.S. Experiments Flown on the Soviet Biosatellite Cosmos 2044, Volume I: Experiments K-7-01-K-7-15*, pp. 491–510. NASA TM 108802, NASA Ames Research Center, Moffett Field.

Lavrova, E.A., Natochin, I.V., Serova, L.V. and Snetkova, E.V. (1993) Body fluid and electrolyte content in rat tissues after space flight in the "Space-2044" spacecraft. Aviakosm. Ekolog. Med. 27, 43–47.

Loginov, V.I. (1993) an Immunocytochemical study of the c-cell function of the thyroid in rats exposed on the Kosmos-2044 biosatellite. Aviakosm. Ekolog. Med. 27, 71–76.

Loginov, V.I. (1997) Immunocytochemical and Histomorphometrical Study of Calcitonin Producing Cells of the Thyroid Gland in Rats Exposed in Weightlessness and Ground-Based Experiments, Institute for Biomedical Problems, Moscow.

Machida, S. and Booth, F.W. (2004) Regrowth of skeletal muscle atrophied from inactivity. Med. Sci. Sports Exerc. 36, 52–59.

Macho, L., Fickova, M., Nemeth, S., Svabova, E., Serova, L. and Popova, I. (1991) The effect of space flight on the board of the satellite Cosmos 2044 on plasma hormone levels and liver enzyme activities of rats. Acta. Astronaut. 24, 329–332.

Martin, E.A., Ritman, E.L. and Turner, R.T. (2003) Time Course of Epiphyseal Growth Plate Fusion in Rat Tibiae. Bone 32, 261–267.

Maurel, D., Ixart, G., Barbanel, G., Mekaouche, M. and Assenmacher (1996) I. Effects of acute tilt from orthostatic to head-down antiorthostatic restraint and of sustained restraint on the intra-cerebroventricular pressure in rats. Brain Res. 736, 165–173.

Merrill, A.H., Jr., Wang, E., LaRocque, R., Mullins, R.E., Morgan, E.T., Hargrove, J.L., Bonkovsky, H.L. and Popova, I.A. (1992a) Differences in glycogen, lipids, and enzymes in livers from rats flown on Cosmos 2044. J. Appl. Physiol. 73, 142S–147S.

Merrill, A.H., Jr., Wang, E., Mullins, R.E., Grindeland, R.E. and Popova, I.A. (1992b) Analyses of plasma for metabolic and hormonal changes in rats flown aboard Cosmos 2044. J. Appl. Physiol. 73, 132S–135S.

Miller, E.S., Bates, R.A., Koebel, D.A. and Sonnenfeld, G. (1999) Antiorthostatic suspension stimulates profiles of macrophage activation in mice. Neuroimmunomodulation 6, 160–167.

Montufar-Solis, D., Duke, P.J. and Durnova, G. (1992) Spaceflight and age affect tibial epiphyseal growth plate histomorphometry. J. Appl. Physiol. 73, 19S–25S.

Montufar-Solis, D., Duke, P.J. and Morey-Holton, E. (2001) The spacelab 3 simulation: Basis for a model of growth plate response in microgravity in the rat. J. Gravit. Physiol. 8, 67–76.

Morey, E.R. (1979) Spaceflight and bone turnover: Correlation with a new rat model of weightlessness. BioScience 29, 168–172.

Morey-Holton, E.R., Garetto, L.P., Doty, S.B., Halloran, B.P. and Turner, R.T. (2003) Is animal age a factor in the response of bone to spaceflight? J. Bone Miner. Res. 18, S219.

Morey-Holton, E.R. and Globus, R.K. (1998) HU of growing rats: A model for predicting skeletal changes during space flight. Bone 22, 83S–88S.

Morey-Holton, E.R. and Globus, R.K. (2002) HU rodent model: Technical aspects. J. Appl. Physiol. 92, 1367–1377.

Morey-Holton, E.R., Halloran, B.P., Garetto, L.P. and Doty, S.B. (2000) Animal housing influences the response of bone to spaceflight in juvenile rats. J. Appl. Physiol. 88, 1303–1309.

Mueller, P.J., Cunningham, J.T., Patel, K.P. and Hasser, E.M. (2003) Proposed role of the paraventricular nucleus in cardiovascular deconditioning. Acta. Physiol. Scand. 177, 27–35.

Musacchia, X.J. (1985) The use of suspension models and comparison with true weightlessness: "A resume". Physiologist. 28, S237–S240.

Musacchia, X.J. (1992) An assessment of suspension systems: Models that reproduce responses to weightlessness. Physiologist. 35, S92–S95.

Musacchia, X.J., Deavers, D.R., Meininger, G.A. and Davis, T.P. (1980) A model for hypokinesia: Effects on muscle atrophy in the rat. J. Appl. Physiol. 48, 479–486.

Musacchia, X.J. and Fagette, S. (1997) Weightlessness simulations for cardiovascular and muscle systems: Validity of rat models. J. Gravit. Physiol. 4, 49–59.

Musacchia, X.J., Steffen, J.M., Fell, R.D., Dombrowski, M.J., Oganov, V.W. and Ilyina-Kakueva, E.I. (1992) Skeletal muscle atrophy in response to 14 days of weightlessness: Vastus medialis. J. Appl. Physiol. 73, 44S–50S.

Nash, P.V., Konstantinova, I.V., Fuchs, B.B., Rakhmilevich, A.L., Lesnyak, A.T. and Mastro, A.M. (1992) Effect of spaceflight on lymphocyte proliferation and interleukin-2 production. J. Appl. Physiol. 73, 186S–190S.

Ohira, Y., Jiang, B., Roy, R.R., Oganov, V., Ilyina-Kakueva, E., Marini, J.F. and Edgerton, V.R. (1992) Rat soleus muscle fiber responses to 14 days of spaceflight and hindlimb suspension. J. Appl. Physiol. 73, 51S–57S.

Ohira, Y., Kawano, F., Wang, X.D., Sudoh, M. and Ishihara, A. (2003) Changes of bone morphology in response to hindlimb suspension of rats. Biol. Sci. Space. 17, 225–226.

Ohira, Y., Tanaka, T., Yoshinaga, T., Kawano, F., Nomura, T., Nonaka, I., Allen, D.L., Roy, R.R. and Edgerton, V.R. (2001) Ontogenetic, gravity-dependent development of rat soleus muscle. Am. J. Physiol. Cell. Physiol. 280, C1008–C1016.

Ojala, B.E., Page, L.A., Moore, M.A. and Thompson, L.V. (2001) Effects of inactivity on glycolytic capacity of single skeletal muscle fibers in adult and aged rats. Biol. Res. Nurs. 3, 88–95.

Oganov, V.S., Murashko, L.M., Kabitskaya, O.E., Szilagyi, T. and Rapcsak, M. (1991) Physiological characteristics of rat skeletal muscles after the flight on board "Cosmos-2044" biosatellite. Physiologist. 34, S174–S176.

Pecaut, M.J., Nelson, G.A., Peters, L.L., Kostenuik, P.J., Bateman, T.A., Morony, S., Stodieck, L.S., Lacey, D.L., Simske, S.J. and Gridley, D.S. (2003) Genetic models in applied physiology: Selected contribution: Effects of spaceflight on immunity in the C57bl/6 mouse. I. Immune Population Distributions. J. Appl. Physiol. 94, 2085–2094.

Pedrini-Mille, A., Maynard, J.A., Durnova, G.N., Kaplansky, A.S., Pedrini, V.A., Chung, C.B. and Fedler-Troester, J. (1992) Effects of microgravity on the composition of the intervertebral disk. J. Appl. Physiol. 73, 26S–32S.

Racine, R.N. and Cornier, S.M. (1992) Effect of spaceflight on rat hepatocytes: a morphometric study. J. Appl. Physiol. 73, 136S–141S.

Riley, D.A. (1999) Is Skeletal muscle ready for long-term spaceflight and return to gravity? Adv. Space Biol. Med. 7, 31–48.

Riley, D.A., Ellis, S., Giometti, C.S., Hoh, J.F., Ilyina-Kakueva, E.I., Oganov, V.S., Slocum, G.R., Bain, J.L. and Sedlak, F.R. (1992) Muscle sarcomere lesions and thrombosis after spaceflight and suspension unloading. J. Appl. Physiol. 73, 33S–43S.

Rivera, C.A., Tcharmtchi, M.H., Mendoza, L. and Smith, C.W. (2003) Endotoxemia and hepatic injury in a rodent model of HU. J. Appl. Physiol. 95, 1656–1663.

Rykova, M.P., Sonnenfeld, G., Lesnyak, A.T., Taylor, G.R., Meshkov, D.O., Mandel, A.D., Medvedev, A.E., Berry, W.D., Fuchs, B.B. and Konstantinova, I.V. (1992) Effect of spaceflight on natural killer cell activity. J. Appl. Physiol. 73, 196S–200S.

Saiki, H., Nakaya, M., Sugita, Y. and Kamachi, M. (1976) Metabolic and hormonal mechanisms of mineral metabolic adaptation to induced hypokinetics in rats. Aviat. Space Environ. Med. 47, 846–852.

Sandmann, M.E., Shoeman, J.A. and Thompson, L.V. (1998) The fiber-type-specific effect of inactivity and intermittent weight-bearing on the gastrocnemius muscle of 30-month-old rats. Arch. Phys. Med. Rehabil. 79, 658–662.

Sawchenko, P.E., Arias, C., Krasnov, I., Grindeland, R.E. and Vale, W. (1992) Effects of spaceflight on hypothalamic peptide systems controlling pituitary growth hormone dynamics. J. Appl. Physiol. 73, 158S–165S.

Sawyer, H.R., Moeller, C.L., Phillips, R.W. and Smirnov, K.L. (1992) Proliferation of jejunal mucosal cells in rats flown in space. J. Appl. Physiol. 73, 148S–150S.

Serova, L. (1996) Weighltessness and Adaptive Capacities of Animals, Institute for Biomodeical Problems, Moscow.

Serova, L. (1998) Simulation models of weightlessness in mammalian's developmental program. J. Gravit. Physiol. 5, P127–P128.

Simmons, D.J., Grazman, B., Russell, J.E., Walker, W.V., Bikle, D.D. and Morey, E.R. (1983) Simulating certain aspects of hypogravity: effects on bone maturation in the non-weight bearing skeleton. Aviat. Space Environ. Med. 54, 1080–1084.

Simmons, D.J., Russell, J.E., Winter, F., Tran Van, P., Vignery, A., Baron, R., Rosenberg, G.D. and Walker, W.V. (1983) Effect of spaceflight on the non-weight-bearing bones of rat skeleton. Am. J. Physiol. 244, R319–R326.

Simske, S.J., Luttges, M.W., Allen, K.A. and Gayles, E.C. (1994) The role of sex and genotype on antiorthostatic suspension effects on the mouse peripheral skeleton. Aviat. Space Environ. Med. 65, 123–133.

Simske, S.J., Luttges, M.W. and Wachtel, H. (1990) Age dependent development of osteopenia in the long bones of tail-suspended mice. Biomed. Sci. Instrum. 26, 87–94.

Sonnenfeld, G. (2003) Animal models for the study of the effects of spaceflight on the immune system. Adv. Space Res. 32, 1473–1476.

Sonnenfeld, G., Mandel, A.D., Konstantinova, I.V., Berry, W.D., Taylor, G.R., Lesnyak, A.T., Fuchs, B.B. and Rakhmilevich, A.L. (1992) Spaceflight alters immune cell function and distribution. J. Appl. Physiol. 73, 191S–195S.

Stauber, W.T., Fritz, V.K., Burkovskaya, T.E. and Ilyina-Kakueva, E.I. (1992) Effect of spaceflight on the extracellular matrix of skeletal muscle after a crush injury. J. Appl. Physiol. 73, 74S–81S.

Stelzer, J.E. and Widrick, J.J. (2003) Effect of hindlimb suspension on the functional properties of slow and fast soleus fibers from three strains of mice. J. Appl. Physiol. 95, 2425–2433.

Stump, C.S., Overton, J.M. and Tipton, C.M. (1990) Influence of single hindlimb support during simulated weightlessness in the rat. J. Appl. Physiol. 68, 627–634.
Stump, C.S., Tipton, C.M. and Henriksen, E.J. (1997) Muscle adaptations to hindlimb suspension in mature and old fischer 344 rats. J. Appl. Physiol. 82, 1875–1881.
Sudoh, M., Waki, H., Kawakami, K., Sakai, O. and Saiki, H. (1994) Changes of blood volume distribution by postural changes in rats. J. Gravit. Physiol. 1, P139–P140.
Tash, J.S., Johnson, D.C. and Enders, G.C. (2002) Long-term (6-Wk) hindlimb suspension inhibits spermatogenesis in adult male rats. J. Appl. Physiol. 92, 1191–1198.
Thomason, D.B., Morrison, P.R., Oganov, V., Ilyina-Kakueva, E., Booth, F.W. and Baldwin, K.M. (1992) Altered actin and myosin expression in muscle during exposure to microgravity. J. Appl. Physiol. 73, 90S–93S.
Tidball, J.G. and Spencer, M.J. (2002) Expression of a calpastatin transgene slows muscle wasting and obviates changes in myosin isoform expression during murine muscle disuse. J. Physiol. 545, 819–828.
Tipton, C.M. (1996) Animal models and their importance to human physiological responses in microgravity. Med. Sci. Sports Exerc. 28, S94–S100.
Tipton, C.M., Overton, J.M., Joyner, M.J. and Hargens, A.R. (1987) Local fluid shifts in humans and rats: Comparison of simulation models with actual weightlessness. Physiologist. 30, S117–S120.
Tou, J.C., Grindeland, R.E. and Wade, C.E. (2004) Effects of diet and exposure to hindlimb suspension on estrous cycling in sprague-dawley rats. Am. J. Physiol. Endocrinol. Metab. 286, E425–E433.
Vailas, A.C., Vanderby, R., Jr., Graf, B., Sears, J., Ashman, R., Grindeland, R., Martinez, D.A., Thielke, R., Ulm, M., Kolis, S., Griffith, P., Choy, A., Kaplansky, A., Durova, G. and Krasnov, I.B. (1994) Biomechanical, biochemical and morphological alterations of intramuscular and dense fibrous connective tissues after 14 days in spaceflight. Part I: Connective Tissue Studies. *In* Connolly, J.P., Grindeland, R.E. and Ballard, R.W. (eds.). *Final Reports of the U.S. Experiments Flown on the Soviet Biosatellite Cosmos 2044, Volume I: Experiments K-7-01-K-7-15*, pp. 105–152. NASA TM 108802, NASA Ames Research Center, Moffett Field.
Vailas, A.C., Vanderby, R., Jr., Martinez, D.A., Ashman, R.B., Ulm, M.J., Grindeland, R.E., Durnova, G.N. and Kaplansky, A. (1992) Adaptations of young adult rat cortical bone to 14 days of spaceflight. J. Appl. Physiol. 73, 4S–9S.
Vico, L., Lafage-Proust, M.H. and Alexandre, C. (1998) Effects of gravitational changes on the bone system in vitro and in vivo. Bone. 22, 95S–100S.
Walton, K. (1998) Postnatal Development under conditions of simulated weightlessness and space flight. Brain Res. Brain Res. Rev. 28, 25–34.
Wilkerson, M.K., Colleran, P.N. and Delp, M.D. (2002) Acute and chronic head-down tail suspension diminishes cerebral perfusion in rats. Am. J. Physiol. Heart Circ. Physiol. 282, H328–H334.

Wronski, T.J. and Morey-Holton, E.R. (1987) Skeletal response to simulated weightlessness: a comparison of suspension techniques. Aviat. Space Environ. Med. 58, 63–68.

Yamauchi, K., Hales, N.W., Robinson, S.M., Niehoff, M.L., Ramesh, V., Pellis, N.R. and Kulkarni, A.D. (2002) Dietary nucleotides prevent decrease in cellular immunity in ground-based microgravity analog. J. Appl. Physiol. 93, 161–166.

Zhang, L.F. (1994) Experimental studies on effects of simulated weightlessness on myocardial function and structure. J. Gravit. Physiol. 1, 133–136.

Experimentation with Animal Models in Space
G. Sonnenfeld (editor)

DOI: 10.1016/S1569-2574(05)10003-3

International Collaboration on Russian Spacecraft and the Case for Free Flyer Biosatellites

Richard E. Grindeland,[1] Eugene A. Ilyin,[2] Daniel C. Holley[3] and Michael G. Skidmore[1]

[1]*Life Sciences Division, NASA – Ames Research Center, Moffett Field, 94035, USA*
[2]*Russian State Research Center – Institute of Biomedical Problems, Russian Academy of Sciences, Moscow, Russia*
[3]*Department of Biological Sciences, San Jose State University, 1 Washington Square, San Jose, CA, 95192, USA*

Abstract

Animal research has been critical to the initiation and progress of space exploration. Animals were the original explorers of "space" two centuries ago and have played a crucial role by demonstrating that the space environment, with precautions, is compatible with human survival. Studies of mammals have yielded much of our knowledge of space physiology. As spaceflights to other planets are anticipated, animal research will continue to be essential to further reveal space physiology and to enable the longer missions. Much of the physiology data collected from space was obtained from the Cosmos (Bion) spaceflights, a series of Russian (Soviet)-International collaborative flights, over a 22 year period, which employed unmanned, free flyer biosatellites. Begun as a Soviet-only program, after the second flight the Russians invited American and other foreign scientists to participate. This program filled the 10 year hiatus between the last US biosatellite and the first animal experiments on the shuttles. Of the 11 flights in the Cosmos program nine of them were international; the flights continued over the years regardless of political differences between the Soviet Union and the Western world. The science evolved from sharing tissues to joint international planning and development, and from rat postmortem tissue analysis to *in vivo* measurements of a host of monkey physiological parameters during flight. Many types of biological specimens were carried on the modified Vostok spacecraft, but only the mammalian studies are discussed herein. The types of studies done encompass the full range of physiology and have begun to answer "critical" questions of space physiology posed by various *ad hoc* committees. The studies have not only yielded a prodigious and significant body of data, they have also introduced some new perspectives in physiology. A number of the physiological insights gained are relevant to physiology on Earth. The Cosmos flights also added significantly to flight-related technology, some of which also has application on our planet. In summary, the Cosmos biosatellite flights were extremely productive and of low cost. The Bion vehicles are versatile in that they can be placed into a variety of orbits and altitudes, and can carry radiation sources or other hazardous material which cannot be carried on manned vehicles. With recent advances in sensor, robotic, and data processing technology, future free flyers will be even more productive, and will largely preclude the need to fly animal experiments on manned vehicles. Currently, mammalian researchers do not have access to space for an unknown time, seriously impeding the advancement and understanding of space physiology during long duration missions. Initiation of a new, international program of free flyer biosatellites is critical to our further understanding of space physiology, and essential to continued human exploration of space.

Introduction

From perusal of this volume and related literature, it is evident that our ability to perform human space missions and our understanding of space physiology depend on animal experiments in space. A majority of published space flight physiological data was garnered from animal experiments on-board unmanned (free flyer) Soviet/Russian biosatellites. Eleven animal research flights took place over nearly a quarter century under the aegis of the Cosmos program (Ilyin, 2000). This program has yielded a wealth of science data and provided a space platform for technological development. Many of the Cosmos-related publications report original observations of some aspects of space physiology, while others have verified observations from both humans and animals in space and have provided insights into the physiological mechanisms underlying those observations. At the termination of the Cosmos program in 1997, before all studies had been published, more than one half of all American space biology studies and more than 200 American biomedical publications had accrued from the Cosmos program experiments. The numbers of Cosmos flight studies performed and publications produced by scientists from other participating countries are undoubtedly even higher. Well over 51 peer-reviewed publications by American and Russian scientists resulted from the Cosmos 2044 flight alone. The Cosmos flights also yielded data that were not obtainable from animal studies carried on other space platforms—Apollo, Skylab 3, or shuttle (e.g., most monkey physiology data). Acquiring human physiological data, comparable to those obtained from some of the Cosmos studies, would have required invasive procedures on human subjects, an ethical and logistical impossibility. Collectively, the numbers of studies and publications and the types of studies possible on biosatellites offer cogent evidence for the importance of animal research in space, the significance of the Cosmos program, the value of free flyer biosatellites, and the critical need to reinitiate a free flyer biosatellite program.

Significance of the cosmos program

The Cosmos program has played a major and pioneering role in the understanding of space physiology and has contributed materially to the development of space technology. First, during the early years of the Cosmos program, the United States (US) and other Western countries had no other significant access to the space environment for animal experiments. Second, results from the Cosmos program provide answers to some of the "critical questions" regarding microgravity physiology which have been posed by a number of *ad hoc* committees over the years (Assembly of Mathematical and Physical Science, 1979; Committee on Space Biology and Medicine, 1987, 1991, 1998, 2000; Task Group on Life Sciences, 1988; Aerospace Medicine Advisory Committee, 1992). Physiological data obtained from humans during flight are often provocative and of clinical significance, but usually do not yield scientifically valid data due to the limited numbers of subjects, variable crew regimens in flight which can obscure the significance of the measurements, and the limitations on the measurements that can be made on humans. The large numbers of animals that can be flown on Cosmos flights and with better controlled dietary regimen, activity–rest cycles, and environmental conditions impart scientific rigor to the observations. More than just expanding our knowledge of space physiology, data from these flights changed some of our physiological concepts; a few of these will be considered later in the review of science findings below. A third, important aspect of international participation in the Cosmos program was the ongoing training and development of a cadre of space life scientists who continued to participate in the Cosmos program and the shuttle (space transportation system: STS) life sciences program as they matured. Fourth, a productive and cordial international collaboration developed in Space Life Sciences Research that hopefully will continue into the era of extended human exploration of space and other planets. Fifth, the free flyer biosatellite has been, and will continue to be, the least expensive and most versatile means of gaining access to space for mammalian research purposes. Sixth, broad-based, working-level scientific contact, such as that which occurred in the Cosmos program, can be a critical element in fostering and maintaining congenial relations between nations of different political philosophies.

Objectives and limitations

This chapter has several objectives. We present a thumbnail sketch of the history of ballistic and orbital mammalian space flights to emphasize the importance of animal research in (1) safely sending humans into space, (2) understanding space physiology, and (3) to put the Cosmos flights into the context of other biological space flights. We also briefly describe the Cosmos program, development of the international collaboration, and the Vostok spacecraft and its payloads. Selected science findings and technology developments resulting from the Cosmos program are summarized and the significance of some studies discussed. Last, we present the case that to meet the objectives of the National Aeronautics and Space Administration's (NASA's) space exploration initiatives it is essential that a new international program of free flyer biosatellite flights be developed. A new free flyer program would most certainly further human exploration of space by enabling the identification and development of countermeasures against the deleterious effects of space flight, providing a low risk platform for rapid technological developments, and elucidating the physiology of microgravity from a fundamental biology perspective. As will be seen, these studies could have important ramifications and benefits for humans on Earth as well as in space.

The original and preferred term for the series of international biological space flights conducted on free flyers is Cosmos. However, for the last (eleventh) flight in the series, and the only one in which Russia and the US had a contractual agreement, a Cosmos number was not assigned so the name Bion (11) was used.

Although the Cosmos vehicles have carried a broad range of biospecimens, which have yielded interesting data, in this chapter we restrict our consideration of science results to mammalian studies. To do this we have largely, but not exclusively, used English language publications as primary sources and English language translations of foreign language original papers or reviews.

Animal balloon and early rocket flights

The first recorded "space flight" of animals was in 1783 by the Montgolfier brothers who sent up a hot air balloon that carried a rooster, a duck, and a sheep to 1500 feet (457 m) (Callahan, 1966; Tipton, 2003a). All animals survived the flight in apparently good condition, thereby stimulating later animal and human balloon flights. Some 170 years later, a series of high altitude balloon flights was used to determine the effects of cosmic radiation on animals, to study the physical environment of space, and to develop equipment needed for animal and human excursions into space (Henry, 1962; Peyton, 1968).

The first animal ventures into space, as usually conceived, employed ballistic rocket flights (Table 1). In 1948, the US began ballistic flights of animals which continued until 1967, whereas the Soviet scientists initiated life sciences rocket flights in 1951 which ended in 1960. The Soviets flew many dogs on ballistic flights over this period and 15 dogs participated in multiple flights. A number of these flights failed, but sufficient physiological data were obtained to convincingly demonstrate that with appropriate precautions the space environment is not antithetical to life (Peyton, 1968). Inspection of Table 1 shows that

Table 1

List of rodent, canine, and non-human primate ballistic* and orbital flights 1948–2003

Species (n)	Misson/Country	Launch Date	Altitude (km)	Duration
Rhesus monkey	V2 Rocket/US	6/11/48	*to 62	
Rhesus monkey	Blossom 3/US	6/18/48	*to 62	
Rhesus monkey	Blossom 4/US	10/48	*to 134	
Cynomologous monkey	Blossom 5/US	5/49	*N/A	
Cynomologous monkey	Blossom 6/US	5/49	*N/A	
Rhesus monkey	V2 Rocket/US	6/14/49	*to 134	
Cynomologous monkey	V2 Rocket/US	9/16/49	*to 11	
Cynomologous monkey	V2 Rocket/US	12/8/49	*to 61	
Mouse	Blossom 7/US	7/50	*N/A	
Mouse	V2 Rocket/US	10/31/50	*to 136	
Cebus monkey; mice (11) (2 per flight)	Aerobee Rocket/US	4/18/51	*to 61	
Cebus monkey; mice (11)	Aerobee Rocket/US	9/20/51	*72	
Dogs (2 per flight)	Rocket R-2A, USSR	1951 6 launches	*100–110	
Cebus monkey (2); mice (2)	Aerobee Rocket/US	5/21/52	*to 62	
Dogs (2 per flight)	Rocket R-2A, USSR	1954 3 launches	*100–110	
Dogs (2 per flight)	Rocket R-2A, USSR	1955 3 launches	*100–110	
Dogs (2 per flight)	Rocket R-2A, USSR	1956 3 launches	*100–110	
Dog (Laika)	Earth's Artificial Satellite, (ISZ-2)/USSR	11/3/57	167–225	
Dogs (2 per flight)	Rocket R-2, USSR	1957 5 launches	*212	
Mouse	Thor-Able/US	4/23/58	*N/A	20 min
Mouse	Thor-Able/US	7/9/58	*N/A	20 min
Mouse	Thor-Able/US	7/23/58	*N/A	20 min
Squirrel monkey	Jupiter IRBM/US	12/13/58	*N/A	
Dogs (2 per flight)	Rocket R-2, USSR	1958 2 launches	*212	
Dogs (2 per flight)	Rocket R-5, USSR	1958 3 launches	*450–473	
Rhesus monkey; squirrel monkey	Jupiter IRBM/US	5/28/59	*N/A	
Mice	Discoverer III/US	6/3/59	*N/A	
Mice (12)	Jupiter/US	9/16/59	*N/A	

Continued

Table 1

Continued

Species (n)	Misson/Country	Launch Date	Altitude (km)	Duration
Rhesus monkey	Hermes 1/US	12/4/59	*to 88	13 min
Dogs (2 per flight)	Rocket R-2, USSR	1959 2 launches	*212	
Rhesus monkey	Hermes 2/US	1/21/60	*to 12	8 min
Dogs (2); white rats (2)	Spaceship-satellite-2	8/19/60	306–339	27 h
Black C57 mice (21); white mice (19)	(KKS-2)/USSR			
Black C57 mice (3)	Atlas/US	10/13/60	*N/A	
Dogs (2) white mice (5)	Spaceship-satellite-3	12/1/60	187–265	27 h
white Rats (2); guinea-pigs (2); black C57, mice (14); CBA/C57 hybrid black mice (7)	(KKS-3)/USSR			
Dogs (2 per flight)	Rocket R-2, USSR	1960 2 launches	*212	
Chimpanzee	Redstone-Mercury2/US	1/31/61	*to 253	18 min
Dog	Spaceship-satellite-4 (KKS-4)/USSR	3/9/61	184–249	1 h 45 min
Dog	Spaceship-satellite-5 (KKS-5)/USSR	3/25/61	179–248	1 h 45 min
Chimpanzee	Atlas-Mercury 5/US	11/29/61	159–235	3 h 20 min
Dogs (2)	Spaceship Cosmos 110/USSR	2/22/66	187–904	22 d
White rat	Aerobee Rocket/US	12/5/67	*N/A	460 sec
Pig tailed monkey	Biosatellite III/US	6/28/69	Orbital	8.8 d
Pocket mice (6)	Apollo 17/US	12/7/72	Orbital/moon	12.0 d
Pocket mice (5)	Skylab 3/US	7/28/73	Orbital	59 d
Rats	**Bion 1/Cosmos605/USSR**	**10/31/73**	**221–424**	**21.5 d**
Rats	**Bion 2/Cosmos690/USSR**	**10/22/74**	**223–389**	**20.5 d**
Rats, wistar (25)	**Bion 3/Cosmos782/USSR**	**11/25/75**	**226–405**	**19.5 d**
Rats, wistar (30)	**Bion 4/Cosmos936/USSR**	**8/3/77**	**224–419**	**18.5 d**
Rats, wistar (37)	**Bion 5/Cosmos1129/USSR**	**9/25/79**	**226–406**	**18.5 d**
Rhesus monkeys (2); rats, wistar (10)	**Bion 6/Cosmos1514/USSR**	**12/14/83**	**226–288**	**5.0 d**
Rats, wistar (6)	STS-8, SSIP/US	8/30/83	Orbit to 354	**6 d**
Rats, wistar (6)	STS-41B, SSIP/US	2/3/84	Orbit to 350	**8 d**
Squirrel monkeys (2); rats (24)	STS-51B, Spacelab3/US	4/29/85	Orbit to 411	**7 d**
Rhesus monkey	**Bion 7/Cosmos1667/USSR**	**7/10/85**	**211–270**	**7.0 d**
Rhesus monkeys (2); rats, wistar (10)	**Bion 8/Cosmso1887/USSR**	**9/29/87**	**222–403**	**12.5 d**

Rats, Long-Evans (4)	STS-29, SSIP/US	3/13/89	Orbit to 341	5 d
Rhesus monkeys (2); rats (10)	**Bion 9/Cosmso2044/USSR**	**9/15/89**	**216–294**	**14 d**
Rats (8)	STS-41, PSE-1/US	10/6/90	Orbit to 296	4 d
Rats, SD (29)	STS-40, SLS-1/US	6/5/91	Orbit to 291	9 d
Rats (8)	STS-48, PARE.01/US	9/12/91	Orbit to 580	5 d
Rats (12)	STS-52, PSE-2	10/22/92	Orbit to 302	10 d
Rhesus monkeys (2)	**Bion 10/Cosmos2229USSR**	**12/29/92**	**226–397**	**11.5 d**
Rats, SD (6)	STS-54, PARE.02/US	1/13/93	Orbit to 306	6 d
Rats, SD (16)	STS-56, PARE.03/US	4/8/93	Orbit to 296	9 d
Rats, fischer 344 (12)	STS-57, PSE-3/US	6/21/93	Orbit to 467	10 d
Rats, SD (48)	STS-58, SLS-2/US	10/18/93	Orbit to 287	14 d
Rats, SD (12)	STS-60, IMMUNE.01/US	2/3/94	Orbit to 354	8 d
Rats, fischer 344 (12)	STS-62, PSE-4/US	3/4/94	Orbit to 302	14 d
Rats, SD (10)	STS-66, NIH.R-1/US	11/3/94	Orbit to 304	11 d
Rats, SD (12)	STS-63, IMMUNE.02/US	2/3/95	Orbit to 394	8 d
Rats, SD (10)	STS-70, NIH.R-2/US	7/13/95	Orbit to 296	9 d
Rats (36)	STS-72, NIH.R-3/US	1/11/96	Orbit to 463	9 d
Rats (12)	STS-77, IMMUNE.03/US	5/19/96	Orbit to 283	10 d
Rats (12)	STS-78, LMS/US	6/20/96	Orbit to 320	17 d
Rats (14)	STS-80, NIH.R-4/US	11/19/96	Orbit to 404	17.5 d
Rhesus monkeys (2)	**Bion 11/Russia**	**12/4/96**	**225–401**	**14 d**
Rats (152); mice (18)	STS-90, Neurolab/US	4/17/98	214–248	16 d
Mice (12)	STS-108	12/5/01		12 d
Rats	STS-107	1/16/03		16 d

Abbreviations are N/A = not available, SD = sprague-dawley, STS = shuttle flight.

the Soviets began orbital flights of animals with the flight of the female dog, Laika, in 1957. With the advent of orbital flights, the space age had truly begun. After the Laika flight, the Soviets flew five additional orbital animal flights before beginning the Cosmos program with the launch of Cosmos 605 in October 1973.

American–Soviet/Russian Cosmos collaboration

The American Biosatellite program ended in 1969. Thereafter, the only access to space for American biology experiments was on the Apollo 17 or Skylab 3 flights for mammals that did not require an independent life support system (e.g., pocket mice). When these flights were completed, there was no access to space for Western biologists. Although shuttles began to fly in 1981, they did not carry non-human mammals until 1983, 10 years after the flight of Skylab 3. During this hiatus, the only space platform that could support mammalian experiments was the Cosmos vehicle. In 1971, the Americans and Soviets formally established a collaboration in space exploration with the signing of the US/USSR Science and Applications Agreement. The following year, the "Intergovernmental Agreement on Cooperation in the Exploration and Use of Outer Space for Peaceful Purposes" was entered into, which led to the 1975 Apollo-Soyuz test project and created a joint working group in space biology and medicine (Edwards, 2000). This latter group provided a context for the US and other Western countries to participate in the Cosmos biosatellite program beginning with the flight of Cosmos 782 in 1975. Cosmos 605 and Cosmos 690 had included scientists only from the Soviet Union and Soviet affiliated countries. With the issuing of the historic invitation to participate in Cosmos 782 by the Soviet Union the Cosmos program bridged the East–West political chasm and became a truly international program.

Various aspects of the Cosmos program have been published, so our comments on the history of the Cosmos program are essentially limited to the evolution of the program (Ilyin, 1983, 2000; Souza, 1996; Edwards, 2000). In the nine flights of the Cosmos series, in which the US was involved, the complexity of US experiments evolved from passive experiments or postflight analysis of shared biological samples to complex inflight physiological measurements, which reached their ultimate development on Bion 11 (Ilyin, 1983) (Table 2). Similarly, the equipment, instrumentation, and procedures developed by scientists and engineers from various countries resulted in increasingly complex inflight monitoring equipment and data systems, more demanding procedures pre-, in-, and postflight, and additional control studies. In later monkey flights, the numbers and types of 1 ***g*** (Earth's normal gravity) controls were increased to include the flight monkeys in simulation of the flight using a flight-type capsule. The more extensive controls greatly facilitated interpretation of flight data. The use of monkeys required extensive training, testing, and careful selection of the candidate animals for flight. Involvement of international scientists

Table 2

Cosmos flight characteristics

Bio-Satellite	Bion 1 Cosmos -605	Bion 2 Cosmos -690	Bion 3 Cosmos -782	Bion 4 Cosmos -936	Bion 5 Cosmos -1129	Bion 6 Cosmos -1514	Bion 7 Cosmos -1667	Bion 8 Cosmos -1887	Bion 9 Cosmos -2044	Bion 10 Cosmos -2229	Bion 11
Launch date	Oct 31, 1973	Oct 22, 1974	Nov 25, 1975	Aug 3, 1977	Sept 25, 1979	Dec 14, 1983	July 10, 1985	Sept 29, 1987	Sept 15, 1989	Dec 29, 1992	Dec 24, 1996
Landing date	Nov 22, 1973	Nov 12, 1974	Dec 15, 1975	Aug 22, 1977	Oct 14, 1979	Dec 19, 1983	July 17, 1985	Oct 12, 1987	Sept 29, 1989	Jan 10, 1993	Jan 7, 1997
Flight duration (days)	21.5	20.5	19.5	18.5	18.5	5	7	12.5	14	11	13.7
Apogee (km)	424	389	405	419	406	288	297	406	294	396.8	401
Perigee (km)	221	223	227	224	226	226	222	224	216	226	225
Orbit inclination (deg)	62.8	62.8	62.8	62.8	62.8	82.3	82.3	62.8	82.3	62.8	62.8
Orbit duration (min)	90.7	90.4	90.5	90.7	90.5	89.3	89	90.5	89.3	90.4	90.5

Adapted from Ilyin, E.A. J. Grav. Physiol. 7(1):51–8, 2000.

progressed from passive guest investigators to active participants in the planning, development and execution of all phases of the experiments, truly collaborative ventures.

Cosmos vehicle

In line with the Soviet/Russian philosophy of reusing and incrementally developing complex systems, the Cosmos spacecraft is a direct descendant of the Vostok vehicle used for the world's first manned space flight on April 12, 1961. Yuri Gagarin's flight required a full environmental support system (maintaining temperature, humidity, pressure, and atmospheric composition) and a human-rated reentry/recovery system. The Cosmos vehicle retains these capabilities and is thus, able to support experimental specimens ranging from cells in culture, plants, aquatic forms to mammalian species including non-human primates. Cosmos vehicles 782, 936, and 1129 carried centrifuges for exposing various specimens to artificial gravity. The ability to monitor biological and other parameters of interest during space flight has also evolved over time. During the Bion 11 space flight, more than 45 parameters were recorded that included relevant environmental measurements as well as monkey muscle contractile force, head motion, electromyograph (EMG), electrocardiograph (ECG), electroencephalograph (EEG), skin, deep body, and brain temperatures, and firing of afferent vestibular neurons.

The payload capsule of the Cosmos biosatellite consists of a 2.5 m diameter sphere with attached battery pack and support modules. The total weight of the entire assemblage at launch is 6300 kg including the internal and external payload masses of 700 kg and 200 kg, respectively. When launched from the Plesetsk Cosmodrome, the modified "Soyuz" launch vehicle used for Cosmos placed the vehicle in a 62.8° inclination orbit between 200 and 400 km altitude. When placed into an 82.3° inclination orbit, the perigee is the same but the apogee is about 300 km. Characteristics of the Cosmos vehicle are summarized in Table 3.

Recovery and parachute landing of Cosmos/Bion vehicles took place up to 22 days later in Kazakhstan near the Caspian Sea. The maximal forces encountered during flight were 5 g linear acceleration during ascent and a transient exposure up to 16 ***g*** (10 to 15 s) during the ballistic reentry. In order to control the landing impact forces, the Cosmos capsule has a "soft landing" system. This system senses ground proximity and fires a small rocket attached to the parachute shroud lines to slow the capsule descent. The differences from the space shuttle reentry ***g*** loads are due mainly to the different atmospheric effects of reentry in a sphere (ballistic) versus that of a controlled lifting body. On-orbit microgravity values in the range of 10^{-3} to 10^{-5} ***g*** are typical for Cosmos. Previous joint American/Russian developmental work has confirmed that, in addition to other smaller research hardware, up to 8 animal enclosure modules (AEM) or an equivalent number of middeck locker payloads can be

Table 3

Cosmos biosatellite characteristics

Acceleration at launch	Not to exceed 4 *g*
Deceleration at de-orbiting	Not to exceed 8 *g* for no longer than 60 s
Impact acceleration	Up to 90 *g* for no longer than 40 ms (Bion 1–5) Soft landing (Bion 6–11)
Maximum flight duration	30 days
Microgravity level	10^{-6} to 10^{-4} *g*
Payload	700 kg inside the recovery module 200 kg outside the recovery module
Placement of biocontainers into the module	2 days before launch Some containers can be installed As late as 8 h before launch
Environmental parameters inside the recovery module:	
Total barometric pressure	730–790 mm Hg
pO_2	140–180 mm Hg
PCO_2	Not to exceed 7 mm Hg
Temperature	18–28°C
Relative Humidity	30–85%

From Ilyin, E.A. J. Grav. Physiol. 7(11):S1–S8, 2000.

accommodated on a Cosmos biosatellite. Table 4 shows the types of specimens that have been flown aboard the Cosmos biosatellite over the years.

Significant findings of selected Cosmos experiments

A number of responses to space flight are discussed in detail in other chapters of this volume. Those discussions cover both Russian and American space flights employing various vehicles and Earth-based simulations. In contrast, the present discussion is limited to findings from the Cosmos flight series and is meant to give a panoramic, rather than an exhaustive view of research in particular areas and to touch on the significance of some of those results (Souza and Ilyin, 1996; Ilyin *et al.*, 2000).

Muscle

Orbital flights of dogs prior to the initiation of the Cosmos program provided the first direct evidence of skeletal muscle wasting in space (Portugalov *et al.*, 1976). Early Cosmos flights of rats revealed that muscle atrophy occurred in this species (Gazenko *et al.*, 1978). These studies also disclosed the loss of motor endplates in the soleus muscle (Ilyina-Kakueva *et al.*, 1976) and changes in microcirculation beds of skeletal muscles (Kaplansky, 1978; Riley *et al.*, 1992). Later studies found motor endplate loss in the adductor longus (Riley *et al.*,

Table 4

Cosmos biospecimens

Bio-Satellite	Bion 1 Cosmos -605	Bion 2 Cosmos -690	Bion 3 Cosmos -782	Bion 4 Cosmos -936	Bion 5 Cosmos -1129	Bion 6 Cosmos -1514	Bion 7 Cosmos -1667	Bion 8 Cosmos -1887	Bion 9 Cosmos -2044	Bion 10 Cosmos -2229	Bion 11
Wistar rats (n)	45	35	25	30	37	10	10	10	10		
Rhesus monkeys (n)						2	2	2	2	2	2
Reptiles (turtles)	+										
Fish (guppy)						+		+			
Amphibians							+	+	+	+	+
Insects	+	+	+	+	+		+	+	+	+	+
Worms								+			
Bacteria	+										
Unicellular Organisms			+		+				+	+	+
Fundulus fish eggs			+								
Japanese quail eggs					+			+			
Animal and plant cultured cells and tissues			+		+			+	+	+	
Plants, seeds, and seedlings	+	+	+	+	+	+	+	+	+	+	+

+ = number unknown

From Ilyin, E.A. J. Grav. Physiol. 7(1):S1–S8, 2000.

1990; Hyde *et al.*, 1992). The nine Cosmos flights carrying rodents included a wide variety of studies on muscles of differing predominant fiber type composition and functions. The extent of atrophy as judged by weight, muscle fiber cross-sectional area, and protein content varied with the muscle, the greatest atrophy typically being shown by the slow extensors, lesser loss by fast extensors, and the least wasting by fast flexor muscles (Castleman *et al.*, 1978; Baldwin *et al.*, 1990; Miu *et al.*, 1990; Musacchia *et al.*, 1992). On Cosmos 936, five rats were kept on a centrifuge (32 cm radius, operated at a force of 1 ***g***) for the duration of flight; in these animals the soleus muscle mass, but not oxidative metabolism was maintained (Ilyina-Kakueva and Portugulov, 1979; Gurovsky *et al.*, 1980; Kotovskaya *et al.*, 1980).

In flight rats, cross-sectional areas of cardiac left ventricular and papillary muscles were 22% smaller than those of control rats (Philpott *et al.*, 1990; Goldstein *et al.*, 1992) and the protein content of jejunal smooth muscle was decreased, indicating that cardiac and smooth muscles also atrophy in space (Mednieks *et al.*, 1989; Baranska *et al.*, 1990; Thomason *et al.*, 1992; Weisbrodt *et al.*, 1994). Protein synthesis, especially of contractile proteins, was decreased in both skeletal and cardiac muscles (Thomason *et al.*, 1992). Skeletal muscle histological studies also disclosed evidence of increased protein catabolism (Riley *et al.*, 1990, 1992). Conversion of slow to fast twitch or to hybrid fibers occurred in flight (Ilyina-Kakueva *et al.*, 1976). That microgravity can affect muscle metabolism was demonstrated by the increased or decreased activities of 16 enzymes in single fibers of soleus and tibialis anterior muscles (Manchester *et al.*, 1990; Chi *et al.*, 1992).

Several rodent models of microgravity exposure have been developed and are commonly used for Earth simulation studies. The Cosmos 2044 flight was the first to include a ground control group of rats exposed to the Morey (1979) model of space flight utilizing hindlimb unloading (thus facilitating comparison of the effects of space flight and hindlimb unloading). Two generalizations seem to apply to skeletal muscle responses in the two "microgravity" systems (Ohira *et al.*, 1992), with one exception, given muscles showed similar degrees of atrophy in space flight and during hindlimb unloading; the adductor longus (AL) showed edema, probably as a result of injury incurred during reentry from space, masking the weight loss due to atrophy, whereas the hindlimb-unloaded rat AL showed a significantly decreased weight and fiber cross-sectional area (Jiang *et al.*, 1992a). The conversion of slow fibers to hybrid or fast fiber type was appreciably slower in the hindlimb-unloaded rat (Jiang *et al.*, 1992a).

Rat gastrocnemius muscles subjected to crush injury before flight showed a "disorganized" and probably nonfunctional pattern of repair and slower healing than 1 ***g*** control animals (Ilyina-Kakueva and Burkovskaya, 1991; Stauber *et al.*, 1992). Similarly, injured soleus muscle responded to microgravity with disorganized and slower healing (Ilyina-Kakueva and Burkovskaya, 1991). The Cosmos 2044 flight provided the first indication that repair of injured muscle in microgravity may result in scar formation and possibly irreversible muscle loss. Rats subjected to hindlimb unloading healed better than flight animals, but not as well as ambulatory control rats.

Similar to dogs and rats, the skeletal muscles of rhesus monkeys responded to microgravity by atrophy of the four leg muscles biopsied: soleus, gastrocnemius, tibialis anterior, and vastus lateralis—and one of the arm muscles biopsied, the triceps (Bodine-Fowler *et al.*, 1992; Carnino *et al.*, 2000; Chopard *et al.*, 2000; Mayet-Sornay *et al.*, 2000; Mounier *et al.*, 2000; Roy *et al.*, 2000a; Shenkman *et al.*, 2000). The muscles showed significant loss of strength due to loss of mass and/or to decreased specific tension (Fitts *et al.*, 2000a). Conversion of slow to hybrid or fast twitch fibers occurred in both arm and leg muscles. Fitts *et al.*, (2000b) reported increased contractile velocity of soleus fibers after flight but were unable to find any fast myosin isoforms in those fibers. Despite the 30% decrease in cross-sectional areas of soleus muscles, the number of myonuclei per millimeter fiber length was unchanged after flight, thus yielding significantly smaller myonuclear domains without evidence of apoptosis (Roy *et al.*, 2000b). The myotendinous junctions of soleus muscle fibers exhibited atrophic changes as a result of space flight (Carnino *et al.*, 2000).

Neuromuscular studies

A sophisticated system for evaluating monkey neuromuscular function pre-, in-, and postflight was developed for the Cosmos flights. A combination of indwelling EMG electrodes in four leg and two arm muscles, a force buckle on a medial gastrocnemius tendon to determine the force exerted, and video taping capability enabled the simultaneous measurement of the various parameters. This system was developed by J.P. Connolly and colleagues at NASA-Ames Research Center, Dr. V.R. Edgerton and colleagues, University of California, Los Angeles and E. Konigsberg, Konigsberg Instruments, Pasadena, CA. The force buckle was the major new development in this technology, but coupling the force measurement with the assessment of muscle recruitment as determined by intramuscular EMG recordings and video taping of a specific motor task performance offered a composite picture of neuromuscular function. Muscle biopsies were taken before and after space flight or 1 ***g*** simulations. Single fibers from the biopsies were analyzed by fiber type histmorphometrically, biochemically, and physiologically. Data from the foregoing tests *vis-à-vis* muscle biopsy analysis yielded an unprecedented synopsis of muscle status.

This system is now being used in spinal cord injury patients for both diagnostic and rehabilitation purposes (Edgerton *et al.*, 2000a). In addition to the adaptation of space flight neuromuscular monitoring systems for the treatment of spinal cord injury (SCI) patients, the Cosmos studies established certain concepts and developed procedures that are also being used to treat SCI patients (Edgerton *et al.*, 2000b). The notion that weight bearing is essential to normal neuromuscular function is being integrated into treatment of SCI patients. In addition, other hardware and software programs originally developed for flight are being utilized in SCI patient therapy.

Exposure to microgravity resulted in altered muscle control (Edgerton *et al.*, 2000a, 2000b; Falempin *et al.*, 2000; Hodgson *et al.*, 2000). An important finding in monkeys was that the motor system "learns" to perform tasks differently in space. After 2–6 days in microgravity, two synergistic extensors (soleus and gastrocnemius) were activated out of normal (1 ***g***) sequence. Significant plasticity was found in the neuronal strategies to select specific groups of motor neurons to perform simple motor tasks (i.e., foot pedal pressing) in microgravity and ambulation postflight. Bion 11 monkeys were taught to press a foot lever in response to a signal and were rewarded with a treat (juice) if they responded appropriately. Unfortunately, the animals did not perform this task very often which was thought to be due to the "floating" animals being unable to press on the foot lever because the restraint system was designed to loosen during flight. Twenty four hours postflight the monkeys demonstrated significant problems with equilibrium while trying to locomote quadrupedally (Recktenwald *et al.*, 2000). The leg muscles in microgravity show oscillatory movement (clonus), a condition similar to that seen in Parkinson Disease, emphasizing the importance of muscle sensory information arising from opposing gravity. After two weeks in space, flight-related neuromuscular changes returned to preflight control level in two to three weeks.

Data gleaned from the Cosmos muscle studies on both rats and monkeys are consistent with biopsy and functional neuromuscular tests performed on humans in space or after landing. These results have not only provided significant insight into muscle responses to the space environment but have also verified, as expected, the relevance of animal studies to human muscle responses in space.

Skeletal system

In Cosmos rat experiments an overall reduced rate of bone growth was found and there appeared to be an actual cessation of growth after 11 days in flight (Yagodovsky *et al.*, 1976; Morey and Baylink, 1978; Popova and Tigranyan, 1981; Cann and Adachi, 1983; Kaplansky *et al.*, 1990). In long bones of the hind legs, periosteal growth was inhibited whereas endosteal bone resorption was not significantly affected (Cann and Adachi, 1983; Wronski and Morey, 1983). Consistent with reduced bone growth were decreases in growth plate thickness (Simmons *et al.*, 1983; Duke *et al.*, 1990; Simmons *et al.*, 1990; Montufar-Solis *et al.*, 1992) rarefaction of the femur cancellous bone (Jee *et al.*, 1983) and reduced bone mineralization. Osteoblast precursor cells did not mature to osteoblasts in space whereas after re-exposure to 1 ***g*** a rapid activation of the osteoblastic cell line ensued (Garetto *et al.*, 1990, 1992). Long bone strength was decreased in younger rats after three weeks in space but not in older rats in space for about two weeks, suggesting that both the bone turnover rate and flight duration are important considerations in terms of bone strength. On Cosmos 936 one group of rats was centrifuged for the duration of the flight as described above. Artificial gravity did not restore periosteal growth in flight but it did appear to counter the decrease in femur strength and to expedite recovery of bone mass and strength over 25 days after flight (Morey and Baylink, 1978; Gurovsky *et al.*, 1980). Vertebral (T7) compression strength was significantly reduced in microgravity (Kazarian, 1981, 1983; Doty *et al.*, 1990; Zernicke *et al.*, 1990; Eurell and Kazarian, 1983; Foldes *et al.*, 1996). Humerus strength was decreased in these rats but the reduction was attributed to changes in bone geometry, not material properties (Vailas *et al.*, 1990, 1992). The calcium-regulating hormones, parathyroid hormone and calcitonin, had decreased concentrations in plasma after flight (Arnaud *et al.*, 1992).

A high-resolution method to localize mineral in cross-sections of rat bones revealed not only lower total mineral content but also a different distribution within a given bone relative to 1 ***g*** controls (Mechanic *et al.*, 1990). The regional nature of the adaptive changes in bone to microgravity exposure has made researchers more aware of the highly localized nature of bone responses to biomechanical stresses or lack of stresses. The changes in mineral content of whole bones seem trivial compared to the marked decreases found in the bone strength of younger rats. This apparent disparity may be a reflection of the within-bone variation in mineral concentration.

The fibulas of space flight rats were cut just prior to flight and the extent of repair was studied immediately after flight (Kaplansky *et al.*, 1991). The healing process of the long bone was markedly inhibited, resulting in a greatly reduced callus size. An increase in the relative osteoid volume of callus trabeculae and a simultaneous decrease in the number of active osteoblasts point to disorders of bone repair in microgravity.

Formation and mineralization of iliac crest bone in space flight monkeys were 35% less than in 1 ***g*** monkeys (Zerath *et al.*, 1991, 2000). Other studies showed that the iliac crest and long bones of the leg developed similar qualitative changes but that the long bone changes were appreciably greater (Kaplansky and Durnova, 1997). This relationship of long bone to iliac crest responses suggests that the long bones sustained significant losses in flight. Electron microscopy (Rodionova *et al.*, 2000) and densitometric studies (Oganov *et al.*, 2000) of monkey bone samples from all six non-human primate flights yielded data consistent with light microscopy studies of the iliac crest biopsies described above. Monkey plasma concentrations of the calcium regulatory hormones, parathyroid hormone, calcitonin, and 1,25-dihydroxy-vitamin D3, were all suppressed at the end of a 14 day flight (Arnaud *et al.*, 2000). Further evidence of connective tissue responses to space were shown by the increased excretion of collagen cross-links (Martinez *et al.*, 2000).

In view of the apparent disparity between bone mineral density and bone strength in space flight rats, a means of measuring bone strength *in vivo* in space flight humans and monkeys is potentially very important. To that end,

Dr. Donald Young and colleagues, NASA-Ames Research Center and Dr. Charles Steele, Stanford University developed a device (Mechanical Response Tissue Analyzer; MRTA) to measure bone bending stiffness, as an index of bone strength in experimental monkeys, rather than inferring strength from indirect measurements such as X-ray densitometry or bone mineral concentration (Young *et al.*, 1979, 1983). Dr. Sara Arnaud, NASA-Ames Research Center, Dr. Steele and the GaitScan Company, have continued the development and testing of the MRTA in humans and monkeys (Hutchinson *et al.*, 2001). The *in vivo* and *in vitro* strengths of a given bone were shown to yield excellent correspondence. Although mineral density and bone strength are usually thought to parallel each other, space flight bones have clearly shown that a dichotomy can exist, and that direct strength measurements are needed. The MRTA has been extensively tested on humans on Earth but has been used thus far only for space flight monkeys. It would appear to have a great value for assessment of bone strength of humans during long duration space flights.

The patellar, but not the Achilles, tendons showed significant reductions in collagen, DNA, and mature collagen cross-link concentrations following 12 days in space (Doty *et al.*, 1992). In rat hindlimb unloading studies, similar changes in tendons and ligaments were found, but there was also a decreased adhesion of the ligament or tendon to the bone.

A particularly noteworthy finding was that proteoglycans in lumbar intervertebral discs were replaced with collagen, resulting in a loss of water content and cushioning ability of the discs (Pedrini-Mille *et al.*, 1992). Whether the collagen could or would be replaced with proteoglycans after return to Earth is unknown and remains an important question.

Together, these findings emphasize how susceptible bone is to effects of microgravity, particularly bone strength, the difficulty injured bone has in repairing properly in space, and the importance of measuring bone strength directly in space flight animals or humans.

Performance, neural, and neurosensory studies

The Psychomotor Test System (PTS) was developed by Drs. Duane Rumbaugh, David Washburn and colleagues, Georgia State University for use in the Rhesus Project at the NASA Ames Research Center, as a forerunner to use in space flights of monkeys. Further development of the PTS was done in conjunction with the Cosmos-Bion 11 space flights in collaboration with Russian scientists at the Institute of Biomedical Problems, Moscow (Rumbaugh *et al.*, 1989; Washburn *et al.*, 2000). The development of the PTS was to provide monkeys with an enriched environment and to test their psychomotor performance before and after space flight or exposure to models simulating some conditions of space flight at 1 ***g***. It was planned to also use the PTS system during the flight of Bion 12. Monkeys were trained to use a PTS at 1 ***g***. The monkeys readily learned to use the system and would use it voluntarily without food reward.

Before and after the Bion 11 flight and 1 *g* simulations, the animals were tested on the PTS. One monkey died 36 h after landing during anesthesia for tissue biopsies; the surviving animal exhibited significantly reduced accuracy and productivity (Washburn *et al.*, 2000). A lesser reduction in performance was seen after 1 *g* simulation in a flight capsule. The death of one monkey and the cancellation of the planned companion flight, Bion 12, compromised the possibility of a rigorous conclusion, but the limited data are certainly suggestive and important. These results argue compellingly, we believe, for more PTS type testing in flight. Inflight tests to evaluate eye–hand coordination revealed reduced precision in the monkeys' hand movements early in flight. By the end of the flight, the precision improved but the tasks took longer (Antsiferova *et al.*, 2000; Shlyk *et al.*, 2000). The PTS is now being used successfully to teach and to test children with learning problems. Some of these children have difficulty with language and the PTS eliminates the need for them to communicate verbally.

Electron and light microscopic studies of rat brains after returning from space flights ranging in length from 5 to 22 days (seven flights) revealed alterations in structures of nerve cell bodies and processes, number and structure of interneuronal contacts, glial cells, and internuclear and intercortical connections (Belichenko *et al.*, 1990; Krasnov and Dyachkova, 1990; Krasnov *et al.*, 1992; Krasnov, 1994). The histological changes observed appeared to correspond to the duration of flight. In rats flown on the Cosmos 2044 mission, motor neurons in the lumbosacral enlargement showed a shift to smaller soma sizes, consistent with the reduced motor activity; succinate dehydrogenase tended to increase in the flight motor neurons (Jiang *et al.*, 1992b). However, neurons from hindlimb-unloaded rats, used to simulate space flight conditions, did not show these changes.

Interestingly, histomorphometric measurements of lumbar dorsal root ganglia of space flight rats offered clear evidence of decreased activity, primarily proprioceptive input (Polyakov *et al.*, 1991). This observation, along with the reduced levels of growth hormone regulatory factors found in the hypothalamus after space flight, led to the suggestion of proprioception as a regulator of bioassayable growth hormone secretion (Sawchenko *et al.*, 1992). Subsequent ground-based and human flight studies verified that relationship (McCall *et al.*, 1997, 1999; Gosselink *et al.*, 1998; Edgerton *et al.*, 2000b) (see Endocrine section). This finding demonstrates an important connection between skeletal muscle activity or inactivity, and one of the mechanisms mediating its effect on metabolism.

In early Cosmos rat flights it was found that neurosecretory cells of the hypothalamic paraventricular nuclei had lowered activity and that the Herring bodies were also reduced (Savina and Alekseev, 1979). These findings are consonant with the later measurement of vasopressin and oxytocin in the posterior pituitary, in which there were marked reductions in both of these hormones (Keil *et al.*, 1989, 1992).

In rats there was little change in neurotransmitters, their receptors, or metabolism in the hippocampus (Lowry *et al.*, 1994). Several areas of rat brain were surveyed for GABA (benzodiazepine) and muscarinic cholinergic receptors that are thought to be involved in motor function. The striatum showed a significant decrease in muscarinic receptors after space flight, but none of the other areas investigated revealed an effect on either type of receptor (Hyde *et al.*, 1992).

Cardiovascular studies did not show any substantive differences in cerebral blood flow or pressure between space flown and control monkeys (Krotov and Nosovsky, 2000). Overall, pO_2 in the brains of monkeys was also unaffected by space flight. These observations certainly suggest that any neurophysiological or neuroendocrine changes observed in space flight animals or humans were not attributable to inadequate cerebral blood flow or oxygen supply.

Space motion sickness has been a besetting problem of cosmonauts and astronauts, especially for short flights, and has been studied extensively in humans (Kozlovskaya *et al.*, 1989). For various reasons, the results of human investigations have not been particularly fruitful (Kozlovskaya *et al.*, 1989; Sirota *et al.*, 1992). Although much remains to be investigated, animal investigations have contributed materially to our understanding of vestibular function in space. Monkey studies, in which scleral search coils have been placed on the eyes and electrodes placed in medial vestibular nucleus, the vestibular afferent nerves, and the cerebellar flocculus to record electrical responses, have been particularly informative (Cohen *et al.*, 1992; Correia *et al.*, 1992a, b, 1994; Sirota *et al.*, 1992; Tomko *et al.*, 1997; Badakva and Miller, 2000; Kozlovskaya *et al.*, 2000).

Collectively the studies have shown that: (1) exposure to microgravity was followed by an almost immediate increase in canal sensitivity or motor system sensitivity to canal stimulation (the horizontal vestibulo-ocular gain increased up to 1.5–1.8 in the first few days of flight in all eight monkeys studied in the Cosmos flights); (2) the increase in gain was closely linked in time to the enhanced responses of the vestibular nuclei neurons to canal stimulation; (3) the coefficient of the vestibulo-ocular reaction (cVOR) gradually decreased during adaptation to microgravity, starting after 2 to 5 days in space and approaching 1.0 (normal for 1 ***g***) at the end of flight on days 10 to 14; and (4) the adaptational suppression of cVOR seemed to be related to activities of the vestibulo-cerebellar loops because, throughout the space flights, activity of vestibulo-cerebellar neurons remained greatly increased (Kozlovskaya *et al.*, 1989, 1990, 2000).

Cardiovascular system

After two weeks Cosmos missions, as noted in the discussion of rat muscle, rats showed decreased fiber cross-sectional areas of both left ventricular and papillary muscles and a mRNA profile consistent with reduced synthesis of cardiac contractile proteins (Baranski *et al.*, 1980; Philpott *et al.*, 1990; Goldstein *et al.*, 1992; Thomason *et al.*, 1992).

Early studies on postflight rats showed altered differential blood cell counts but a more recent investigation did not demonstrate any significant change in the cell counts nor in circulating levels of erythropoietin (Lange *et al.*, 1994). However, studies of monkey blood revealed decreased hemoglobin, hematocrit, and reticulocytes (Burkovskaya and Korolkov, 2000). The reason for the differing responses in red blood cell (RBC) masses of rats and monkeys is unknown, but may be a species phenomenon. It is unknown, but of importance, if the RBC population continues to decrease with a lengthened duration of flight. The RBC mass of humans in space was reported to decrease, particularly after long duration missions, but the mechanisms leading to the reduced RBC are still unclear. The half-life of rat RBCs was decreased by about 20% after three weeks space flights (Leon *et al.*, 1978, 1980). Interestingly, rats subjected to centrifugation during flight did not show a reduced half-life of their RBCs. There are reports of reduced RBC production, suggesting that both a shortened half-life and diminished formation of RBCs may have contributed to the decreased mass. One of the changes found in livers of space flight rats was a decrease in D-levulinic acid synthase, the rate-limiting enzyme in heme synthesis (Merrill *et al.*, 1990, 1992a). Whether the reduced heme synthesis resulted in decreased corpuscular hemoglobin or contributed to a decreased RBC production is unknown.

In monkeys, the volumes of total body water and of the several fluid compartments, including plasma volume, were all reduced in response to space flight (Lobachik *et al.*, 2000). Dehydration is a universal finding after flight (Ushakov *et al.*, 1980; Pitts *et al.*, 1983; Ilyin, 2000).

The facial puffiness experienced by cosmonauts/astronauts in space has been ascribed to a headward movement of blood in the absence of gravity, resulting in an increased central venous pressure. Cardiovascular studies in monkeys, and a human measurement on a shuttle flight, have not supported that hypothesis (Korolkov *et al.*, 1992; Sandler *et al.*, 1986, 1994). Central venous pressure, central blood volume, cardiac output, arterial pressure, and ECG of flight monkeys did not differ appreciably from those measurements in 1 ***g*** control animals. As mentioned, cerebral blood flow and pressure and pO_2 did not differ significantly from control values (Krotov and Nosovsky, 2000).

Endocrine studies

Growth hormone (GH) concentrations in the rat pituitary gland, measured by radioimmunoassay (immunoreactive growth hormone; IGH), were normal or elevated postflight, but plasma titers of IGH were reduced, suggesting that the secretory process was suppressed (Grindeland *et al.*, 1978, 1990a, b; Merrill *et al.*, 1992b). Pituitary GH, measured by bioassay (bioassayable growth hormone, BGH), was invariably decreased by 50 or 60% relative to 1 ***g*** controls.

Anterior pituitary cells from flight rats, when contained in an Amicon hollow fiber and placed into the cerebral ventricles of hypophysectomized rats, did not promote growth of the recipient rats whereas cells from control rats did, supporting the view that space flight depletes pituitary BGH (Hymer *et al.*, 1992). Space flight caused marked decreases in peptide regulators of growth hormone secretion: growth hormone releasing factor (GRF) and growth hormone release inhibiting factor (somatostatin; SS). In the hypophysiotropic nuclei of the hypothalamus both of these regulatory peptides, and the mRNAs for their precursor forms, were significantly decreased in space flight rats (Grindeland *et al.*, 1990b; Sawchenko *et al.*, 1992). In contrast, two weeks of hindlimb unloading resulted in only marginal decreases in GRF and SS peptides and their mRNAs. In postflight monkeys, plasma IGH was either marginally detectable or undetectable, and remained depressed for up to 11 days after flight (Grindeland *et al.*, 2000).

Measurement of posterior pituitary levels of vasopressin and oxytocin revealed markedly decreased levels in space flight rats (Keil *et al.*, 1992). In the latter flights of the Cosmos series, plasma thyroxine (T4) concentrations were within normal limits and triiodothyronine (T3) was consistently decreased (Merrill *et al.*, 1992b), but in earlier flights Russian investigators found reduced plasma levels of thyrotropic hormone. Space flight monkeys exhibited significant decreases in T3 but no change in T4. During 1 ***g*** simulations the T4 was still unaffected and the T3 showed a smaller decrease than postflight (Grindeland *et al.*, 2000). The significant reduction in circulating T3 and normal T4 after flight suggests a reduction in T4 deiodinase activity, the site of which remains unknown, but the decreases appear to be related, at least in part, to hypokinesia.

Male rats (Grindeland *et al.*, 1990a; Merrill *et al.*, 1992b) and monkeys (Grindeland *et al.*, 2000) have typically shown large decreases in circulating and/or testicular testosterone (T) levels after space flight. Testicular luteinizing hormone (LH) receptor populations did not differ from those of Earth controls in flight-type cages, suggesting that deficient LH secretion might be the cause of the diminished T levels (Amann *et al.*, 1992). Plasma LH levels have not been measured in either postflight rats or monkeys due to competition for the limited volumes of plasma. However, male astronauts also showed diminished T levels but, surprisingly, had elevated LH titers (Strollo *et al.*, 1997). Hindlimb-unloaded rats had decreased levels of T despite circulating LH titers comparable to or greater than those of controls (Deaver *et al.*, 1992). The low T and normal LH levels in hindlimb-unloaded rats were found regardless of whether the testicles were prevented from moving into the body cavity or not. It is of interest that the low T found in space flight astronauts did not appear to be due to a general stress reaction as cortisol levels were unchanged from control, suggesting a more specific effect of microgravity. It is known that pineal melatonin is anti-gonadal, and a study by Holley *et al.* (1994) found increased melatonin precursors (5-HT and 5-HIAA) in pineal glands of space flight and suspended rats, suggesting a possible mechanism for the decreased testosterone levels.

Reproductive function

Although circulating T was reduced in rats during flights of up to three weeks there was little or no effect on spermatogonia counts (Sapp *et al.*, 1990; Amann *et al.*, 1992). Following a flight of 13 days, male rats successfully mated with 1 ***g*** females, yielding normal offspring (Serova *et al.*, 1989). On a flight of 19 days, male and female rats were flown together to see if they would mate and if normal gestational development would ensue (Gurovsky *et al.*, 1980; Serova *et al.*, 1984; Keefe, 1985). At the end of the flight, none of the rats was pregnant and there was no clear evidence that they had been. Moreover, it is unknown whether the animals mated in space. Female rats flown in space during days 13 to 18 of pregnancy have given birth postflight to smaller and fewer, but otherwise normal, pups (Serova *et al.*, 1984). Rats that were subjected to hindlimb unloading for 28 days had lengthened estrous cycles; they also had reduced plasma concentrations of estradiol (E2) during the estrus phase of the cycle, but normal LH levels (Tou *et al.*, 2004). It remains unclear how exposure to actual microgravity or models simulating some conditions of microgravity inhibits T and E2 secretion, but does not appear to affect LH levels. The apparently normal LH but low E2 concentrations observed in hindlimb-unloaded rats raises an interesting question of whether a mammal would ovulate while in space.

In support of space physiological studies, including reproductive studies, sensors have been developed. One device, dubbed "the fetal biotelemetry system", was originally developed by NASA-Ames Research Center engineers to monitor various physiological parameters in space flight animals. The device was redesigned to pass through a 10 mm surgical trocar for potential use in women whose babies required surgical interventions performed *in utero* and who were thus in danger of premature birth. Due to intersecting logistical and technical issues the device was never used in humans. However, it was successfully used to monitor temperature and pressure *in utero* in rats during pregnancy and in rats, giving birth in space-related studies on Earth.

Circadian system/temperature regulation

Primate circadian rhythms persisted in microgravity when normal light dark cycles were in effect. Activity rhythms appeared to be maintained with a 24 h period, while thermoregulatory rhythms were more variable in period and showed greater variability of phase (time of peak), usually showing phase delays (Sulzman, 1986; Fuller *et al.*, 1996; Alpatov *et al.*, 2000). Skin and deep body temperature, but not brain temperature, were reduced in microgravity relative to ground controls (Klimovitsky *et al.*, 2000).

Metabolism and body composition

Prior non-human primate studies showed decreases in monkey metabolic rate following exposure to microgravity but the Bion 11 study did not (Popova and Grigoriev, 1994; Dotsenko *et al.*, 2000; Hoban-Higgins *et al.*, 2000). However, in the latter study, monkeys did exhibit a 40% decrease in their circulating T3, consistent with reduced metabolic rates (Grindeland *et al.*, 2000). The reason for the disparity between metabolic rate and T3 measurements is unknown. Studies of skeletal muscles, liver, and plasma revealed that space flight elicited significant changes in carbohydrate and fat metabolism, with markedly increased levels of plasma glucose and liver glycogen (Abraham *et al.*, 1978; Merrill *et al.*, 1990, 1992a; Racine and Cormier, 1992; Dotsenko *et al.*, 2000). Cosmos flights provided the first clear evidence that pharmacokinetics may be altered, since hepatic enzymes involved in xenobiotic metabolism were inhibited during two biosatellite space flights (Merrill *et al.*, 1990, 1992a). The anesthesia-related death of a Bion 11 monkey, following a postflight biopsy procedure, may have been associated with reduced xenobiotic metabolism and manifested as increased anesthetic/drug sensitivity. As mentioned, the liver also showed reduced heme biosynthesis.

Comparison of body compositions of rats after an 18.5 day of space flight and 1 ***g*** simulation revealed a 7% decrease in total body water and a 36% decrease in extracellular water volume in the flight rats (Ushakov *et al.*, 1980; Pitts *et al.*, 1983). There was also a shift of water from skin to the viscera. Whole body calcium and bone mineral were decreased 22% by exposure to space flight. Body fat was similar for flight and simulation control rats but both groups had almost twice as much total fat as did vivarium animals, suggesting that the increased fat was due to hypokinesia in the single animal flight-type cages. Perhaps the most surprising finding, in view of wasting of muscles previously discussed in this chapter, was the net increase in muscle mass as adduced from creatine, protein, and potassium measurements.

Immune system

Decreases in weights of the thymus, spleen, and inguinal lymph nodes were found in space flight rats. Lymphocytes from bone marrow or spleens of flight rats differed from 1 ***g*** controls in their abundance of various subpopulations, responsiveness to granulocyte–monocyte colony stimulating factor, and toxicity of natural killer cells for a particular cell line (Mandel and Balish, 1977; Sonnenfeld *et al.*, 1990, 1992; Nash *et al.*, 1992; Rykova *et al.*, 1992). Space flight appears to selectively affect various elements of the immune system with differing time courses. The results of Cosmos experiments may be significant harbingers, since immune system deficiencies may not become evident in the short term, but may result in human immune problems during lengthy space missions.

The case for a new free flyer biosatellite program

Cosmonauts and astronauts have accrued extensive experience in low-Earth orbit, with one cosmonaut remaining in space for as long as 14 months. This record of successful space flights belies a number of obstacles, environmental and physiological, which may limit longer duration human voyages in space beyond the Earth's protective magnetic field such as journeys to another planet. Radiation and meteorite showers are other environmental hazards of concern. Some of the physiological concerns that have arisen from space flight experience to date, such as bone and muscle atrophy and metabolic changes, may compromise crew health and performance. Other concerns are unanswered "critical" questions formulated by *ad hoc* committees over the years and cited earlier. Finding answers to these questions will necessitate animal studies. As stated, animal studies have provided the backbone of our present understanding of space physiology.

There has not been an opportunity to expose animals to the deep space environment beyond the Van Allen Belts, except for the Apollo 17 flight, which carried pocket mice. The longest duration animal flight to date has been Skylab 3 (84 days); all other animal flights have been 22 days or less. Thus, there is a critical need to perform longer duration mammalian flights beyond the Earth's protective atmosphere. In addition to supporting human space exploration, animal research in space is important for clarifying the role of gravity in life processes, and by direct extension, furthering our understanding of fundamental biological processes on Earth (Tipton, 2003b).

Choice of animals for the primary payload

The physiological changes resulting from exposure to microgravity appear to be systems changes. For example, skeletal muscle changes reflect not only direct effects of unloading on the muscle but also the indirect effects mediated by the nervous and endocrine systems. Thus, species whose physiology and genomes have been well studied, whose physiology mimics that of humans, and which are large enough to accommodate physiological sensors are the species of choice for many of the primary experiments. Small rodents, such as mice, or other species might be the choice for particular studies, such as radiation biology experiments where larger numbers and tissue analysis are needed. For certain types of experiments, e.g., neuromuscular and neurosensory physiology, the monkey is certainly the optimal species. If the primary objective of these studies is to support human space exploration, then species meeting the above criteria are essential. The laboratory rat will continue to be the most useful species for most experiments. Lower forms and tissue cultures could also be accommodated as secondary payloads, but the primary payload species has to be amenable to careful evaluation of physiological systems as suggested in this review.

The species of animal chosen and the inflight facilities needed will obviously determine the type of spacecraft to be flown.

Free flyer biosatellite vehicles

It bears repeating that the Cosmos spacecraft is a research platform with no humans on board. This is a critical distinction when research requires the use of radiation exposure (either ambient or from an onboard source), toxic compounds, virulent disease vectors, odoriferous animals or orbital inclinations/altitudes not achievable in crewed research platforms. Uncrewed platforms can be launched, recovered, and operated in response to research priorities and without regard for the operational imperatives imposed on crewed space missions. The absence of crew also removes the possibility of unintended disturbances that may perturb research that require continuous microgravity or confound circadian rhythm research.

As mentioned, the present Cosmos spacecraft can stay on orbit for periods of up to 30 days or be returned as may be required to meet scientific imperatives. An important point is that the existing Cosmos spacecraft could accommodate up to approximately 50 rats; employing a single cohort of rats to investigate a specific system, e.g., neuromuscular on a given flight using a variety of approaches, would yield a prodigious amount of data. Future "Cosmos" spacecrafts are expected to carry larger payloads for longer periods and with even more electrical power available. Additionally, the Cosmos platform offers a developmental environment where science and technology concepts can be rapidly applied to space flight research activities with less rigorous safety and verification requirements than must be imposed when humans are present. This provides an opportunity to develop and test novel sensors, instrumentation, technologies, and operational concepts before embarking on the lengthy and expensive process required prior to use on crewed vehicles. For example, the effects of reentry on physiology can be obviated for many studies by using a variety of biosensors. Animals implanted with biopotential, biophysical, biochemical, and physiological telemetric sensors complemented by video observation, data recording, and robust data downlinking capabilities, largely preclude the need for human manipulation of animals in flight. In fact, avoiding handling of conscious animals or the need to use anesthetics will probably yield more pristine data on the physiological responses to the various stages of flight. The implanted animals can be followed during the interval between loading and launch, during launch and insertion into orbit, while in microgravity, and during reentry. With those capabilities in place, it appears that the great majority of mammalian experiments could be flown on a free flyer research platform such as Cosmos. Tissues of these animals would be available after landing and, conceivably, the animals could be killed and preserved robotically in space to avoid the confounding effects of reentry g loads on the tissues.

In addition to those functions that it is uniquely suited to perform, the Cosmos research platform can serve as an important adjunct to research conducted onboard the Space Shuttle (STS) and International Space Station (ISS) where human safety concerns or operational conflicts may limit research opportunities. This platform can also be a bridge to those crewed platforms that will be developed to support space flight research when the STS and ISS are no longer available.

The current lack of access to space for mammalian researchers is reminiscent of the hiatus between the Skylab 3 and STS flights. Not only does the present lack of opportunity seriously impede progress in space physiology but an aging cadre of experienced investigators will also soon retire, further compounding the deficit. Loss of this experience or knowledge base, unfortunately, often leads to needless duplication of research at a later date. In the absence of access to space, the enthusiasm for space life science studies by younger scientists will also be dampened. Initiation of an international program of regular biosatellite flights would materially advance our understanding of space physiology and provide opportunities to recruit and train the next generation of space physiologists.

Summary and conclusions

In the 220 years since the first venture of mammals into "space" aboard a hot air balloon, through the eras of balloon and ballistic rocket flights to orbital space flights, animals have pioneered space physiology and technology investigations. As we contemplate interplanetary travel, animal studies in space become even more important in understanding the long-term effects of microgravity and in maximizing the well-being and performance of crew members. As illustrated in this review, there are also many areas of fundamental space physiology that merit more thorough investigation.

The invitation by the Soviet government for other countries to participate in orbital animal flights in the Cosmos program made major advances in space physiology and technology possible. An abundance of physiological data on adaptive and maladaptive responses to microgravity has emerged from the Cosmos program and given rise to some of the new physiological concepts described in this review. Many of the observed changes in Cosmos biosatellites have been extended in later STS flights with animals and have also been observed in astronauts and cosmonauts. To date, the Cosmos program has been the single largest source of physiological data concerning microgravity exposure and has been predictive of human responses to microgravity. The data generated from Cosmos flights have provided the discipline of space biology and medicine with a scientific and rational basis for the development of countermeasures to the deleterious aspects of space flight.

Cosmos data have also provided insights into terrestrial mammalian physiology, e.g., the neural plasticity found in space flight monkeys has provided important clues to understanding spinal cord physiology on Earth. These insights are being used in treating spinal cord injury patients. Other significant technologies that have been developed for Cosmos flight experiments and described in this review, are also now being employed for human benefit on Earth.

Advancement of our understanding of space biology cannot occur in the absence of a space platform. With the era of the STS and ISS drawing to a close, there is a need to identify a space platform where research into the effects of space flight on living systems can continue. Free flyer satellites, such as Cosmos, have served and can continue to serve the needs of the space physiology research community. However, in today's constrained budget environments, the animal research needed to develop countermeasures and to explore the basic mechanisms of life's adaptive responses to space flight is at great risk. It is critical that policy makers and their constituents, who are interested in manned space flight, support efforts to develop a robust program for recurrent access to space by the animal research community. With appropriate technology development, nearly all space experiments one can visualize can be done on a free flyer biosatellite. With the advancement of nanotechnology and robotics, the types of animal experiments that can be done on free flyer biosatellites should be nearly limitless. When one considers the low cost and versatility of the Cosmos free flyer spacecraft, it is clear that for the foreseeable future, Cosmos-type free flyers will be a critical tool for physiological research on animals in space. As we have demonstrated, animal research in space reduces the risk entailed in human exploration activities and provides methodologies for palliative treatments on Earth. Like the Cosmos program of the past, a new biosatellite program could be an international collaborative effort that would provide critical information to support our journeys to the Moon, Mars, and beyond. It is to be hoped that the international space physiology community: academic, commercial, and governmental, will rally to the challenge to continue and expand the investigations begun under the aegis of the Cosmos program.

Acknowledgments

Tables 2, 3, and 4 are adapted from tables published by E.A. Ilyin in the Journal of Gravitational Physiology. We thank Dr. C. Fuller and the journal publishers for permission to use the tables. We also thank Dr. P.X. Callahan and Ms. L. Briones of the Ames Research Center Data Archives for providing space flight data.

The authors express their sincere gratitude to Drs. O.G. Gazenko, A.I. Grigoriev, and L.P. Chambers, and many others for their contributions to the Cosmos research program. We remember fondly and with appreciation members of the international Cosmos team that died during the course of these space flights. We want to thank Ms. G. Tverskaya for her work as translator *par excellence*, for her many other very helpful roles during the entire span of the Cosmos program, and for her assistance with this manuscript. We also gratefully acknowledge our debt to Ms. P. Bala and Dr. J. Tou for their help in preparation of this chapter.

References

Aerospace Medicine Advisory Committee (1992) Strategic considerations for support of humans in space and Moon/Mars exploration missions. *In Life Sciences and Technology Programs*. NASA Advisory Council.

Assembly of Mathematical and Physical Sciences (1979) Life beyond Earth's environment. *In Space Science Board, National Research Council, National Academy of Sciences*. National Academy Press, Washington, D.C.

Committee on Space Biology and Medicine (1987) A strategy for space biology and medical space. *In National Research Council, National Academy of Sciences*. National Academy Press, Washington, D.C.

Committee on Space Biology and Medicine (1991) Assessment of programs in space biology and medicine. *In National Research Council, National Academy of Sciences*. National Academy Press, Washington, D.C.

Committee on Space Biology and Medicine (1998) A strategy for research in space biology and medicine in the new century. *In Space Studies Board, National Research Council, National Academy of Sciences*. National Academy Press, Washington, D.C.

Committee on Space Biology and Medicine (2000) Review of NASA's biomedical research program. *In Space Studies Board, National Research Council, National Academy of Sciences*. National Academy Press, Washington, D.C.

Task Group on Life Sciences. (1988) Space science in the twenty-first century. *In National Research Council, National Academy of Sciences*. National Academy Press, Washington, D.C.

Abraham, S., Klein, H.P., Lin, C.Y. and Volkmann, C. (1978) The effects of spaceflight on some liver enzymes concerned with carbohydrate and lipid metabolism in the rat. *In* Souza, K.A. (ed.). *Final Reports of U.S. Experiments Flown on the Soviet Satellite Cosmos 936*, NASA TM-78526, Washington, D.C.

Alpatov, A.M., Hoban-Higgins, T.M., Klimovitsky, V.Y., Tumurova, E.G. and Fuller, C.A. (2000) Circadian rhythms in Macaca mulatta monkeys during Bion 11 flight. J. Gravit. Physiol. 7, S119–S123.

Amann, R.P., Deaver, D.R., Zirkin, B.R., Grills, G.S., Sapp, W.J., Veeramachaneni, D.N., Clemens, J.W., Banerjee, S.D., Folmer, J., Gruppi, C.M., Wolgemuth, D.J. and Williams, C.S. (1992) Effects of microgravity or simulated launch on testicular function in rats. J. Appl. Physiol. 73, 174S–185S.

Antsiferova, L.I., Shlyk, G.G. and Ignatenko, A.V. (2000) Effect of microgravity on the implementation of conditioned reflex skills of rhesus monkeys. J. Gravit. Physiol. 7, S95–S98.

Arnaud, S.B., Fung, P., Popova, I.A., Morey-Holton, E.R. and Grindeland, R.E. (1992) Circulating parathyroid hormone and calcitonin in rats after spaceflight. J. Appl. Physiol. 73, 169S–173S.

Arnaud, S.B., Navidi, M., Deftos, L., Buckendahl, P., Dotsenko, M.A., Bengtson, S., Bigbee, A. and Grindeland, R.E. (2000) Calcium metabolism in Bion 11 monkeys. J. Gravit. Physiol. 7, S153.

Badakva, A.M. and Miller, N.V. (2000) The effect of microgravity on the autonomic nervous system of rhesus monkeys. J. Gravit. Physiol. 7, S113–S118.

Baldwin, K.M., Herrick, R.E., Ilyina-Kakueva, E.I. and Oganov, V.S. (1990) Effects of zero gravity on myofibril content and isomyosin distribution in rodent skeletal muscle. FASEB J. 4, 79–83.

Baranska, W., Shopinski, P. and Kaplansky, A. (1990) Morphometrical evaluation of myocardium of rats flown on biosatellite Kosmos 1887. Materica. Medica. Polona. 22, 255–257.

Baranski, S., Kujawa, M. and Kaplansky, A. (1980) Ultrastructural qualitative and quantitative evaluation of cytoplasmatic structures of heart muscle of rats living aboard biosputnik Kosmos 936. Physiologist. 23, S38–S40.

Belichenko, P.V., Machanov, M.A., Fedorov, A.A., Krasnov, I.B. and Leontovich, T.A. (1990) Effects of space flight on dendrites of the neurons of the rat's brain. Physiologist. 33, S12–S15.

Bodine-Fowler, S.C., Roy, R.R., Rudolph, W., Haque, N., Kozlovskaya, I.B. and Edgerton, V.R. (1992) Spaceflight and growth effects on muscle fibers in the rhesus monkey. J. Appl. Physiol. 73, 82S–89S.

Burkovskaya, T.E. and Korolkov, V.I. (2000) Hemopoiesis in bone marrow of monkeys after spaceflight. J. Gravit. Physiol. 7, S129–S134.

Callahan, P.X. (1966) *The Use of Animals in Space Research*. Seventeenth Annual Animal Care Panel, Chicago.

Cann, C.E. and Adachi, R.R. (1983) Bone resorption and mineral excretion in rats during spaceflight. Am. J. Physiol. 244, R327–R331.

Carnino, A., Roffino, S., Chopard, A. and Marini, J.F. (2000) Effects of a 14-day spaceflight on soleus myotendinous junction ultrastructure in the rhesus monkey. J. Gravit. Physiol. 7, S65–S68.

Castleman, K.R., Chui, L.A. and Van Der Meulen, J.P. (1978) Spaceflight effects on muscle fibers. *In* Souza, K.A. (ed.). *Final Reports of U.S. Experiments Flown on the Soviet Satellite Cosmos 936*, NASA TM-78526, Washington, D.C.

Chi, M.M., Choksi, R., Nemeth, P., Krasnov, I., Ilyina-Kakueva, E., Manchester, J.K. and Lowry, O.H. (1992) Effects of microgravity and tail suspension on enzymes of individual soleus and tibialis anterior fibers. J. Appl. Physiol. 73, 66S–73S.

Chopard, A., Leclerc, L., Pons, F., Leger, J.J. and Marini, J.F. (2000) Effects of 14-day spaceflight on myosin heavy chain expression in biceps and triceps muscles of the rhesus monkey. J. Gravit. Physiol. 7, S47–S49.

Cohen, B., Kozlovskaya, I., Raphan, T., Solomon, D., Helwig, D., Cohen, N., Sirota, M. and Yakushin, S. (1992) Vestibuloocular reflex of rhesus monkeys after spaceflight. J. Appl. Physiol. 73, 121S–131S.

Correia, M.J., Perachio, A.A., Dickman, J.D. and Kozlovskaya, I.B. (1992a) Sensitivity changes in semicircular canal following microgravity (abstract). *43rd Congress of the International Astronautical Federation*, 541 p. World Space Congress, Washington, D.C.

Correia, M.J., Perachio, A.A., Dickman, J.D., Kozlovskaya, I.B., Sirota, M.G., Yakushin, S.B. and Beloozerova, I.N. (1992b) Changes in monkey horizontal semicircular canals afferent responses after spaceflight. J. Appl. Physiol. 73, 112S–120S.

Correia, M.J. (1994) Studies of vestibular neurons in normal, hyper- and hypogravity. *In* Conolly, J.P., Skidmore, M.G. and Helwig, D. (eds.). *Final Reports of U. S. Experiments Flown on Soviet Biosatellite Cosmos 2229*, 16 p. NASA CR-194734, Washington, D.C.

Deaver, D.R., Amann, R.P., Hammerstedt, R.H., Ball, R., Veeramachaneni, D.N. and Musacchia, X.J. (1992) Effects of caudal elevation on testicular function in rats. Separation of effects on spermatogenesis and steroidogenesis. J. Androl. 13, 224–231.

Dotsenko, M.A., Rudneva, R.I., Bengtson, S.G. and Korolkov, V.I. (2000) Metabolic parameters of Bion 11 monkeys. J. Gravit. Physiol. 7, S139–S142.

Doty, S.B., Morey-Holton, E.R., Durnova, G.N. and Kaplansky, A.S. (1990) Cosmos 1887: morphology, histochemistry, and vasculature of the growing rat tibia. FASEB J. 4, 16–23.

Doty, S.B., Morey-Holton, E.R., Durnova, G.N. and Kaplansky, A.S. (1992) Morphological studies of bone and tendon. J. Appl. Physiol. 73, 10S–13S.

Duke, P.J., Durnova, G. and Montufar-Solis, D. (1990) Histomorphometric and electron microscopic analyses of tibial epiphyseal plates from Cosmos 1887 rats. FASEB J. 4, 41–46.

Edgerton, V.R., Roy, R.R., Hodgson, J.A., Day, M.K., Weiss, J., Harkema, S.J., Dobkin, B., Garfinkel, A., Konigsberg, E. and Koslovskaya, I. (2000a) How the science and engineering of spaceflight contribute to understanding the plasticity of spinal cord injury. Acta. Astronaut. 47, 51–62.

Edgerton, V.R., Roy, R.R., Recktenwald, M.R., Hodgson, J.A., Grindeland, R.E. and Kozlovskaya, I. (2000b) Neural and neuroendocrine adaptations to microgravity and ground-based models of microgravity. J. Gravit. Physiol. 7, 45–52.

Edwards, K.E. (2000) The US-Soviet/Russian Cosmos biosatellite program. Quest. 73, 20–35.

Eurell, J.A. and Kazarian, L.E. (1983) Quantitative histochemistry of rat lumbar vertebrae following spaceflight. Am. J. Physiol. 244, R315–318.

Falempin, M., Canu, M.H., Langlet, C. and Kozlovskaya, I.B. (2000) Effect of microgravity on the electromyographic activity of two upperlimb muscles during a goal-directed movement and during locomotion. J. Gravit. Physiol. 7, S69–S70.

Fitts, R.H., Romatowski, J.G., Blaser, C., De La Cruz, L., Gettelman, G.J. and Widrick, J.J. (2000a) Effect of spaceflight on the isotonic contractile properties of single skeletal muscle fibers in the rhesus monkey. J. Gravit. Physiol. 7, S53–S54.

Fitts, R.H., Romatowski, J.G., De La Cruz, L., Widrick, J.J. and Desplanches, D. (2000b) Effect of spaceflight on the maximal shortening velocity, morphology, and enzyme profile of fast- and slow-twitch skeletal muscle fibers in rhesus monkeys. J. Gravit. Physiol. 7, S37–S38.

Foldes, I., Kern, M., Szilagyi, T. and Oganov, V.S. (1996) Histology and histochemistry of intervertebral discs of rats participated in spaceflight. Acta Biol. Hung. 47, 145–156.

Fuller, C.A., Hoban-Higgins, T.M., Klimovitsky, V.Y., Griffin, D.W. and Alpatov, A.M. (1996) Primate circadian rhythms during spaceflight: results from Cosmos 2044 and 2229. J. Appl. Physiol. 81, 188–193.

Garetto, L.P., Gonsalves, M.R., Morey, E.R., Durnova, G. and Roberts, W.E. (1990) Preosteoblast production 55 hours after a 12.5-day spaceflight on Cosmos 1887. FASEB J. 4, 24–28.

Garetto, L.P., Morey, E.R., Durnova, G.N., Kaplansky, A.S. and Roberts, W.E. (1992) Preosteoblast production in COSMOS 2044 rats: short-term recovery of osteogenic potential. J. Appl. Physiol. 73, 14S–18S.

Gazenko, O.G. (1976) Preliminary results of investigations on-board Kosmos 782 physiological experiments with rats. *Ministry of Health USSR, Institute of Biomedical Problems and Academy of Sciences USSR Report*, TTF-17288 National Aeronautics and Space Administration, Washington, D.C.

Gazenko, O.G., Genin, A.M., Ilyin, E.A., Portugalov, V.V., Serova, L.V. and Tigranyan, R.A. (1978) Basic results of an experiment with mammals on the Kosmos-782 biosatellite. Kosm. Biol. Aviakosm Med. 6, 43–49.

Goldstein, M.A., Edwards, R.J. and Schroeter, J.P. (1992) Cardiac morphology after conditions of microgravity during COSMOS 2044. J. Appl. Physiol. 73, 94S–100S.

Gosselink, K.L., Grindeland, R.E., Roy, R.R., Zhong, H., Bigbee, A.J., Grossman, E.J. and Edgerton, V.R. (1998) Skeletal muscle afferent regulation of bioassayable growth hormone in the rat pituitary. J. Appl. Physiol. 84, 1425–1430.

Grindeland, R.E., Keil, L.C., Ellis, S., Parlow, A.F., Kendall, J.W., Gaudette, D. and Geschwind, I.I. (1978) Effects of spaceflight on plasma and glandular concentrations of pituitary hormones. *In* Souza, K.A. (ed.). *Final Reports of U.S. Experiments Flown on the Soviet Satellite Cosmos 782*, NASA TM-78525, Washington, D.C.

Grindeland, R.E., Popova, I.A., Vasques, M. and Arnaud, S.B. (1990a) Cosmos 1887 mission overview: effects of microgravity on rat body and adrenal weights and plasma constituents. FASEB J. 4, 105–109.

Grindeland, R.E., Vale, W., Hymer, W.C., Sawchenko, P.E., Vasques, M., Krasnov, I., Kaplansky, A. and Popova, I.A. (1990b) Growth hormone regulation, synthesis and secretion in microgravity. *In* Connolly, J.P., Grindeland, R.E. and Ballard, R.W. (eds.). *Final Reports of U.S. Experiments Flown in Soviet Biosatellite Cosmos 1887*, NASA Technical Memorandum 102254.

Grindeland, R.E., Dotsenko, M.A., Mukku, V.R., Bigbee, A.J. and Bengtson, S.G. (2000) Rhesus monkey hormonal responses to microgravity. J. Gravit. Physiol. 7, S143.

Gurovsky, N.N., Gazenko, O.G., Adamovich, B.A., Ilyin, E.A., Genin, A.M., Korolkov, V.I., Shipov, A.A., Kotovskaya, A.R., Kondratyeva, V.A., Serova, L.V. and Kondratyev, Yu I. (1980) Study of physiological effects of weightlessness and artificial gravity in the flight of the biosatellite Cosmos-936. Acta Astronaut. 7, 113–121.

Henry, J.P. (1962) Effects of weightlessness in ballistic and orbital flight. A progress report. Aeromed. Acta 33, 1056–1068.

Hoban-Higgins, T.M., Stein, T.P., Dotsenko, M.A., Korolkov, V.I. and Fuller, C.A. (2000) Energy metabolism of Macaca mulatta during spaceflight. J. Gravit. Physiol. 7, S145–S148.

Hodgson, J.A., Riazansky, S.N., Goulet, C., Badakva, A.M., Kozlovskaya, I.B., Recktenwald, M.R., McCall, G., Roy, R.R., Fanton, J.W. and Edgerton, V.R. (2000) Rhesus leg muscle EMG activity during a foot pedal pressing task on Bion 11. J. Gravit. Physiol. 7, S87.

Holley, D.C., Soliman, M.R.I., Krasnov, I. and Asadi, H. (1994) Pineal physiology after spaceflight: relation to rat gonadal function. *In* Connolly, J.P., Grindeland, R.E. and Ballard, R.W. (eds.). *Final Reports of U.S. Experiments Flown on the Soviet Biosatellite Cosmos 2044*, NASA Technical Memorandum 108802.

Hutchinson, T.M., Bakulin, A.V., Rakhmanov, A.S., Martin, R.B., Steele, C.R. and Arnaud, S.B. (2001) Effects of chair restraint on the strength of the tibia in rhesus monkeys. J. Med. Primatol. 30, 313–321.

Hyde, T.M., Wu, L.C., Krasnov, I.B., Sigworth, S.K., Daunton, N.G. and D'Amelio, F. (1992) Quantitative autoradiographic analysis of muscarinic cholinergic and GABA (benzodiazepine) receptors in the forebrain of rats flown on the Soviet Biosatellite Cosmos 2044. Brain Res. 593, 291–294.

Hymer, W.C., Grindeland, R.E., Krasnov, I., Victorov, I., Motter, K., Mukherjee, P., Shellenberger, K. and Vasques, M. (1992) Effects of spaceflight on rat pituitary cell function. J. Appl. Physiol. 73, 151S–157S.

Ilyin, E.A. (1983) Investigations on biosatellites of the Cosmos series. Aviat. Space Environ. Med. 54, S9–S15.

Ilyin, E.A. (2000) Historical overview of the Bion project. J. Gravit. Physiol. 7, S1–S8.

Ilyin, E.A., Korolkov, V.I., Skidmore, M.G., Viso, M., Kozlovskaya, I.B., Grindeland, R.E., Lapin, B.A., Gordeev, Y.V., Krotov, V.P., Fanton, J.W., Bielitzki, J.T., Golov, V.K., Magedov, V.S. and Hines, J.W. (2000) Bion 11 mission: primate experiments. J. Gravit. Physiol. 7, S9–S17.

Ilyina-Kakueva, E.I., Portugalov, V.V. and Krivenkova, N.P. (1976) Space flight effects on the skeletal muscles of rats. Aviat. Space Environ. Med. 47, 700–703.

Ilyina-Kakueva, E.I. and Portugulov, V.V. (1979) Effects of artificial gravity on the skeletal musculature of rats in spaceflight conditions. Archives d'Anatomie d' Histologie et Embryologie 76, 22–27.

Ilyina-Kakueva, E.I. and Burkovskaya, T.E. (1991) The microgravity effect on a repair process in m. soleus of the rats flown on Cosmos-2044. Physiologist. 34, S141–S142.

Jee, W.S., Wronski, T.J., Morey, E.R. and Kimmel, D.B. (1983) Effects of spaceflight on trabecular bone in rats. Am. J. Physiol. 244, R310–R314.

Jiang, B., Ohira, Y., Roy, R.R., Nguyen, Q., Ilyina-Kakueva, E.I., Oganov, V. and Edgerton, V.R. (1992a) Adaptation of fibers in fast-twitch muscles of rats to spaceflight and hindlimb suspension. J. Appl. Physiol. 73, 58S–65S.

Jiang, B., Roy, R.R., Polyakov, I.V., Krasnov, I.B. and Edgerton, V.R. (1992b) Ventral horn cell responses to spaceflight and hindlimb suspension. J. Appl. Physiol. 73, 107S–111S.

Kaplansky, A.S. (1978) Distinctive features of microcirculatory bed in gastrocnemius of rats in weightlessness. Arkhiv. Anatomiy, Gistologiy, i Embryologiy. 75, 27–30.

Kaplansky, A.S., Durnova, G.N., Ilyina-Kakueva, E.I., Sakharova, Z.F. and Vorotnikova, E.V. (1990) Histomorphometric analysis of bones of Cosmos-1887 rats. Physiologist. 33, S20–S22.

Kaplansky, A.S., Durnova, G.N., Burkovskaya, T.E. and Vorotnikova, E.V. (1991) The effect of microgravity on bone fracture healing in rats flown on Cosmos-2044. Physiologist. 34, S196–S199.

Kaplansky, A.S. and Durnova, G.N. (1997) Development of osteoporosis in primates in hypokinesia with head-down tilt. J. Gravit. Physiol. 4, P133–P134.

Kazarian, L.E. (1981) Strength characteristics of rat spinal columns: Cosmos-1129. *In* Gazenko, O.G. (ed.). *Advances in Physiological Sciences. Gravitational Physiology*, pp. 129–138. Pergamon Press, New York.

Keefe, J.R. (1985) Vertebrate development in space. *In* Halstead, T.W. (ed.). *NASA Developmental Biology Workshop. NASA TM-86756*, Ames Research Center, Moffett Field, CA.

Keil, L., Evans, J., Grindeland, R. and Krasnov, I. (1989) The effect of spaceflight on pituitary oxytocin and vasopressin content of rats. *In* Connolly, J.P., Grindeland, R.E. and Ballard, R.W. (eds.). *Final Reports of U. S. Experiments Flown on Soviet Biosatellite Cosmos 1887*, NASA Technical Memorandum 102254.

Keil, L., Evans, J., Grindeland, R. and Krasnov, I. (1992) Pituitary oxytocin and vasopressin content of rats flown on Cosmos 2044. J. Appl. Physiol. 73, 166S–168S.

Klimovitsky, V.Y., Alpatov, A.M., Hoban-Higgins, T.M., Utekhina, E.S. and Fuller, C.A. (2000) Thermal regulation in Macaca mulatta during space flight. J. Gravit. Physiol. 7, S149–S152.

Korolkov, V.T., Krotov, V.P., Badakva, A.M., Lobachik, V.T., Truzhennikov, A.N., Yushin, V.A., Vacek, A. and Sandler, H. (1992) Circulation in primates exposed to microgravity aboard the "Cosmos" biosatellites. Physiologist. 35, S245–S247.

Kotovskaya, A.R., Ilyin, E.A., Korolkov, V.I. and Shipov, A.A. (1980) Artificial gravity in space flight. Physiologist. 23, S27–S29.

Kozlovskaya, I.B., Babayev, B.M., Barmin, V.A., Beloozerova, I.N., Kornilova, L.N., Sirota, M.G. and Yakushin, S.B. (1990) Human and animal results on vestibular research in space—VERIS. *Fourth European Symposium on Life Sciences Research in Space, Nov. 1990*, pp. S353–S357. Trieste, Italy.

Kozlovskaya, I.B., Ilyin, E.A., Sirota, M.G., Korolkov, V.I., Babayev, B.M., Beloozerova, I.N. and Yakushin, S.B. (1989) Studies of space adaptation syndrome in experiments on primates performed on board of Soviet biosatellite "Cosmos-1887". Physiologist. 32, S45–S48.

Kozlovskaya, I.B., Grindeland, R.E., Viso, M. and Korolkov, V.I. (2000) Bion 11 science objectives and results. J. Gravit. Physiol. 7, S19–S25.

Krasnov, I. (1994) Gravitational neuromorphology. *In* Bonting, S.L. (ed.). *Advances in Space Biology and Medicine*, Vol. 4, pp. 85–110. IAI Press Inc., England.

Krasnov, I.B. and Dyachkova, L.N. (1990) The effect of space flight on the ultrastructure of the rat cerebellar and hemisphere cortex. Physiologist. 33, S29–S30.

Krasnov, I.B., Polyakov, I.V. and Drobyshev, V.I. (1992) Neuron-glia-capillary system in spinal cord of rats after 14-day space flight. Physiologist. 35, S218–S219.

Krotov, V.P. and Nosovsky, A.M. (2000) Oxygen tension in the somatosensory cortex of rhesus monkeys in space. J. Gravit. Physiol. 7, S125–S127.

Lange, R.D., Driscoll, T.B., Khrushchov, N.G., Serova, L.V. and Vacek, A. (1994) Hematology studies in rats flown on Soviet biosatellite Cosmos 2044. *In* Connolly, J.P., Grindeland, R.E. and Ballard, R.W. (eds.). *Final Reports of U.S. Experiments Flown on the Soviet Biosatellite Cosmos 2044*, NASA Technical Memorandum 108802.

Leon, H.A., Serova, L.V., Cummins, J. and Landaw, S.A. (1978) Alterations in erythrocyte survival parameters in rats after 19.5 days aboard Cosmos 782. Aviat. Space Environ. Med. 49, 66–69.

Leon, H.A., Serova, L.V. and Landaw, S.A. (1980) Effect of weightlessness and centrifugation on red cell survival in rats subjected to space flight. Aviat. Space Environ. Med. 51, 1091–1094.

Lobachik, V.I., Chupushtanov, S.A., Pischulina, G.N. and Nosovsky, A.M. (2000) The effect of a 14-day flight on body fluids in primates. J. Gravit. Physiol. 7, S135–S137.

Lowry, O.H., Krasnov, I.B., Ilyina-Kakueva, E.I., Nemeth, P., McDougal, D.B., Jr., Choksi, R., Carter, J.G., Chi, M.M.Y., Manchester, J.K. and Pusteri, M.E. (1994) Effect of microgravity: The distribution of selected enzymes and amino acids in the hippocampal formation. *In* Connolly, J.P., Grindeland, R.E. and Ballard, R.W. (eds.). *Final Reports of U.S. Experiments Flown on the Soviet Biosatellite Cosmos 2044*, NASA Technical Memorandum 108802.

Manchester, J.K., Chi, M.M., Norris, B., Ferrier, B., Krasnov, I., Nemeth, P.M., McDougal, D.B., Jr. and Lowry, O.H. (1990) Effect of microgravity on metabolic enzymes of individual muscle fibers. FASEB J. 4, 55–63.

Mandel, A.D. and Balish, E. (1977) Effect of space flight on cell-mediated immunity. Aviat. Space Environ. Med. 48, 1051–1057.

Martinez, D.A., Buckendahl, P.E., Grindeland, R.E., Dotsenko, M.A., Bigbee, A.J. and Vailas, A.C. (2000) Evaluation of bone and collagen metabolism by assessing urinary biomarkers in non-human primates. J. Gravit. Physiol. 7, S169–S170.

Mayet-Sornay, M.H., Hoppeler, H., Shenkman, B.S. and Desplanches, D. (2000) Structural changes in arm muscles after microgravity. J. Gravit. Physiol. 7, S43–S44.

McCall, G.E., Goulet, C., Grindeland, R.E., Hodgson, J.A., Bigbee, A.J. and Edgerton, V.R. (1997) Bed rest suppresses bioassayable growth hormone release in response to muscle activity. J. Appl. Physiol. 83, 2086–2090.

McCall, G.E., Goulet, C., Roy, R.R., Grindeland, R.E., Boorman, G.I., Bigbee, A.J., Hodgson, J.A., Greenisen, M.C. and Edgerton, V.R. (1999) Spaceflight suppresses exercise-induced release of bioassayable growth hormone. J. Appl. Physiol. 87, 1207–1212.

Mechanic, G.L., Arnaud, S.B., Boyde, A., Bromage, T.G., Buckendahl, P., Elliott, J.C., Katz, E.P. and Durnova, G.N. (1990) Regional distribution of mineral and matrix in the femurs of rats flown on Cosmos 1887 biosatellite. FASEB J. 4, 34–40.

Mednieks, M.I., Popova, I.A., Kato, K., Stevenson, J. and Philpott, D.E. (1989) Cellular distribution of cyclic AMP dependent protein kinase regulatory subunits in heart muscle of rats flown on Cosmos 1887. *In* Connolly, J.P., Grindeland, R.E. and Ballard, R.W. (eds.). *Final Reports of U.S. Experiments Flown on Soviet Biosatellite Cosmos 1887*, NASA Technical Memorandum 102254.

Merrill, A.H., Jr., Hoel, M., Wang, E., Mullins, R.E., Hargrove, J.L., Jones, D.P. and Popova, I.A. (1990) Altered carbohydrate, lipid, and xenobiotic metabolism by liver from rats flown on Cosmos 1887. FASEB J. 4, 95–100.

Merrill, A.H., Jr., Wang, E., LaRocque, R., Mullins, R.E., Morgan, E.T., Hargrove, J.L., Bonkovsky, H.L. and Popova, I.A. (1992a) Differences in glycogen, lipids, and enzymes in livers from rats flown on Cosmos 2044. J. Appl. Physiol. 73, 142S–147S.

Merrill, A.H., Jr., Wang, E., Mullins, R.E., Grindeland, R.E. and Popova, I.A. (1992b) Analyses of plasma for metabolic and hormonal changes in rats flown aboard Cosmos 2044. J. Appl. Physiol. 73, 132S–135S.

Miu, B., Martin, T.P., Roy, R.R., Oganov, V.S., Ilyina-Kakueva, E.I., Marini, J.F., Leger, J.J., Bodine-Fowler, S.C. and Edgerton, V.R. (1990) Metabolic and morphologic properties of single muscle fibers in the rat after spaceflight Cosmos 1887. FASEB J. 4, 64–72.

Montufar-Solis, D., Duke, P.J. and Durnova, G.N. (1992) Spaceflight and age affect tibial epiphyseal growth plate histomorphometry. J. Appl. Physiol. 73, 19S–25S.

Morey, E.R. and Baylink, D.J. (1978) Inhibition of bone formation during space flight. Science 201, 1138–1141.

Morey, E.R. (1979) Spaceflight and bone turnover: correlation with a new rat model of weightlessness. Bioscience 29, 168–172.

Mounier, Y., Stevens, L., Shenkman, B.S., Kischel, P., Lenfant, A.M., Montel, V., Catinot, M.P., Toursel, T. and Picquet, F. (2000) Effect of spaceflight on single fiber function of triceps and biceps muscles in rhesus monkeys. J. Gravit. Physiol. 7, S51–S52.

Musacchia, X.J., Steffen, J.M., Fell, R.D., Dombrowski, M.J., Oganov, V.S. and Ilyina-Kakueva, E.I. (1992) Skeletal muscle atrophy in response to 14 days of weightlessness: vastus medialis. J. Appl. Physiol. 73, 44S–50S.

Nash, P.V., Konstantinova, I.V., Fuchs, B.B., Rakhmilevich, A.L., Lesnyak, A.T. and Mastro, A.M. (1992) Effect of spaceflight on lymphocyte proliferation and interleukin-2 production. J. Appl. Physiol. 73, 186S–190S.

Oganov, V.S., Bakulin, A.V., Novikov, V.E. and Murashko, L.M. (2000) Change and recovery of bone mass and acoustic properties of rhesus monkeys after Bion 11 spaceflight. J. Gravit. Physiol. 7, S163–S168.

Ohira, Y., Jiang, B., Roy, R.R., Oganov, V., Ilyina-Kakueva, E., Marini, J.F. and Edgerton, V.R. (1992) Rat soleus muscle fiber responses to 14 days of spaceflight and hindlimb suspension. J. Appl. Physiol. 73, 51S–57S.

Pedrini-Mille, A., Maynard, J.A., Durnova, G.N., Kaplansky, A.S., Pedrini, V.A., Chung, C.B. and Fedler-Troester, J. (1992) Effects of microgravity on the composition of the intervertebral disk. J. Appl. Physiol. 73, 26S–32S.

Peyton, G. (1968) Fifty years of aerospace medicine:its evolution since the founding of aerospace medicine in January 1918. AFSC historical publication no. 67–180. Air Force Systems Command Aerospace Medical Division. Brooks Air Force Base, San Antonio, TX.

Philpott, D.E., Popova, I.A., Kato, K., Stevenson, J., Miquel, J. and Sapp, W. (1990) Morphological and biochemical examinations of Cosmos 1887 rat heart tissue: Part I—Ultrastructure. FASEB J. 4, 73–78.

Pitts, G.C., Ushakov, A.S., Pace, N., Smith, A.H., Rahlmann, D.F. and Smirnova, T.A. (1983) Effects of weightlessness on body composition in the rat. Am. J. Physiol. 244, R332–R337.

Polyakov, I.V., Drobyshev, V.I. and Krasnov, I.B. (1991) Morphological changes in the spinal cord and intervertebral ganglia of rats exposed to different gravity levels. Physiologist. 34, S187–S188.

Popova, I.A. and Tigranyan, R.A. (1981) Phosphatase activity in the humerus of rats flown in biosatellite Kosmos-936. Kosm. Biol. Aviakosm Med. 15, 82–84.
Popova, I.A. and Grigoriev, A.I. (1994) Metabolic effects of spaceflight: Cosmos missions overview. Advances in Space Biology and Medicine 4, 69–83.
Portugalov, V.V., Savina, E.A., Kaplansky, A.S., Yakovleva, V.I., Plakhuta-Plakutina, G.I., Pankova, A.S., Katunyan, P.I., Shubich, M.G. and Buvailo, S.A. (1976) Effect of space flight factors on the mammal: experimental-morphological study. Aviat. Space Environ. Med. 47, 813–816.
Racine, R.N. and Cormier, S.M. (1992) Effect of spaceflight on rat hepatocytes: A morphometric study. J. Appl. Physiol. 73, 136S–141S.
Recktenwald, M.R., Hodgson, J.A., Roy, R.R., Riazansky, S.N., McCall, G., Kozlovskaya, I.B., Washburn, D.A., Fanton, J.W. and Edgerton, V.R. (2000) Quadrupedal locomotion in rhesus monkeys after 14 days of spaceflight. J. Gravit. Physiol. 7, S71.
Riley, D.A., Ilyina-Kakueva, E.I., Ellis, S., Bain, J.L., Slocum, G.R. and Sedlak, F.R. (1990) Skeletal muscle fiber, nerve, and blood vessel breakdown in space-flown rats. FASEB J. 4, 84–91.
Riley, D.A., Ellis, S., Giometti, C.S., Hoh, J.F., Ilyina-Kakueva, E.I., Oganov, V.S., Slocum, G.R., Bain, J.L. and Sedlak, F.R. (1992) Muscle sarcomere lesions and thrombosis after spaceflight and suspension unloading. J. Appl. Physiol. 73, 33S–43S.
Rodionova, N.V., Shevel, I.M., Oganov, V.S., Novikov, V.E. and Kabitskaya, O.E. (2000) Bone ultrastructural changes in Bion 11 rhesus monkeys. J. Gravit. Physiol. 7, S157–S161.
Roy, R.R., Zhong, H., Bodine, S.C., Pierotti, D.J., Talmadge, R.J., Barkhoudarian, G., Kim, J., Fanton, J.W., Kozlovskaya, I.B. and Edgerton, V.R. (2000a) Fiber size and myosin phenotypes of selected rhesus lower limb muscles after a 14-day spaceflight. J. Gravit. Physiol. 7, S45.
Roy, R.R., Zhong, H., Bodine, S.C., Talmadge, R.J., Fanton, J.W., Kozlovskaya, I.B. and Edgerton, V.R. (2000b) Myonuclear number and size of rhesus soleus fibers after 14 days of actual or simulated flight. J. Gravit. Physiol. 7, S63.
Rumbaugh, D.M., Richardson, W.K., Washburn, D.A., Savage-Rumbaugh, E.S. and Hopkins, W.D. (1989) Rhesus monkeys (Macaca mulatta), video tasks, and implications for stimulus-response spatial contiguity. J. Comp. Psy. 103, 32–38.
Rykova, M.P., Sonnenfeld, G., Lesnyak, A.T., Taylor, G.R., Meshkov, D.O., Mandel, A.D., Medvedev, A.E., Berry, W.D., Fuchs, B.B. and Konstantinova, I.V. (1992) Effect of spaceflight on natural killer cell activity. J. Appl. Physiol. 73, 196S–200S.
Sapp, W.J., Philpott, D.E., Williams, C.S., Kato, K., Stevenson, J., Vasquez, M. and Serova, L.V. (1990) Effects of spaceflight on the spermatogonial population of rat seminiferous epithelium. FASEB J. 4, 101–104.

Sandler, H., Stone, H.L., Hines, J.W., Benjamin, B., Halpryn, B.M., Krotov, V., Kulaev, B., Krilov, V., Magedov, V. and Nazin, A.N. (1986) Cardiovascular results from a rhesus monkey flown aboard the Cosmos 1514 spaceflight and ground-based controls. *In* Mains, R.C. and Gomersall, E.W. (eds.). *Final Reports of U.S. Monkey and Rat Experiments Flown on the Soviet Satellite Cosmos 1514*, NASA TM-88223.

Sandler, H., Skidmore, M., Hines, J., Agasid, E., MacKenzie, R., Krotov, V.P., Bazunova, E.G., Begorodsky, A.O., Estratov, Y.A. and Nazin, A.N. (1994) Primate and cardiovascular flight experiment and ground based controls. *In* Hines, J.W. and Skidmore, M.G. (eds.). *Final Reports of U.S. Experiments Flown on the Soviet Biosatellite Cosmos 1667*, NASA TM-10880.

Savina, E.A. and Alekseev, E.I. (1979) Morphological investigation of the hypothalamo-hypophyseal neurosecretory system. *Effects of Spaceflight Dynamic Factors on the Animal Organism*, pp. 79–81. Nauka Press, Moscow.

Sawchenko, P.E., Arias, C., Krasnov, I., Grindeland, R.E. and Vale, W. (1992) Effects of spaceflight on hypothalamic peptide systems controlling pituitary growth hormone dynamics. J. Appl. Physiol. 73, 158S–165S.

Serova, L.V., Denisova, L.A. and Makeeva, V.F. (1984) The effect of microgravity on the prenatal development of mammals. Physiologist 27, S5–S8.

Serova, L.V., Denisova, L.A. and Baikova, O.V. (1989) The effect of microgravity on the reproductive function of male rats. Physiologist 32, S29–S30.

Shenkman, B.S., Beloozerova, I.N., Lee, P. and Nemirovskaya, T.L. (2000) Structural and metabolic characteristics of rhesus monkey m. soleus after spaceflight. J. Gravit. Physiol. 7, S39–S42.

Shlyk, G.G., Ignatenko, A.V., Efimova, M.Y. and Zavadskaya, O.I. (2000) Behavior and performance of rhesus monkeys during Bion 11 flight. J. Gravit. Physiol. 7, S81–S85.

Simmons, D.J., Russell, J.E., Winter, F., Tran Van, P., Vignery, A., Baron, R., Rosenberg, G.D. and Walker, W.V. (1983) Effect of spaceflight on the non-weight bearing bones of rat skeleton. Am. J. Physiol. 244, R319–R326.

Simmons, D.J., Grynpas, M.D. and Rosenberg, G.D. (1990) Maturation of bone and dentin matrices in rats flown on the Soviet biosatellite Cosmos 1887. FASEB J. 4, 29–33.

Sirota, M.G., Babayev, B.M., Belozerova, I.N., Ivanov, A.M., Yakushin, S.B. and Kozlovskaya, I.B. (1992) Vestibulooculomotor interactions in the microgravity environment. *In* Gazenko, O.G. (ed.). *Results of Investigations on Biosatellites*, pp. 25–29. Nauka Press, Moscow.

Sonnenfeld, G., Mandel, A.D., Konstantinova, I.V., Taylor, G.R., Berry, W.D., Wellhausen, S.R., Lesnyak, A.T. and Fuchs, B.B. (1990) Effects of spaceflight on levels and activity of immune cells. Aviat. Space Environ. Med. 61, 648–653.

Sonnenfeld, G., Mandel, A.D., Konstantinova, I.V., Berry, W.D., Taylor, G.R., Lesnyak, A.T., Fuchs, B.B. and Rakhmilevich, A.L. (1992) Spaceflight alters immune cell function and distribution. J. Appl. Physiol. 73, 191S–195S.

Souza, K.A. and Ilyin, E.A. (1996) Major results from biological experiments in space. *In* Grigoriev, A.I. (ed.). *Space Biology and Medicine*, pp. 31–62. Nauka Press, Moscow.

Stauber, W.T., Fritz, V.K., Burkovskaya, T.E. and Ilyina-Kakueva, E.I. (1992) Effect of spaceflight on the extracellular matrix of skeletal muscles after a crush injury. J. Appl. Physiol. 73, 74S–81S.

Strollo, F., Strollo, G., More, M., Ferretti, C., Mangrossa, N., Casarosa, E., Luisi, M. and Riondino, G. (1997) Changes in human adrenal and gonadal function onboard Spacelab. J. Gravit. Physiol. 4, P103–P104.

Sulzman, F.M. (1986) Synchronization of primate circadian rhythms in space. *In* Mains, R.C. and Gomersall, E.W. (eds.). *Final Reports of U.S. Monkey and Rat Experiments Flown on the Soviet Satellite Cosmos 1514*, NASA TM-88223.

Thomason, D.B., Morrison, P.R., Oganov, V., Ilyina-Kakueva, E.I., Booth, F.W. and Baldwin, K.M. (1992) Altered actin and myosin expression in muscle during exposure to microgravity. J. Appl. Physiol. 73, 90S–93S.

Tipton, C.M. (2003a) Animals and biosatellites in space: a historical perspective. J. Gravit. Physiol. 10, 1–4.

Tipton, C.M. (2003b) Animals and biosatellites in space: recapturing lost opportunities and creating new initiatives. J. Gravit. Physiol. 10, 13–14.

Tomko, D.L., Kozlovskaya, I.B., Paige, G.D. and Badakva, A.M. (1997) Adaptation to microgravity of oculomotor reflexes amor:otolith-ocular reflexes. *In* Connolly, J.P., Skidmore, M.G. and Helwig, D. (eds.). *Final Reports of U.S. Experiments Flown in Soviet Biosatellite Cosmos 2229*, NASA Technical Memorandum 110439.

Tou, J.C., Grindeland, R.E. and Wade, C.E. (2004) Effects of diet and exposure to hindlimb suspension on estrous cycling in Sprague-Dawley rats. Am. J. Physiol. Endocrinol. Metab. 286, E425–E433.

Ushakov, A.S., Smirnova, T.A., Pitts, G.C., Pace, N., Smith, A.H. and Rahlmann, D.F. (1980) Body composition of rats flown aboard Cosmos-1129. Physiologist 23, S41–S44.

Vailas, A.C., Zernicke, R.F., Grindeland, R.E., Kaplansky, A.S., Durnova, G.N., Li, K.C. and Martinez, D.A. (1990) Effects of spaceflight on rat humerus geometry, biomechanics, and biochemistry. FASEB J. 4, 47–54.

Vailas, A.C., Vanderby, R., Jr., Martinez, D.A., Ashman, R.B., Ulm, M.J., Grindeland, R.E., Durnova, G.N. and Kaplansky, A.S. (1992) Adaptations of young adult rat cortical bone to 14 days of spaceflight. J. Appl. Physiol. 73, 4S–9S.

Washburn, D.A., Rumbaugh, D.M., Richardson, W.K., Gulledge, J.P., Shlyk, G.G. and Vasilieva, O.N. (2000) PTS performance by flight- and control-group macaques. J. Gravit. Physiol. 7, S89–S93.

Weisbrodt, N.W., Booth, F.W., Lai, M. and Thomason, D.B. (1994) mRNA levels in smooth muscle. *In* Connolly, J.P., Grindeland, R.E. and Ballard, R.W. (eds.). *Final Reports of U.S. Experiments Flown on the Soviet Biosatellite Cosmos 2044*, NASA Technical Memorandum 108802.

Wronski, T.J. and Morey, E.R. (1983) Effect of spaceflight on periosteal bone formation in rats. Am. J. Physiol. 244, R305–R309.
Yagodovsky, V.S., Triftanidi, L.A. and Gorokhova, G.P. (1976) Space flight effects on skeletal bones of rats (light and electron microscopic examination). Aviat. Space Environ. Med. 47, 734–738.
Young, D.R., Howard, W.H., Cann, C. and Steele, C.R. (1979) Noninvasive measures of bone bending rigidity in the monkey (m. nemestrina). Calcif Tissue Int. 27, 109–115.
Young, D.R., Niklowitz, W.J. and Steele, C.R. (1983) Tibial changes in experimental disuse osteoporosis in the monkey. Calcif Tissue Int. 35, 304–308.
Zerath, E., Holy, X., Malouvier, A., Caissard, J.C. and Nogues, C. (1991) Rat and monkey bone study in the Biocosmos 2044 space experiment. Physiologist 34, S194–S195.
Zerath, E., Holy, X., Andre, C., Renault, S., Noel, B., Delannoy, P., Hott, M. and Marie, P.J. (2000) Effects of Bion 11 14-day space flight on monkey iliac bone. J. Gravit. Physiol. 7, S155–S156.
Zernicke, R.F., Vailas, A.C., Grindeland, R.E., Kaplansky, A., Salem, G.J. and Martinez, D.A. (1990) Spaceflight effects on biomechanical and biochemical properties of rat vertebrae. Am. J. Physiol. 258, R1327–R1332.

Experimentation with Animal Models in Space
G. Sonnenfeld (editor)

DOI: 10.1016/S1569-2574(05)10004-5

Mouse Infection Models for Space Flight Immunology

Stephen Keith Chapes and Roman Reddy Ganta

Division of Biology and Department of Diagnostic Medicine and Pathobiology, Kansas State University, Manhattan, KS 66506, USA

Abstract

Several immunological processes can be affected by space flight. However, there is little evidence to suggest that flight-induced immunological deficits lead to illness. Therefore, one of our goals has been to define models to examine host resistance during space flight. Our working hypothesis is that space flight crews will come from a heterogeneous population; the immune response gene make-up will be quite varied. It is unknown how much the immune response gene variation contributes to the potential threat from infectious organisms, allergic responses or other long term health problems (e.g. cancer). This article details recent efforts of the Kansas State University gravitational immunology group to assess how population heterogeneity impacts host health, either in laboratory experimental situations and/or using the skeletal unloading model of space-flight stress. This paper details our use of several mouse strains with several different genotypes. In particular, mice with varying MHCII allotypes and mice on the C57BL background with different genetic defects have been particularly useful tools with which to study infections by *Staphylococcus aureus*, *Salmonella typhimurium*, *Pasteurella pneumotropica* and *Ehrlichia chaffeensis*. We propose that some of these experimental challenge models will be useful to assess the effects of space flight on host resistance to infection.

Space flight affects the immune response

The bold plan to return to the Moon and to send humans to Mars has excited people around the world (Lawler, 2004). However, the announcement raises several questions about safety and the maintenance of astronaut health. Although preflight screening, quarantine protocols and limited duration space flights do not allow for long-term decline of the immune system and the manifestation of severe illnesses, there is considerable evidence that astronauts may be more susceptible to disease during space flight (Taylor, 1993). Taylor noted that during long Skylab missions "...the mean scores for gingival inflammation and dental calculus approximately doubled over preflight values during . . . 30–84 days in space . . .'' (Taylor, 1993). The observation that delayed-type hypersensitivity reactions (a complex process involving multiple components of the immune system) are depressed during even short-term space flight (Taylor and Janney, 1992) can only mean that astronaut health would eventually be compromised. Even innate immune responses are reported to be affected by short-term space flights (Mehta *et al.*, 2001). When the immune system does not function normally, such as in the normal function of interferon-γ (Dalton *et al.*, 1993; Huang *et al.*, 1993), host resistance dramatically declines. Therefore, the loss of immune function during space flight cannot be taken lightly.

Space flight suppresses hematopoietic differentiation of macrophages as well as other blood cells (Vacek *et al.*, 1983; Sonnenfeld *et al.*, 1990, 1992; Ichiki *et al.*, 1996). It also decreases the number of circulating blood monocytes (Taylor *et al.*, 1986), induces the release of abnormal monocytes that lack the expression of insulin growth factor receptors (Meehan *et al.*, 1992), or alters the leukocyte subpopulations in the bone marrow and spleen (Pecaut *et al.*, 2003). A murine model used to simulate some of the physiological changes associated with space flight, antiorthostatic suspension (hindlimb unloading), also diminishes the number of macrophage progenitor cells in the bone marrow and affects hematopoiesis (Dunn *et al.*, 1983, 1985; Armstrong *et al.*, 1993, 1994, 1995b; Sonnenfeld *et al.*, 1992). The effects of space flight or skeletal unloading on macrophage hematopoiesis may explain why human peripheral blood lymphocyte blastogenic responses are generally suppressed post flight (Taylor and Dardano, 1983; Taylor *et al.*, 1986) and delayed-type hypersensitivity responses are suppressed (Taylor and Janney, 1992). Lower numbers or defective monocytes may not be able to effectively work as accessory cells. This hypothesis is supported by the observation that *in vitro* T cell blastogenesis is depressed by space flight only when monocytes are not adhered to a substrate and cannot effectively secrete IL-1 (Cogoli *et al.*, 1993). Moreover, it appears that space flight can disrupt the normal immune system not only in humans but also in rat (Ichiki *et al.*, 1994, 1996; Chapes *et al.*, 1999a; Lesnyak *et al.*, 1996), mouse (Gridley *et al.*, 2003; Pecaut *et al.*, 2003) and rhesus monkey (Sonnenfeld *et al.*, 1996).

Skeletal unloading using hindlimb unloading affects the immune response

The first question that arises when the term ''antiorthostatic suspension'' is mentioned is, ''what is antiorthostatic suspension?'' It is a technique developed in response to the need to study physiological systems that change in a microgravity environment, now commonly referred to as hindlimb unloading. The original model described by Morey and subsequent variations are often referred to as hindlimb unloading (Morey, 1979; Morey *et al.*, 1979). However, other names are commonly used including, antiorthostatic suspension, spinal traction, tail traction, harness suspension, and skeletal unloading with head down tilt. These techniques have been used to investigate several kinds of physiological changes from stress hormones to musculoskeletal changes (Musacchia and Steffen, 1983; Morey-Holton and Arnaud, 1991). The technique allows for hindlimb unloading while forelimbs remain weight bearing. The rodent is still able to move due to its ability to grasp the floor grid. Animals are able to move at will and have reduced stress levels (Wronski and Morey-Holton, 1987; Chapes *et al.*, 1993). We have reviewed the use and development of the hindlimb unloading model and its use in immunological studies (Chapes *et al.*, 1993). Aspects of the hindlimb unloading model have been revisited more recently (Morey-Holton and Globus, 2002); therefore, it is not the aim of this current work to review the usefulness of hindlimb unloading as a model for space flight. However, it is important to point out that hindlimb unloading is probably the most widely used model to induce changes in host physiology that can also be seen during space flight; and in particular immune function. For example, direct comparison of splenic lymphocyte proliferation and lymph node lymphocyte proliferation shows that hindlimb unloading has a lymphoid organ-dependent effect in mice (Armstrong *et al.*, 1993) and rats (Nash *et al.*, 1991) and there appears to be some consistency between space flight and hindlimb unloading with regard to mitogen-induced cellular proliferation (Chapes *et al.*, 1993). There is also some suggestion that hematopoiesis can be impacted in both systems (Sonnenfeld *et al.*, 1990, 1992). Pecaut *et al.* (2000) compared hindlimb unloading combined with simulations of the enhanced gravitational effects of launch and compared several immunological parameters to space flight. Those investigators were disappointed with the lack of correlation in spleen cell phenotype changes. However, there are even inconsistencies in changes in rat hematopoiesis when measured after space flight where some see a space flight effect (Vacek *et al.*, 1983; Ichiki *et al.*, 1994, 1996; Sonnenfeld *et al.*, 1990, 1992) and others do not (Chapes *et al.*, 1999a, b) Therefore, there are aspects of the "unloading" experience that are not understood and hard to duplicate, even in space flight. This inconsistency is not limited to immunophysiology. Changes in bone are space-flight as well as equipment dependent (Wronski *et al.*, 1998).

Pathogenesis studies

Even though immunological measurements have become routine for astronauts and other animals during space flight, there is little evidence to suggest that flight-induced immunological deficits lead to illness. The facilities and technology also have not allowed for pathogen challenge experiments to test that hypothesis. Recent data from several space flight experiments done *in vitro* and *ex vivo* post flight suggest that macrophage cytokine secretion may be exacerbated (Chapes *et al.*, 1992; Chapes and Beharka, 1998; Cogoli *et al.*, 1993). The over production of macrophage cytokines and the subsequent development of immunopathologies, such as cachexia (Tracey, 1992) could be just as devastating as immunosuppression to host health. Therefore, immunological disregulation and its manifestations during space flight need to be understood.

Although pathogen challenge studies have not been done during space flight, some challenges have been done in hindlimb-unloaded mice by the Sonnenfeld group. For example, mice that were resistant to infection with the encephalomyocarditis virus D variant showed decreased resistance to infection when antiorthostatically unloaded (Gould and Sonnenfeld, 1987). Pathogenesis was associated with depressed interferon-γ concentrations in the mice. This same group found that enhanced interferon-γ transiently induced at the start of hindlimb unloading was protective against an intracellular pathogen, *Listeria monocytogenes* (Miller and Sonnenfeld, 1993, 1994). Therefore, the timing and innate host resistance at the time of infection may play an important role in host defense and survival. More recent studies have found that *Klebsiella pneumoniae* infections are exacerbated by hindlimb unloading (Belay and Sonnenfeld, 2002). Although the investigators have not identified the immune system components that were compromised, stress hormones elevated during hindlimb unloading could enhance bacterial growth *in vitro* (Aviles *et al.*, 2003; Sonnenfeld *et al.*, 2002).

Although the *K. pneumoniae* challenges were not done by the natural route of infection (Belay and Sonnenfeld, 2002) the data raise a concern about the impact of space flight on host resistance to disease. One only need to look at the impact the loss of part of the immune response to see what might happen during flight. For example, slight alterations in the balance between T_H1 and T_H2 cytokines produced could compromise the cell-mediated immune response (Kaufmann and Ladel, 1994; Forsthuber *et al.*, 1996; Romagnani, 1997; Wakeham *et al.*, 1998) and weaken host resistance. The development of disease and pathogenesis in immunologically compromised mice that lack single or multiple immune response genes also exemplify possible astronaut health outcomes of long-term space flight (Mak and Simard, 1998). Moreover, immunocompromise can be complicated by the reactivation of latent virus infections (Stowe *et al.*, 2001a, b; Mehta *et al.*, 2004) and the microbes normally isolated from space craft (Ott *et al.*, 2004; Pierson, 2001).

The gravitational immunology group at Kansas State University has completed several studies designed to understand the effects of space flight or hindlimb unloading on macrophage differentiation, function and secretion products (Fleming *et al.*, 1990, 1991; Chapes *et al.*, 1992, 1993; Kopydlowski *et al.*, 1992; Armstrong *et al.*, 1993; Armstrong and Chapes, 1994; Woods and Chapes, 1994). For example, bone marrow macrophages were tested for their ability to produce cytokines in space during shuttle flights STS-37 and STS-43. Lipopolysaccharide-activated bone marrow macrophages secreted significantly more interleukin-1 and tumor necrosis factor-α when stimulated in space compared to cells stimulated on Earth (Chapes *et al.*, 1992). The enhanced TNF-α secretory response was subsequently confirmed by Cogoli's group

(Cogoli *et al.*, 1993) and demonstrates that macrophage function is significantly affected by space flight. We also have completed some investigations that indicate that signal transduction is significantly affected by space flight. We found that space flight abrogated TNF-mediated killing of LM929 cells (Woods and Chapes, 1994). Experiments conducted on space shuttle missions STS-54 and STS-57 indicated that inhibitors of protein kinase-C activation restored TNF-mediated cytotoxicity to normal levels during space flight and suggested that space flight affects protein kinase-C. Similar suggestions have been made by others (deGroot *et al.*, 1991; Limouse *et al.*, 1991; Schmitt *et al.*, 1996), and may be isoform dependent (Hatton *et al.*, 1999, 2002). We have also found that whole animal immunological responsiveness can be affected negatively and positively based on rats flown on three shuttle flights (Chapes *et al.*, 1999a and b; Bateman *et al.*, 1998). These space studies have been complemented by KC-135 flights that have demonstrated that macrophages and inflammatory cells are affected directly by microgravity (Armstrong *et al.*, 1995a; Fleming *et al.*, 1991).

Using hindlimb unloading modeling, we have investigated neutrophil function and killing of bacteria (Fleming *et al.*, 1990), examined peritoneal macrophage function (Kopydlowski *et al.*, 1992) and used hindlimb unloading to induce depressed macrophage hematopoiesis, which occurs during space flight (Armstrong *et al.*, 1993; Chapes *et al.*, 1993). We have used this latter system to investigate mechanisms for the suppression. For example, bone marrow macrophage progenitor numbers were depressed in hindlimbs that were unloaded (femora) and the forelimbs that were not unloaded (humeri); suggesting the systemic nature of the unloading event on bone marrow macrophage differentiation (Armstrong *et al.*, 1994). The depression of macrophage progenitor cells correlated with a depressed ability of bone marrow cells to produce both colony stimulating factor-1 and interleukin-6, but not transforming growth factor-β (Armstrong *et al.*, 1994, 1995b). These studies have helped to define some therapies that might be useful as countermeasures.

We have made significant progress in attempting to elaborate the effects of space flight, microgravity or hindlimb unloading on leukocytes and animal immune responses. However, our studies have not addressed changes in host resistance that are altered by space flight. Therefore, one of our goals has been to define models that may be useful for examination of host resistance during space flight. Our working hypothesis has been that individuals subjected to space flight come from a heterogeneous population; the immune response gene make-up of the astronauts is quite varied. It is unknown how much the immune response gene variation contributes to the potential threat from infectious organisms, allergic responses or other long term health problems (e.g., cancer). It would be valuable to have quantitative data to show how population heterogeneity impacts host health either in laboratory experimental situations and/or during space flight. Because of the complexity of heterogeneous populations, we chose to develop simple models to begin to acquire these data. The availability of mice with functional and defective alleles, in several combinations of innate and acquired immune components offers the opportunity to examine the level of host response necessary for resistance to a number of microorganisms. In addition, we wanted a model that would allow us the opportunity to examine whether increasing the number of defective immune response genes along with the physiological changes induced by space flight would impact host resistance. The additive pressure of increasing the number of defective immune response genes may trigger immunological problems that may not be manifested on Earth.

The mouse model

We have used mouse strains with several different genotypes. Mice with varying MHCII allotypes and mice on the C57BL background with different genetic defects have been particularly useful tools with which to study infections. The mouse strains span the range of immunocompetence and allow for the assessment of risk of different pathogens (Tables 1 and 3). Stock-Abb^{tm1}, MHCII$^{-/-}$, *Tlr4*$^{LPS\text{-}del/LPS\text{-}del}$, *Slc11a1*$^{s/s}$ (B10×C2D) mice lack critical components necessary for the development of both acquired and innate immunity and are susceptible to the most benign of opportunistic pathogens, including *Pasteurella pneumotropica* and *Pneumocystis carinii* (Wright and Chapes, 1999). C57BL/10ScN mice lack a functional *Tlr4* gene (Poltorak *et al.*, 1998), but do not carry the IL-12 defect reported in C57BL/10 ScCr mice (Merlin *et al.*, 2001). They have functional acquired immunity and have proven useful in the monitoring of pathogens that require immediate activation of the macrophage system in the lung for complete host resistance (Chapes *et al.*, 2001; Hart *et al.*, 2003). In contrast, B6.129-*Abb*^{tm1} (C2D) mice have functional innate immunity but lack CD4^{+} T cells because of the absence of MHCII molecules. They are susceptible to pathogens that need the activation of helper-T-cells for host resistance (Grusby *et al.*, 1991; Grusby and Gilmcher,

Table 1

Mouse strains

Mouse	Abbreviation	Genotype
C57BL/6J	B6	MHCII$^{+/+}$, *Tlr4*$^{n/n}$
C57BL/10ScN	B10	MHCII$^{+/+}$, *Tlr4*$^{d/d}$
B6.129-*Abb*^{tm1} N5F20	C2D	MHCII$^{-/-}$, *Tlr4*$^{n/n}$
Stock-*Abb*^{tm1}, MHCII$^{-/-}$, *Tlr4*$^{LPS\text{-}del/LPS\text{-}del}$, *Slc11a1*$^{s/s}$	B10×C2D	MHCII$^{-/-}$, *Tlr4*$^{d/d}$
B6.129S6-*Cd4*^{tm1Knw}	CD4D	MHCII$^{+/+}$, *Tlr4*$^{n/n}$

1995). B6.129S6-*Cd4*^{tm1Knw} (CD4D) mice express MHCII but do not express CD4 molecules (McCarrick *et al.*, 1993). They also lack $CD4^+$ helper T-cells. C57BL/6J mice represent the wild-type genotype of mice on the C57BL background. Balb/cJ ($MHCII^{+/+}$, IA^d, *Tlr4*$^{LPS-n}$), C3HeB/FeJ ($MHCII^{+/+}$, IA^k, *Tlr4*$^{LPS-n}$, FeJ) and C3H/HeJ ($MHCII^{+/+}$, *Tlr4*$^{LPS-d}$, HeJ) mice carry different MHCII alloztypes from C57BL mice. FeJ mice were embryo-derived from HeJ mice and have the same genetic background as HeJ mice (Festing, 1994). The subsequent spontaneous mutation of the *Tlr4* gene (Vogel *et al.*, 1979; Vogel, 1992) at Jackson Laboratories between 1960 and 1965 allows for congenic comparisons between FeJ and HeJ mice. Balb/c mice have a functional *Tlr4* gene but carry nonfunctional *Slc11a1* alleles. These mice have been described in detail previously with the genetic recombinant B10×C2D mouse having been created at Kansas State University (Wright and Chapes, 1999; Chapes *et al.*, 2001; Hart *et al.*, 2003).

Staphylococcus aureus and Salmonella typhimurium

We examined the pathogenesis of the pyogenic bacterium, *Staphylococcus aureus* and the facultative, intracellular bacterium, *Salmonella typhimurium* in C2D and wild-type, C57BL/6J mice. *S. aureus* is commonly isolated from the space craft (Pierson, 2001) and *Salmonella* always pose a potential risk where food may need to be stored for long periods of time (e.g., a flight to Mars). There was only a narrow dose range where the MHCII knock-out affected mouse survival after *S. aureus* injection compared to wild-type mice (Table 2). Using cyclophosphamide-induced neutropenic mice, we confirmed that neutrophils were the critical host resistance mechanism to *S. aureus* (Devalon *et al.*, 1987) and the lack of $CD4^+$ T cells or the absence of MHCII played

Table 2

Effect of injecting live *Staphylococcus aureus* (ISP 479C) into C2D and B6 mice

Mouse strain	CFU injected[a]	# sick at 6–8 h	# died at 12 h	# died total
C57BL/6J	1×10^7	0/10	0/10	0/10
C57BL/6J	1×10^8	0/15	0/15	0/15
C57BL/6J	1×10^9	3/15	0/15	0/15
C57BL/6J	5×10^9	3/5	1/5	3/5
C57BL/6J	1×10^{10}	10/15	9/15	14/15
C2D ($MHCII^{-/-}$)	1×10^7	0/10	0/10	0/10
C2D ($MHCII^{-/-}$)	1×10^8	2/15	0/15	0/15
C2D ($MHCII^{-/-}$)	1×10^9	10/15[b]	0/15	5/15[b]
C2D ($MHCII^{-/-}$)	5×10^9	5/5	1/5	4/5
C2D ($MHCII^{-/-}$)	1×10^{10}	14/15	12/15	15/15

[a]Bacteria injected i.p. and mice were observed for 72 h.
[b]Significantly different from wild-type mice, $p < 0.05$.

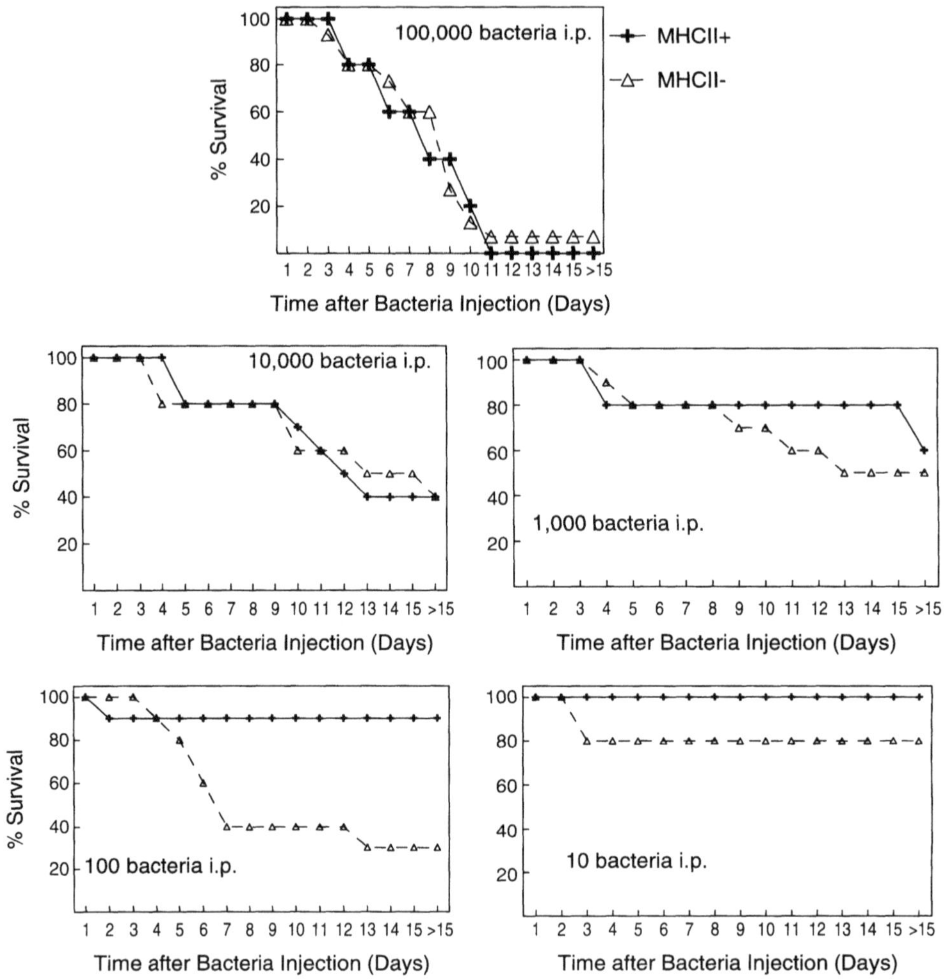

Fig. 1. Survival kinesics of C2D (MHCII$^{-/-}$) and B6 (MHCII$^{+/+}$) mice after treatment with 150 mg/kg of 2H-1,3,2-oxazaphosphorine (cyclophosphamide) and injection of various doses of *S. aureus*; 10–15 mice per treatment group.

a significantly lesser role (Fig. 1). This experimental system would be an excellent model with which to test the effects of space flight on neutrophil function. Transient neutrophilias have been reported in animals and men in postflight analyses (Allebban *et al.*, 1994; Kimzey, 1977; Taylor, 1988). However, analysis of rats during flight reveals that the neutrophilia was probably due to the stresses of launch and landing (Ichiki *et al.*, 1996). There is controversy about the effects of hindlimb unloading on PMN function (Fleming *et al.*, 1990; Miller *et al.*, 1994; Smolen *et al.*, 2000); however, nothing is known about how neutrophil function would be affected over long periods of space flight. Therefore, this model would allow for the definition of neutrophil function in space.

The MHCII knock-out significantly reduced the dose of *Salmonella typhimurium* needed to kill mice (Chapes and Beharka, 1998) (Fig. 2). Notably, MHCII knock-out mice had significantly higher serum IL-10 concentrations than B6 mice and their macrophages secreted significantly more IL-10 and less NO and O_2 after LPS stimulation *in vitro* than wild-type macrophages (Chapes and Beharka, 1998).

Obviously, we will not be sending astronauts into space with such dramatic MHCII defects. However, these data suggest that if space flight disrupts immunological functions controlled by inappropriate MHCII expression, $CD4^+$ T cell function or the balance of cytokines produced, there could be health consequences after infection. B6 mice are generally considered "immunocompetent" and have a complex immune response gene haplotype. Similar complexity would be typical of human astronauts. However, B6 mice are at increased risk from *Salmonella* because of their genotype (Nauciel *et al.*, 1988). We speculated that part of that susceptibility may have been because of a poor interferon-γ response following infection. We tested that hypothesis by injecting B6 mice with the potent IFN-γ inducer, *P. acnes*

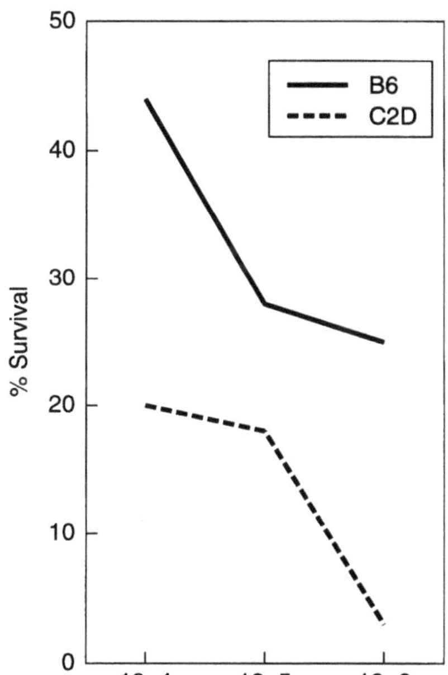

Fig. 2. Survival kinesics of C2D ($MHCII^{-/-}$) and B6 ($MHCII^{+/+}$) mice after i.p. injection of various doses of *S. typhimurium*. Survival determined one month after experimental challenge.

(Fantuzzi and Dinarello, 1996). We found Balb/cJ and FeJ mice challenged with a lethal dose of *Salmonella* 4 days after the injection of *P. acnes* survived longer and in higher numbers than mice not primed with *P. acnes* (Table 3). B6 mice did not exhibit any benefits from a *P. acnes* injection. In fact, their survival rate was relatively low; FeJ mice had over 95% survival compared to <30% survival for B6 mice.

C3HeB/FeJ mice are *Slc11a1*$^{r/r}$ and IA^{k} and B6 mice are *Slc11a1*$^{s/s}$ and IA^{b}. Innate resistance to *Salmonella* has been linked to the *Slc11a1*^{r} allele (the *Slc11a1* gene was formerly referred to as the *ity* or *Nramp1* gene) (Govoni *et al.*, 1996; Plant and Glynn, 1976). Since IA^{b} mice have been found to be more sensitive to *Salmonella* infections (Nauciel *et al.*, 1988) than other haplotypes, it was unclear if the ability to respond successfully to *Salmonella* challenge was more dependent on IA gene expression or to differences in *Slc11a1* alleles. When Balb/c, C57BL/6J, C3HeB/FeJ and C3H/HeJ mice were challenged with *Salmonella* 4 days following an i.p. injection of *P. acnes*, Balb/c mice were comparable in their resistance to *Salmonella* challenge after priming with *P. acnes*, as FeJ mice (Table 3). These data suggested that resistance after *P. acnes* therapy was not as strongly linked to the *Slc11a1*^{s} expression as it was to the IA^{b} expression. We found that resistance required a potent IFN-γ inducer to provide protection because thioglycollate injection 4 days before *Salmonella* challenge provided no protection (Table 3). We also found that HeJ mice, without functional *Tlr4* genes also did not acquire increased resistance to *Salmonella* after *P. acnes* therapy (Table 3); this is indicative of the fact that host resistance to *Salmonella* is under complex genetic regulation. Nevertheless, the data illustrate the usefulness in using congenic mice in evaluating host immune responsiveness and would be powerful tools for use in determining impact of space flight on a particular genotype.

Table 3

Impact of MHCII, *Tlr4* and *Slc11a1* genotype on host resistance to *Salmonella typhimurium*[a]

Mouse strain	Genotype	Therapy[a]	Survival #[b]	Survival %[b]
Balb/cJ	I-A$^{d/d}$, MHCII$^{+/+}$, *Tlr4*$^{n/n}$,*Slc11a1*$^{s/s}$	Thioglycollate	7/18	39
		P. acnes	23/30	77
C57BL/6J	I-A$^{b/b}$, MHCII$^{+/+}$, *Tlr4*$^{Lps\text{-}n/Lps\text{-}n}$, *Slc11a1*$^{s/s}$	Thioglycollate	5/18	28
		P. acnes	8/30	27
C3HeB/FeJ	I-A$^{k/k}$, MHCII$^{+/+}$, *Tlr4*$^{Lps\text{-}n/Lps\text{-}n}$, *Slc11a1*$^{r/r}$	Thioglycollate	8/18	44
		P. acnes	26/30	87
C3H/HeJ	I-A$^{k/k}$, MHCII$^{+/+}$, *Tlr4*$^{Lps\text{-}d/Lps\text{-}d}$, *Slc11a1*$^{r/r}$	Thioglycollate	0/12	0
		P. acnes	4/24	17

[a]Mice challenged with 5×10^{5} bacteria either i.p. or s.c. 4 days after i.p. injection of 700 μg of heat-killed *P. acnes*.

[b]Survival measured ≥30 days after *Salmonella* injection.

Pasteurella pneumotropica

The respiratory tract is a unique system with a characteristic branched morphology, much like an upside-down tree. The lung is different from many other host organs in that it is exposed to the external environment, it is highly vascularized and it can be considered a mucosal immune organ. Interestingly, many species use the lung as the primary filtering apparatus superceding the liver (Chitko-McKown *et al.*, 1991, 1992). Inflammatory cells are vital to the resolution of lung infections (Lukacs *et al.*, 1995). CD18 is important for leukocyte recruitment (Yu and Limper, 1997; Mizgerd *et al.*, 1997; Doherty *et al.*, 1994) and their cognate receptor, ICAM-1, appears to be necessary on pulmonary endothelium for a successful host response (Yu and Limper, 1997). The inflammatory cells make both oxidative and nonoxidative products to help resolve lung infections (Aratani *et al.*, 1999; Jensen *et al.*, 1998). The lung was also identified as a potential target of space flight (Todd *et al.*, 1993, 1994). Particulates that normally fall on Earth remain suspended and can be inhaled. Changes in pressure because of extra vehicular activities, fluid shifts, risks from radiation and stress can also contribute to impaired pulmonary function.

Studies with knock-out and immune-deficient mice have provided insights into some lung immune mechanisms. IL-12-knock-out mice fail to produce both type 1 and type 2 cytokines following infection with *Mycobacteria* (Wakeham *et al.*, 1998) leading to impaired host clearance of the organism. This is consistent with observations that IFN-γ plays an important role in immune clearance of *Chlamydia pneumoniae* (Penttila *et al.*, 1998) and other pathogens (Neumann *et al.*, 1998). IL-12 regulates the production of IFN-γ (Sutterwala and Mosser, 1999). Therefore, the classical T-cell-macrophage delayed-type hypersensitivity response appears to be active in the lung. Indeed, macrophages are critical for the removal of *P. carinii* and other pathogens (Hickman-Davis *et al.*, 1997; Limper *et al.*, 1997).

In addition to IL-12 and IFN-γ, other protective cytokines are produced in the lung. For example, lung fibroblasts constitutively produce several cytokines and chemokines that can maintain immune cells *in situ*, in particular, monocytes and macrophages (Koyama *et al.*, 1998). Cytokines and chemokines are also important to the recruitment of inflammatory neutrophils and monocytes during infection (Lukacs *et al.*, 1995; Debs *et al.*, 1988; Sutterwala and Mosser, 1999). T-cell-dependent granuloma formation appears to be particularly dependent on monocyte-chemoattractant protein-1 (MCP-1), which is produced in high concentrations in response to both TNFα and IL-1 (Ichiyasu *et al.*, 1999). Therefore, components of both innate and acquired immunity play a role in host lung protection.

Because the lung may be a vulnerable organ during space flight, we think it is important to have *in vivo* models that allow us to monitor the impact of space flight on pulmonary function and immune responses. In recent years we have characterized four mouse strains on the C57BL background (Table 1) with different genetic defects to study lung infections by the opportunistic pathogen *P. pneumotropica*. We propose this as a space model system because *P. pneumotropica* does not pose a significant threat to humans unless they are severely immunocompromised (Cuadrado-Gomez *et al.*, 1995; Campos *et al.*, 2000). Even long-term Mir missions have not resulted in such immunocompromise. Therefore, we would not anticipate any human risks. Moreover, because the four mouse strains span the range of immunocompetence, experimental *P. pneumotropica* challenges during space flight will allow for the assessment of risk to different components of the immune system. For example, because C57BL/10ScN mice lack a functional *Tlr4* gene but have functional acquired immunity, most infections in this mouse strain do not result in death because the cell-mediated immune response can respond. If space flight impairs the cell-mediated immunity, experimental *P. pneumotropica* infections would become more severe. Alternatively, C2D mice (B6.129-Abb^{tm1}) can serve as sentries of the impact of space flight on innate immunity. These mice do not develop pneumonia after a respiratory challenge with *P. pneumotropica*. Therefore, if these mice develop pneumonia, it would indicate an effect of space flight on host resistance mediated by immune components. If B6 mice develop pneumonia after experimental *P. pneumotropica* infection, it would warn of a potentially devastating effect of space flight on the respiratory immune response. B10×C2D mice, lacking both helper-T-cells as well as early innate immunity mediated by Tlr4 receptors, critical components necessary for the development of both acquired and innate immunity, have served as the benchmark of what can happen under circumstances when even opportunistic pathogens pose a health risk (Wright and Chapes, 1999; Chapes *et al.*, 2001; Hart *et al.*, 2003). They have been valuable in studies looking at the therapeutic potential of countermeasures (Hart *et al.*, 2003).

To test the value of this mouse model for space flight research, we have initiated studies on B10 mice that have been hindlimb-unloaded for 8 days (see Armstrong *et al.*, 1993, 1994 for technical details of hindlimb unloading). Mice were challenged with approximately 5×10^9 CFU of *P. pneumotropica* on the day of suspension. Mice were euthanized at day 8 and lungs were visually assessed for pneumonic lesions and for the presence of bacteria by PCR assay of *P. pneumotropica* DNA isolated from lung tissue using the *P. pneumotropica*-specific primers. Densitometric comparisons between *P. pneumotropica* DNA and the mammalian housekeeping gene S14 DNA were performed on ethidium bromide-stained agarose gels as previously described (Chapes *et al.*, 2001). We found that there were over 100% more bacteria in lungs from hindlimb-unloaded mice compared to tail restraint controls (Table 4). This correlated with the severity of the bacterial lesions that were observed in the pathological assessment. These data indicate that hindlimb unloading can delay or inhibit the helper-T-cell response that is necessary for B10 mice to clear the infection. Furthermore, these data exemplify how this model can be useful for assessing changes in host resistance.

Table 4

Effect of hindlimb unloading on *Pasteurella pneumotropica* infection in C57BL/10ScN *Tlr4*-gene deficient mice

Treatment	Assay	Lung assessment after hindlimb unloading for 8 days
Hindlimb unloading	Lesion score[a]	2.1
	Bacterial load[b]	0.5[c]
Tail restraint	Lesion score[a]	1.5
	Bacterial load[b]	0.2

[a]Lesion scores were assessed by examining lungs for evidence of bacterial-induced pneumonia. Scores were determined after 8 days of hindlimb unloading. Mice were infected immediately prior initiation of the suspension period. See Hart *et al.*, 2003.
[b]Bacterial load was determined by using a densitometric ratio of *P. pneumotropica* DNA/S14 DNA and imaged with ethidium bromide after agarose gel electrophoresis. See Hart *et al.*, 2003.
[c]Statistically different from tail restraint mice using a Mann–Whitney test ($P < 0.02$); $n = 8$ mice per treatment group (2 experiments combined).

Ehrlichia chaffeensis

Experiments using *S. aureus*, *S. typhimurium* or *P. pneumotropica* or even pathogens used by other investigators (e.g., *K. pneumonia*) to infect mice during space flight would yield novel and informative data on the impact of space flight on host resistance. The use of congenic and knock-out mice would allow us to directly determine what components of the immune response are impacted. However, there may be containment issues about facilities that are available for the use of mice (Dalton *et al.*, 2003). Until facilities are available that will completely isolate animals from astronauts during shuttle flight or on space station, there may be some concerns about the use of potential human pathogens. The possible spread of these microorganisms might be viewed as too high a risk. Although these might be the more appropriate pathogens and model systems with which to test host resistance and immunocompetence, alternative model systems where the spread of the bacteria is not seen as problematic will also be needed. To that end, we have been experimenting with an additional infection model that appears to be compatible with the requirement that the organism not be highly contagious.

We have been working to understand host resistance to the tick-born pathogen *Ehrlichia chaffeensis*. *E. chaffeensis* causes human monocytic ehrlichiosis and impacts mostly immune compromised people (Paddock *et al.*, 1993, 1997). We have used various combinations of knock-out mice to establish the contributions of various immunological components to the clearance of the organism (Ganta *et al.*, 2002, 2004). Although this is a human pathogen, it is only spread by direct contact with blood and is not highly contagious. Therefore, it offers advantages over the other bacterial models we are interested

Table 5

Clearance of *E. chaffeensis* in mice

Mouse	Affected immune component	Time to cure[a]
C57BL/6J (B6)	Wild-type/normal	14–16 days
C57BL/10ScN (B10)	Tlr4 receptor-deficient	> 30 days
B6.129-*Abb*^{tm1} N5F20 (C2D)	MHCII and CD4 T-cell deficient	> 120 days
B6.129S6-*Cd4*^{tm1Knw} (CD4D)	CD4 T-cell deficient	≥27 days

[a]Clearance determined by the presence of *E. chaffeensis* by culture assay. See Ganta *et al.*, 2002 and 2004.

in because it is not a respiratory or intestinal pathogen. Furthermore, although some mouse strains are bacteremic for long periods of times, death is not a final outcome of the infection; even with mice lacking several immune response genes (Ganta *et al.*, 2004). Mice can be injected i.p. with *E. chaffeensis* and immunocompetence can be gauged by the time it takes to clear bacteria. Using this paradigm, we have established the impact of bacterial clearance in the absence of various immune response components (Table 5). Wild type mice clear primary infections in about two weeks. The absence of *Tlr4* gene delays clearance to times longer than 30 days, sometimes 40 days. The absence of $CD4^+$ T-cells results in clearance in about 4 weeks. Mice that lack MHCII molecule expression develop long lasting bacteremia. Therefore, if mice are infected with *E. chaffeensis* before flight, the kinetics of clearance of the bacteria in various mice or the change from a nonlethal to a lethal infection would provide significant insight into the role of various immunological mechanisms of host resistance. These mouse strains could be complemented by the addition and characterization of other immunological knock-out mice such as $CD8^+$ T-cell knock outs (B6.129S2-Cd8^{atm1Mak} {N13}). Moreover, traditional immune responses can be measured on these mice to correlate the disease cure with immune function.

The future of animals in space flight

The ultimate test of whether space flight impacts the immune system is the ability of people and animals to remain healthy or to recover from "routine" illness like colds. In fact, the October 1, 2003 NASA Biological and Physical Research Enterprise Strategic plan (Horne *et al.*, 2003) outlines that a major research priority is "to assure survival of humans traveling far from Earth" (Organizing question 1). In particular, it is important to identify how the human body adapts to space flight and identify effective countermeasures to those changes. Important research targets for the period of 2004–2008 outlined in the strategic plan are to study immune function and determine any increases in number or virulence of pathogens. The strategic plan outlines goals to develop and test new therapies for maintaining or enhancing immune function. It also addresses organizing question 2 in the NASA strategic plan: "How do space environments affect life at molecular and cellular levels" (Question 2a; research target for 2004–2008).

The announcement that a key goal for NASA is the return to the Moon and later prepare for a trip to Mars (Lawler, 2004) only exacerbates the need to understand what happens to the immune system during space flight. Moreover, there will not be enough astronaut hours in space to accomplish all of the tests and experiments needed to accomplish this monumental task. The need to understand space-induced changes in the musculoskeletal systems, along with the immune system, will require the input of vertebrate animal models. Given the powerful genetic, immunologic and experimental systems the mouse offers, NASA will need to move quickly in developing facilities to accomplish these goals.

The AEM has been the major rodent habitat used by NASA. It has flown on approximately two dozen missions. It supports approximately 1250–1500 g of animal mass and it is designed to be self-contained in a mid-deck locker (Bonting *et al.*, 1991). Total animal floor space, with water box installed, is 645 cm^2. This facility has been mostly dedicated for the use of rats and has only been used for mouse studies on STS-90 (Dalton *et al.*, 2003) and STS-108 (Gridley *et al.*, 2003; Pecaut *et al.*, 2003) because of odor issues (Dalton *et al.*, 2003). Therefore, although this facility may be used for some future mouse research, additional habitat facilities will be necessary.

The limitations of the AEM led NASA to authorize the construction of a rodent facility called the Advanced Animal Habitat-Centrifuge. According to NASA (Sarver, 2004) "The Advanced Animal Habitat-Centrifuge (AAH-C) is a research environment for laboratory rats and mice that will be orbiting for up to 90 days. The AAH-C will control temperature, humidity, and lighting, as well as food and water delivery, and waste management. An airflow rate of at least 10 changes per hour will prevent carbon dioxide and ammonia from accumulating in the specimen chamber. Air will be filtered and conditioned before being exchanged with the air in the Space Station environment; this will maintain bio-isolation between the crew and the specimens." In fact, in early 2004, a Life Sciences Advisory Subcommittee was charged to develop recommendations about the first rodent habitats for ISS for consideration by the Director of Fundamental Space Biology. The Task Force recommended that a mouse facility needs to be built and should closely follow the completion of the rat facility. The committee emphasized that some immediate funding should be targeted to facilitate the completion of a mouse facility as soon as possible. In addition, the committee felt that it is not prudent to limit studies on space station to male or female mice and that NASA needs a strategic plan for the research community to be able to use both rats and mice on Shuttle in middeck lockers. Therefore, there is every expectation that mouse infection models will be employed when this equipment comes on line. The models presented here will allow comprehensive analysis of the immune system in space.

Acknowledgments

The KSU gravitational immunology group has been supported past and present by the following grants: NASA NAG2-1274, NAGW-1197 and NCC5-168; American Heart Association grants KS-94-GS-33 and KS-97-GS-02; USDA Animal Health Funds Section 1433 grant 4-81895; the Army Research and Development Command, grant DAMD-17-89-Z-9039; NIH grants CA09418, AI052206, AI55052, AI50785, and RR16475; and the Kansas Agricultural Experiment Station. This is Kansas Agricultural Experiment Station Publication # 04-321-B.

References

Allebban, Z., Ichiki, A., Gibson, L., Jones, J., Congdon, C. and Lange, R. (1994) Effects of spaceflight on the number of rat peripheral blood leukocytes and lymphocyte subsets. J. Leukoc. Biol. 55, 209–213.

Aratani, Y., Koyama, H., Nyui, S., Suzuki, K., Kura, F. and Maeda, N. (1999) Severe impairment in early host defense against *Candida albicans* in mice deficient in myeloperoxidase. Infect. Immun. 67, 1828–1836.

Armstrong, J.A., Balch, S. and Chapes, S.K. (1994) Interleukin-2 therapy reverses some immunosuppressive effects of skeletal unloading. J. Appl. Physiol. 77, 584–589.

Armstrong, J.A. and Chapes, S.K. (1994) Effects of extracellular matrix proteins on macrophage differentiation, growth, and function: Comparison of liquid and agar culture systems. J. Exp. Zool. 269, 178–187.

Armstrong, J., Gerren, R. and Chapes, S.K. (1995a) The effect of space and parabolic flight on macrophage hematopoiesis and function. Exp. Cell Res. 216, 160–168.

Armstrong, J., Kirby-Dobbels, K. and Chapes, S. (1995b) The effects of rM-CSF and rIL-6 therapy on immunosuppressed antiorthostatically suspended mice. J. Appl. Physiol. 78, 968–975.

Armstrong, J.A., Nelson, K., Simske, S., Luttges, M., Iandolo, J.J. and Chapes, S.K. (1993) Skeletal unloading causes organ specific changes in immune cell responses. J. Appl. Physiol. 75, 2734–2739.

Aviles, H., Belay, T., Fountain, K., Vance, M., Sun, B. and Sonnenfeld, G. (2003) Active hexose correlated compound enhances resistance to *Klebsiella pneumoniae* infection in mice in the hindlimb-unloading model of spaceflight conditions. J. Appl. Physiol. 95, 491–496.

Bateman, T.A., Zimmerman, R.J., Ayers, R.A., Ferguson, V.L., Chapes, S.K. and Simske, S.J. (1998) Histomorphometric, physical and mechanical, effects of spaceflight and insulin-like growth factor-1 on rat long bones. Bone 23, 527–535.

Belay, T. and Sonnenfeld, G. (2002) Differential effects of catecholamines on *in vitro* growth of pathogenic bacteria. Life Sci. 71, 447–456.

Bonting, S., Kishiyama, J. and Arno, R. (1991) Facilities for animal research in space. Adv. Space Biol. Med. 1, 279–325.

Campos, A., Taylor, J.H. and Campbell, M. (2000) Hamster bite peritonitis: *Pasteurella pneumotropica* peritonitis in a dialysis patient. Pediatr. Nephrol. 15, 31–32.

Chapes, S.K. and Beharka, A.A. (1998) *Salmonella* infections in the absence of the major histocompatibility complex II. J. Leukoc. Biol. 63, 297–304.

Chapes, S.K., Mastro, A., Sonnenfeld, G. and Berry, W. (1993) Antiorthostatic suspension as a model for the effects of spaceflight on the immune system. J. Leukoc. Biol. 54, 227–235.

Chapes, S., Morrison, D., Guikema, J., Lewis, M. and Spooner, B. (1992) Cytokine secretion by immune cells in space. J. Leukoc. Biol. 52, 104–110.

Chapes, S.K., Mosier, D.A., Wright, A.D. and Hart, M.L. (2001) MHCII, *Tlr4* and *Nramp1* genes control host pulmonary resistance against the opportunistic bacterium *Pasteurella pneumotropica*. J. Leukoc. Biol. 69, 381–386.

Chapes, S.K., Simske, S.J., Forsman, A.D., Bateman, T.A. and Zimmerman, R.J. (1999a) Effects of space flight and IGF-1 on immune function. Adv. Space Res. 23, 1955–1964.

Chapes, S.K., Simske, S.J., Sonnenfeld, G., Miller, E.S. and Zimmerman, R.J. (1999b) Effects of spaceflight and PEG-IL-2 on rat physiological and immunological responses. J. Applied Physiol. 86, 2065–2076.

Chitko-McKown, C. and Blecha, F. (1992) Pulmonary intravascular macrophages: A review of immune properties and functions. Ann. Rech. Vet. 23, 201–214.

Chitko-McKown, C.G., Chapes, S.K., Brown, R.E., Phillips, R.M., Mckown, R.D. and Blecha, F. (1991) Porcine alveolar and pulmonary intravascular macrophages: Comparison of immune functions. J. Leukoc. Biol. 50, 364–372.

Cogoli, A., Bechler, B., Cogoli-Greuter, M., Criswell, S., Joller, H., Joller, P., Hunzinger, E. and Muller, O. (1993) Mitogenic signal transduction in T lymphocytes in microgravity. J. Leukoc. Biol. 53, 569–575.

Cuadrado-Gomez, L.M., Arranz-Caso, J.A., Cuadros-Gonzalez, J. and Albarran-Hernandez, F. (1995) *Pasteurella pneumotropica* pneumonia in a patient with AIDS. Clin. Infect. Dis. 21, 445–446.

Dalton, D., Pitts-Meek, S., Keshav, S., Figari, I., Bradley, A. and Stewart, T. (1993) Multiple defects of immune cell function in mice with disrupted interferon-γ genes. Science 259, 1739–1742.

Dalton, P., Gould, M., Girten, B., Stodieck, L.S. and Bateman, T.A. (2003) Preventing annoyance from odors in spaceflight: A method for evaluating the sensory impact of rodent housing. J. Appl. Physiol. 95, 2113–2121.

Debs, R., Fuchs, H., Philip, R., Montgomery, A., Brunette, E., Liggitt, D., Patton, J. and Shellito, J. (1988) Lung-specific delivery of cytokines induces sustained pulmonary and systemic immunomodulation in rats. J. Immunol. 140, 3482–3488.

deGroot, R., Rijken, P., den Hertog, J., Boonstra, J., Berkleij, A., de Laat, S. and Kruijer, W. (1991) Nuclear responses to protein kinase C signal trasduction are sensitive to gravity changes. Exp. Cell Res. 197, 87–90.
Devalon, M., Elliott, G. and Regelmann, W. (1987) Oxidative response of human neutrophils, monocytes, and alveolar macrophages induced by unopsonized surface-adherent *Staphylococcus aureus*. Infect. Immun. 55, 2398–2403.
Doherty, D., Downey, G., Schwab, B., III, Elson, E. and Worthen, G. (1994) Lipopolysaccharide-induced monocyte retention in the lung. J. Immunol. 153, 241–255.
Dunn, C., Johnson, P. and Lange, R. (1983) Hematopoiesis in antiorthostatic, hypokinesic rats. The Physiologist 26, S133–S134.
Dunn, C., Johnson, P., Lange, R., Perez, L. and Nessel, R. (1985) Regulation of hematopoiesis in rats exposed to antiorthostatic, hypokinetic/hypodynamia: I. Model description. Aviat. Space and Environ. Med. 56, 419–426.
Fantuzzi, G. and Dinarello, C.A. (1996) The inflammatory response in interleukin-1β-deficient mice: Comparison with other cytokine-related knock-out mice. J. Leukoc. Biol. 59, 489–493.
Festing, M.F.W. (1994) Inbred strains of mice. Companion issue to mouse genome 92, 373–374.
Fleming, S.D., Edelman, L. and Chapes, S.K. (1991) Effects of corticosterone and microgravity on inflammatory cell production of superoxide. J. Leukoc. Biol. 50, 69–76.
Fleming, S.D., Rosenkrans, C. and Chapes, S.K. (1990) Test of the antiorthostatic suspension model on mice: Effects on the inflammatory cell response. Aviat. Space and Environ. Med. 61, 327–332.
Forsthuber, T., Yip, H.C. and Lehmann, P.V. (1996) Induction of Th1 and Th2 immunity in neonatal mice. Science 271, 1728–1730.
Ganta, R.R., Cheng, C., Wilkerson, M.J. and Chapes, S.K. (2004) Delayed clearance of *Ehrlichia chaffeensis* infection in CD4(+) T-cell knockout mice. Infect. Immun. 72, 159–167.
Ganta, R.R., Wilkerson, M.J., Cheng, C., Rokey, A.M. and Chapes, S.K. (2002) Persistent *Ehrlichia chaffeensis* infection occurs in the absence of functional MHCII genes. Infect. Immun. I70, 380–388.
Gould, C. and Sonnenfeld, G. (1987) Enhancement of viral pathogenesis in mice maintained in an antiorthostaticmodel: Coordination with effects on interferon production. J. Biol. Reg. Homeostatic Agents 1, 33–36.
Govoni, G., Vidal, S., Gauthier, S., Skamene, E., Malo, D. and Gros, P. (1996) The *Bcg/Ity/Lsh* locus: genetic transfer of resistance to infections in C57BL/6J mice transgenic for the $Nramp1^{Gly169}$ allele. Infect. Immun. 64, 2923–2929.
Gridley, D.S., Nelson, G.A., Peters, L.L., Kostenuik, P.J., Bateman, T.A., Morony, S., Stodieck, L.S., Lacey, D.L., Simske, S.J. and Pecaut, M.J. (2003) Genetic models in applied physiology: Selected contribution: Effects of

spaceflight on immunity in the C57BL/6 mouse. II. Activation, cytokines, erythrocytes, and platelets. J. Appl. Physiol. 94, 2095–2103.
Grusby, M. and Gilmcher, L. (1995) Immune responses in MHC class II-deficient mice. Annu. Rev. Immunol. 13, 417–435.
Grusby, M., Johnson, R., Papaioannou, V. and Glimcher, L. (1991) Depletion of CD4+ T cells in major histocompatibility complex class II-deficient mice. Science 253, 1417–1420.
Hart, M.L., Mosier, D.A. and Chapes, S.K. (2003) Toll-like receptor 4-positive macrophages protect mice from Pasteurella pneumotropica-induced pneumonia. Infect. Immun. 71, 663–670.
Hatton, J.P., Gaubert, F., Cazenave, J.P. and Schmitt, D. (2002) Microgravity modifies protein kinase C isoform translocation in the human monocytic cell line U937 and human peripheral blood T-cells. J. Cell. Biochem. 87, 39–50.
Hatton, J.P., Gaubert, F., Lewis, M.L., Darsel, Y., Ohlmann, P., Cazenave, J.P. and Schmitt, D. (1999) The kinetics of translocation and cellular quantity of protein kinase C in human leukocytes are modified during spaceflight. FASEB J. 13 Suppl., S23–S33.
Hickman-Davis, J.M., Michalek, S.M., Gibbs-Erwin, J. and Lindsay, J.R. (1997) Depletion of alveolar macrophages exacerbates respiratory mycoplasmosis in mycoplasma-resistant C57BL mice but not mycoplasma-susceptible C3H mice. Infect. Immun. 65, 2278–2282.
Horne, W., Gorman, J.J., Guerra, L.A., Kishiyama, J.S., May, L.D., Sunshine, P.L. and Williams, G.J. (eds.). (2003) *National Aeronautics and Space Administration: Biological and Physical Research Enterprise Strategy*, National Aeronautics and Space Administration, Washington, D.C.
Huang, S., Hendriks, W., Althage, A., Hemmi, S., Bluethmann, H., Kamijo, R., Vilcek, J., Zinkernagel, R. and Aguet, M. (1993) Immune response in mice that lack the interferon gamma receptor. Science 259, 1742–1745.
Ichiki, A., Gibson, L. and Allebban, Z. (1994) Effects of spaceflight on rat leukocytes. J. Leukoc. Biol. 56, Supplement 34.
Ichiki, A., Gibson, L., Jago, T., Strickland, K., Johnson, D., Lange, R. and Allebban, Z. (1996) Effects of spaceflight on rat peripheral blood leukocytes and bone marrow progenitor cells. J. Leukoc. Biol. 60, 37–43.
Ichiyasu, H., Suga, M., Matsukawa, A., Iyonaga, K., Mizobe, T., Takahashi, T. and Ando, M. (1999) Functional roles of MCP-1 in *Propionibacterium acnes* induced, T cell-mediated pulmonary granulomatosis in rabbits. J. Leukoc. Biol. 65, 482–491.
Jensen, T., Aliouat, E.M., Lundgren, B., Settnes, O.P., Karlsson, A., Dahlgren, C., Dei-Cas, E. and Obel, N. (1998) Cultured rat and purified human *Pneumocystis carinii* stimulate intra- but not extracellular free radical production in human neutrophils. J. Eukaryot. Microbiol. 45, 544–547.
Kaufmann, S.H. and Ladel, C.H. (1994) Role of T cell subsets in immunity against intracellular bacteria: Experimental infections of knock-out mice with

Listeria monocytogenes and *Mycobacterium bovis* BCG. Immunobiology 191, 509–519.

Kimzey, S. (1977) Hematology and immunology studies. *In* Johnston, R.S. and Dietlin, L.F. (eds.). NASA SP-377, National Aeronautics and Space Administration, Washington, D.C.

Kopydlowski, K., McVey, D., Woods, K., Iandolo, J. and Chapes, S.K. (1992) Effects of antiorthostatic suspension and corticosterone on macrophage and spleen cell function. J. Leukoc. Biol. 52, 202–208.

Koyama, S., Sato, E., Masubuchi, T., Takamizawa, A., Nomura, H., Kubo, K., Nagai, S. and Izumi, T. (1998) Human lung fibroblasts release chemokinetic activity for monocytes constitutively. Am. J. Physiol. Regul. Integr. Comp. Physiol. 275, L223–L230.

Lawler, A. (2004) Remaking NASA: How much space for science? Science 303, 610–612.

Lesnyak, A., Sonnenfeld, G., Avery, L., Konstantinova, I., Rykova, M., Meshkov, D. and Orlova, T. (1996) Effect of SLS-2 spaceflight on immunologic parameters of rats. J. Appl. Physiol. 81, 178–182.

Limouse, M., Manie, S., Konstantinova, I., Ferrua, B. and Schaffar, L. (1991) Inhibition of phorbol ester-induced cell activation in microgravity. Exp. Cell Res. 197, 82–86.

Limper, A.H., Hoyte, J.S. and Standing, J.E. (1997) The role of alveolar macrophages in *Pneumocystis carinii* degradation and clearance from the lung. J. Clin. Invest. 99, 2110–2117.

Lukacs, N., Strieter, R., Chensue, S., Widmer, M. and Kunkel, S. (1995) TNF-α mediates recruitment of neutrophils and eosinophils during airway inflammation. J. Immunol. 154, 5411–5417.

Mak, T.W. and Simard, J.J.L. (1998) *Handbook of Immune Response Genes*, pp. 1–393, Plenum Press, New York.

McCarrick, J.W., 3rd, Parnes, J.R., Seong, R.H., Solter, D. and Knowles, B.B. (1993) Positive-negative selection gene targeting with the diphtheria toxin A-chain gene in mouse embryonic stem cells. Transgenic Res. 2, 183–190.

Meehan, R., Neale, L., Kraus, E., Stuart, C., Smith, M., Cintron, N. and Sams, C. (1992) Alteration in human mononuclear leucocytes following space flight. Immunol. 76, 491–497.

Mehta, S.K., Cohrs, R.J., Forghani, B., Zerbe, G., Gilden, D.H. and Pierson, D.L. (2004) Stress-induced subclinical reactivation of varicella zoster virus in astronauts. J. Med. Virol. 72, 174–179.

Mehta, S.K., Kaur, I., Grimm, E.A., Smid, C., Feeback, D.L. and Pierson, D.L. (2001) Decreased non-MHC-restricted (CD56+) killer cell cytotoxicity after spaceflight. J. Appl. Physiol. 91, 1814–1818.

Merlin, T., Sing, A., Nielsen, P.J., Galanos, C. and Freudenberg, M.A. (2001) Inherited IL-12 unresponsiveness contributes to the high LPS resistance of the *lps*(d) C57BL/10ScCr mouse. J. Immunol. 166, 566–573.

Miller, E., Koebel, D., Davis, S., Klein, J., McLeish, K., Goldwater, D. and Sonnenfeld, G. (1994) Influence of suspension on the oxidative burst by rat neutrophils. J. Appl. Physiol. 76, 387–390.

Miller, E. and Sonnenfeld, G. (1993) Influence of suspension on the expression of protective immunological memory to murine *Listeria monocytogenes* infection. J. Leukoc. Biol. 54, 578–583.

Miller, E. and Sonnenfeld, G. (1994) Influence of antiorthostatic suspension on resistance to murine *Listeria monocytogenes* infection. J. Leukoc. Biol. 55, 371–378.

Mizgerd, J.P., Kubo, H., Kutkoski, G.J., Bhagwan, S.D., Scharffetter-Kochanek, K., Beaudet, A.L. and Doerschuk, C.M. (1997) Neutrophil emigration in the skin, lungs, and peritoneum: Different requirements for CD11/CD18 revealed by CD18-deficient mice. J. Exp. Med. 186, 1357–1364.

Morey, E. (1979) Spaceflight and bone turnover: Correlation with a new rat model. Bioscience 29, 168–172.

Morey, E., Sabelman, E., Turner, R. and Baylink, D. (1979) A new rat model simulating some aspects of space flight. The Physiologist S23–S24.

Morey-Holton, E. and Arnaud, S. (1991) Skeletal responses to space flight. Adv. Space Biol. Med. 1, 37–69.

Morey-Holton, E.R. and Globus, R.K. (2002) Hindlimb unloading rodent model: Technical aspects. J. Appl. Physiol. 92, 1367–1377.

Musacchia, X. and Steffen, J. (1983) The validity of an animal model for experiments related to weightlessness. The Physiologist 26, S37–S40.

Nash, P., Bour, B. and Mastro, A. (1991) Effect of hindlimb suspension simulation of microgravity on *in vitro* immunological responses. Exp. Cell Res. 195, 353–360.

Nauciel, C., Ronco, E., Guenet, J.-L. and Pla, M. (1988) Role of *H-2* and non-*H-2* genes in control of bacterial clearance from the spleen in *Salmonella typhimurium*-infected mice. Infect. Immun. 56, 2407–2411.

Neumann, B., Emmanuilisis, K., Stadler, M. and Holzmann, B. (1998) Distinct functions of interferon gamma for chemokine expression in models of acute lung inflammation. Immunology 95, 512–521.

Ott, C.M., Bruce, R.J. and Pierson, D.L. (2004) Microbial characterization of free floating condensate aboard the Mir space station. Microb. Ecol. 47, 133–136.

Paddock, C.D., Suchard, D.P., Grumbach, K.L., Hadley, W.K., Kerschmann, R.L., Abbey, N.W., Dawson, J.E., Anderson, B.E., Sims, K.G., Dumler, J.S. and Herndier, B.G. (1993) Brief report: Fatal seronegative ehrlichiosis in a patient with HIV infection. N. Engl. J. Med. 329, 1164–1167.

Paddock, C.D., Sumner, J.W., Shore, G.M., Bartley, D.C., Elie, R.C., McQuade, J.G., Martin, C.R., Goldsmith, C.S. and Childs, J.E. (1997) Isolation and characterization of *Ehrlichia chaffeensis* strains from patients with fatal ehrlichiosis. J. Clin. Microbiol. 35, 2496–2502.

Pecaut, M.J., Nelson, G.A., Peters, L.L., Kostenuik, P.J., Bateman, T.A., Morony, S., Stodieck, L.S., Lacey, D.L., Simske, S.J. and Gridley, D.S. (2003)

Genetic models in applied physiology: Selected contribution: Effects of spaceflight on immunity in the C57BL/6 mouse. I. Immune population distributions. J. Appl. Physiol. 94, 2085–2094.

Pecaut, M.J., Simske, S.J. and Fleshner, M. (2000) Spaceflight induces changes in splenocyte subpopulations: Effectiveness of ground-based models. Am. J. Physiol. Regul. Integr. Comp. Physiol. 279, R2072–R2078.

Penttila, J.M., Anttila, M., Puolakkainen, M., Laurila, A., Varkila, K., Sarvas, M., Makela, P.H. and Rautonen, N. (1998) Local immune responses to *Chlamydia pneumoniae* in the lungs of BALB/c mice during primary infection and reinfection. Infect. Immun. 66, 5113–5118.

Pierson, D.L. (2001) Microbial contamination of spacecraft. Gravit. Space Biol. Bull. 14, 1–6.

Plant, J. and Glynn, A. (1976) Genetics of resistance to infection with *Salmonella typhimurium* in mice. J. Infect. Dis. 133, 72–78.

Poltorak, A., He, X., Smirnova, I., Liu, M.Y., Huffel, C.V., Du, X., Birdwell, D., Alejos, E., Silva, M., Galanos, C., Freudenberg, M., Ricciardi-Castagnoli, P., Layton, B. and Beutler, B. (1998) Defective LPS signaling in C3H/HeJ and C57BL/10ScCr mice: mutations in *Tlr4* gene. Science 282, 2085–2088.

Romagnani, S. (1997) The Th1/Th2 paradigm. Immunol. Today 18, 263–266.

Sarver, G. (2004) Habitats: Advanced Animal Habitat-Centrifuge. *http://brp.arc.nasa.gov/GBL/Habitats/aah.html*, National Aeronautics and Space Administration, Washington, D.C.

Schmitt, D.A., Hatton, J.P., Emond, C., Chaput, D., Paris, H., Levade, T., Cazenave, J.-P. and Schaffar, L. (1996) The distribution of protein kinase C in human leukocytes is altered in microgravity. FASEB J. 10, 1627–1634.

Smolen, J.E., Fossett, M.C., Joe, Y., Prince, J.E., Priest, E., Kanwar, S. and Smith, C.W. (2000) Antiorthostatic suspension for 14 days does not diminish the oxidative response of neutrophils in mice. Aviat. Space Environ. Med. 71, 1239–1247.

Sonnenfeld, G., Aviles, H., Belay, T., Vance, M. and Fountain, K. (2002) Stress, suspension and resistance to infection. J. Gravit. Physiol. 9, 199–200.

Sonnenfeld, G., Davis, S., Taylor, G.R., Mandel, A.D., Konstantinova, I.V., Lesnyak, A., Fuchs, B.B., Peres, C., Tkackzuk, J. and Schmitt, D.A. (1996) Effect of space flight on cytokine production and other immunologic parameters of rhesus monkeys. J. Interferon and Cytokine Res. 16, 409–415.

Sonnenfeld, G., Mandel, A., Konstantinova, I., Berry, W., Taylor, G., Lesnyak, A., Fuchs, B. and Rakhmilevich, A. (1992) Space flight alters immune cell function and distribution. J. Appl. Physiol. 73, 191S–195S.

Sonnenfeld, G., Mandel, A., Konstantinova, I., Taylor, G., Berry, W., Wellhausen, S., Lesnyak, A. and Fuchs, B. (1990) Effects of spaceflight on levels and activity of immune cells. Aviat. Space Environ. Med. 61, 648–653.

Stowe, R.P., Mehta, S.K., Ferrando, A.A., Feeback, D.L. and Pierson, D.L. (2001a) Immune responses and latent herpesvirus reactivation in spaceflight. Aviat. Space Environ. Med. 72, 884–891.

Stowe, R.P., Pierson, D.L. and Barrett, A.D. (2001b) Elevated stress hormone levels relate to Epstein-Barr virus reactivation in astronauts. Psychosom. Med. 63, 891–895.

Sutterwala, F.S. and Mosser, D.M. (1999) The taming of IL-12: suppressing the production of proinflammatory cytokines. J. Leukoc. Biol. 65, 543–551.

Taylor, G. (1988) Cell anomalies associated with spaceflight conditions. Adv. Exp. Med. Biol. 225, 259–271.

Taylor, G.R. (1993) Immune changes during short-duration missions. J. Leukoc. Biol. 54, 202–208.

Taylor, G. and Dardano, J. (1983) US/USSR space biology and medicine: Human cellular immune responsiveness following space flight. Aviat. Space and Environ. Med. 54, S55–S59.

Taylor, G. and Janney, R. (1992) *In vivo* testing confirms a blunting of the human cell-mediated immune mechanism during space flight. J. Leukoc. Biol. 51, 129–132.

Taylor, G., Neale, L. and Dardano, J. (1986) Immunological analyses of U.S. Space Shuttle crew members. Aviat. Space Environ. Med. 57, 213–217.

Todd, P., Sklar, M., Ramirez, W.F., Smith, G. and Morgenthaler, G.W. (1993) Physics, chemistry and pulmonary sequelae of thermodegradation events in long-mission space flight. Presented at International Conference on Environmental Systems, Warrendale, PA.

Todd, P., Sklar, V., Ramirez, W.F., Smith, G.J., Morgenthaler, G.W., McKinnon, J.T., Oberdorster, G. and Schulz, J. (1994) Inhalation risk in low-gravity spacecraft. Acta Astronaut 33, 305–315.

Tracey, K. (1992) The acute and chronic pathophysiogic effects of TNF: Mediation of septic shock and wasting (cachexia). *In* Beutler, B. (ed.). *Tumor Necrosis Factors: The Molecules and Their Emerging Role in Medicine*, pp. 255–273, Raven Press, New York.

Vacek, A., Bartonickov, A., Rotkovsk, D., Michurina, T., Damaratskaya, E. and Serova, L. (1983) The effects of weightlessness and increased gravity on hemopoietic stem cells of rats and mice. The Physiologist 26, S131–S132.

Vogel, S.N. (1992) The *Lps* gene: Insights into the genetic and molecular basis of LPS responsiveness and macrophage differentiation. *In* Beutler, B. (ed.). *Tumor Necrosis Factors: The Molecules and Their Emerging role in Medicine*, pp. 485–513, Raven Press, New York.

Vogel, S., Hansen, C. and Rosenstreich, D. (1979) Characterization of a congenitally LPS-resistant, athymic mouse strain. J. Immunol. 122, 619–622.

Wakeham, J., Wang, J., Magram, J., Croitoru, K., Harkness, R., Dunn, P., Zganiacz, A. and Xing, Z. (1998) Lack of both types 1 and 2 cytokines, tissue inflammatory responses, and immune protection during pulmonary infection by *Mycobacterium bovis* bacille Calmette-Guerin in IL-12-deficient mice. J. Immunol. 160, 6101–6111.

Woods, K.M. and Chapes, S.K. (1994) Abrogation of TNF-mediated cytotoxicity by space flight involves protein kinase-C. Exp. Cell Res. 211, 171–174.

Wright, A.D. and Chapes, S.K. (1999) LPS sensitivity in recombinant mice lacking functional alleles at MHCII, *Lps*, and *Nramp1* Genes. J. Endotoxin Res. 5, 297–305.

Wronski, T.J., Li, M., Shen, Y., Miller, S.C., Bowman, B.M., Kostenuik, P. and Halloran, B.P. (1998) Lack of an effect of space flight on bone mass and bone formation in group-housed rats. J. Applied Physiol. 85, 279–285.

Wronski, T. and Morey-Holton, E. (1987) Skeletal response to simulated weightlessness: A comparison ofsuspension techniques. Aviation, Space, and Environmen. Med. 58, 63–68.

Yu, M.L. and Limper, A.H. (1997) *Pneumocystis carinii* induces ICAM-1 expression in lung epithelial cells through a TNF-α-mediated mechanism. Am. J. Physiol. Regul. Integr. Comp. Physiol. 273, L1103–L1111.

Experimentation with Animal Models in Space
G. Sonnenfeld (editor)

DOI: 10.1016/S1569-2574(05)10005-7

Vestibular Experiments in Space

Bernard Cohen,[1] Sergei B. Yakushin,[1] Gay R. Holstein,[1] Mingjia Dai,[1] David L. Tomko,[2] Anatole M. Badakva[3] and Inessa B. Kozlovskaya[3]

[1]*Department of Neurology, Mount Sinai School of Medicine, New York, USA;*
[2]*National Aeronautics and Space Agency;*
[3]*the Institute of Biomedical Problems, Moscow, Russia*

Introduction

Life on Earth is conditioned by the constant downward pull of gravity, which can be considered as an equivalent constant upward linear acceleration (Einstein, 1911). During lateral, fore-aft, and vertical translations as well as during centripetal accelerations generated by turning around distant axes, these linear accelerations summate with the linear acceleration of gravity to form a resultant vector, which we term gravito-inertial acceleration (GIA). These linear accelerations increase the magnitude of the GIA and tilt it relative to the head. The otolith organs and body tilt receptors sense the resultant changes in the GIA and generate compensatory and orienting eye movements through the linear vestibulo-ocular reflex (lVOR) to direct and stabilize vision. Angular head movements are sensed by the semicircular canals, which produce compensatory eye movements over the angular vestibulo-ocular reflex (aVOR). The vestibular system also responds to linear and angular accelerations through vestibulo-spinal reflexes to produce body postural movements that stabilize the body when stationary and support it when one is in motion. Additionally, the vestibular system responds to changes in head and body position relative to the GIA to maintain blood pressure and enhance respiration. Thus, the vestibular system plays a critical role in sensing and responding to linear and angular accelerations in a gravitational environment.

In microgravity, the linear acceleration of gravity is reduced to negligible levels, and a standing posture is no longer maintained. Instead, subjects swim in Space rather than walk, and they no longer experience the vertical linear translations that accompany locomotion. Additionally, gravity no longer contributes to the resultant change in the GIA when the head is translated. Thus, the GIA is always in the direction of the translational and centripetal accelerations due to movement. Angular acceleration, which is in a head coordinate frame, is not markedly altered in microgravity. Because of the ubiquitous nature of gravity on Earth and the difficulty of performing experiments on orbit, we know relatively little about how the nervous system is altered by the prolonged absence of gravity.

From the 1970s, the Russian Space Agency made orbital space flight available to INTERCOSMOS scientists from many countries for animal experimentation, primarily on rats. This included Czechoslovakia, Poland, Hungary, Germany, France, and the United States. Between 1983 and 1995, monkeys and rats flew for approximately 1–2 week periods on a Vostok Space Capsule in experiments supported by the Russian Space Agency and the National Aeronautics and Space Aadministration (NASA). Rodents and invertebrates also flew on NASA Space Shuttle missions, particularly on the Neurolab Mission (STS-90) in 1998. The purpose of this chapter is to provide a summary of these experiments. Experiments on rodents and other animal species in the NASA Space Shuttle flights and the Russian COSMOS/BION flights are summarized in the first section. The second section summarizes inflight experiments performed by the Russian Space Agency on the aVOR and on single unit activity in the vestibular nuclei and flocculus of monkeys. Many of these data appear here for the first time. In the third section, we summarize NASA-supported research on rhesus monkeys on the linear and angular vestibulo-ocular reflex (lVOR and aVOR). The intent was to provide a summary of experimental data on which future experimentation in Space can be expanded. In addition to revealing some new and interesting scientific results, it is hoped that it will become obvious that basic studies, involving subhuman primates as well as rodents should be essential components of any future space-related research. The results of the earliest flights are not included in this chapter, but can be found in references at the end of the bibliography.

SECTION 1: CELLULAR RESPONSES TO ALTERED GRAVITATIONAL ENVIRONMENTS IN ADULT ANIMALS

A number of studies have examined the impact of altered gravity exposure on vestibular system structures at the cellular level. As a whole, they clearly document the modifiability of cellular and subcellular constituents of vestibular pathways in response to altered gravitational stimuli.

Peripheral vestibular system

Some evidence suggests that exposure of adult as well as developing animals to altered gravitational environments can trigger adaptive responses in weight-lending structures. For example, an increase in otoconial mass has been reported in the utricles of adult animals of several species (rats, frogs) following 7 days of exposure to microgravity (Vinnikov *et al.*, 1980; Ross *et al.*, 1985;

Ross, 1987; Lychakov *et al.*, 1989). Conversely, the saccular otolithic membrane volume is reduced in rats exposed to 2.3 ***g*** or 4.15 ***g*** hypergravity through centrifugation (Lim *et al.*, 1974), although otoconial morphology (examined by scanning electron microscopy) and otoconial synthesis (assessed by mRNA expression of the otoconial matrix protein osteopontin) appear to be normal after 2 h to 7 days of hypergravity exposure (Uno *et al.*, 2000).

Increases in otoconial mass have also been observed in the saccular otoliths of newt larvae (Weiderhold *et al.*, 1997) and late-stage embryos of swordtail fish (Weiderhold *et al.*, 2000) reared in Space, although such increases are not observed in older fish. As in adult animals, the saccular otolithic membrane volume is reduced in cichlid fish reared on a centrifuge (Anken *et al.*, 1998), and the volume of statoconia in marine mollusk larvae reared at 2–5 ***g*** is diminished in a graded manner as compared with 1 ***g*** controls (Pedrozo and Wiederhold, 1994). Similarly, there is a reduction in the number of large otoconia that are present in the anterior peripheral portion of the utricle in rats raised in a 2 ***g*** environment for 60 days than in 1 ***g*** controls (Krasnov, 1991). Taken together, these results suggest that otoconial mass adapts to fluctuations in the gravitational stimulus, perhaps to maintain a consistent force on the maculae.

A number of studies have addressed the question of morphologic alterations in the sensory hair cells of adult mammalian otolith organs in response to altered gravitational conditions. These studies indicate that short-term (e.g., 7 day) exposure to space flight does not cause otolith receptor cell degeneration in rats (Ross *et al.*, 1985), although the type I hair cells reportedly exhibit abnormal cytologic features such as increased chromatin, and enlarged perinuclear and intercellular spaces (Krasnov, 1987, 1991). However, more prolonged exposure to microgravity clearly impacts hair cell synaptology. In general, these studies indicate that type II hair cells in particular, but also type I cells evidence adaptive structural modifications in response to alterations in the gravitational environment (Ross, 1993, 1994, 2000). Quantitative ultrastructural analysis of utricular hair cells from adult rats sacrificed immediately after landing from a 9-day space flight demonstrated statistically significant increases in the number of ribbon synapses (Ross, 1994). In addition, the number of synapse pairs and clusters markedly increased in the type II hair cells. Nine days after landing, which was the mission duration, synapse counts remained elevated in the flight rats, but were also elevated in the type II hair cells of ground control animals, suggesting that these receptor cells are particularly vulnerable to stress-induced morphologic changes. In a subsequent study (Ross and Tomko, 1998), the mean number of synapses on type II hair cells doubled, and those on type I cells increased by over 40% by the 13th day of a 14-day space flight. Immediately postflight, these synapse counts were reduced to 67% and 13% increases, respectively, over control values. Comparable studies of vestibular hair cells have been conducted utilizing 2–4 ***g*** hypergravity exposure for periods of time ranging from 14–30 days (Lim *et al.*, 1974;

Lychakov *et al.*, 1988; Ross, 1993). In general, these experiments support Ross' centrifugation study (Ross, 1993) demonstrating that type I hair cell synapse number does not change in response to hypergravity, but there is a decrement of approximately 40% in type II hair cell synapses. Clearly, the otolith organs of adult mammals have the potential for morphologic reorganization at several sites in response to altered gravity conditions (Ross, 1997).

Since studies of molecular changes in peripheral vestibular cells in response to fluctuations in the gravity stimulus have been initiated relatively recently, generalizations about molecular alterations are premature. One recent study investigated glutamate receptor mRNA expression using RT-PCR on cells of the vestibular ganglion, as well as the vestibular nuclei, and vestibulo-cerebellum of rats exposed to hypergravity for 2 h–7 days (Uno *et al.*, 2002). In the vestibular periphery, this experiment demonstrated that synthesis of GluR2 (an AMPA glutamate receptor subunit) receptors in vestibular ganglion cells is reduced in rats exposed to 7 days of hypergravity.

Behavioral, physiological and biochemical correlates of vestibular cellular responses

A spinning movement during swimming, termed "looping," was initially reported in some (but not all) fish at the transition from 1 ***g*** to microgravity during parabolic flight (von Baumgarten *et al.*, 1972), and subsequently described in larval cichlid fish and *Xenopus laevis* immediately following prolonged 3 ***g*** centrifugation (Rahmann *et al.*, 1992). Looping has also been reported in killifish in microgravity (von Baumgarten *et al.*, 1975). Fish that evidence looping behavior exhibit a significantly higher asymmetry in otolith size (Anken *et al.*, 1998) and weight (Beier *et al.*, 2002), in comparison with nonlooping siblings. In addition, although the total number of sensory and supporting cells in the utricular maculae is the same, the cell density is significantly lower in looping than nonlooping fish, due to the atypical presence of enlarged epithelial cells (Bauerle *et al.*, 2004).

Adult mammals may be more profoundly affected by changes in the gravitational environment than are fish. Upon landing following 9 days of space flight, adult rats reportedly show substantially reduced locomotion (Ross, 1994). They maintain a posture with their abdomens and tails flat against the cage floor, with their limbs and digits extended. After 9 days at 1 ***g***, flight rats display normal body posture and movement. Similarly, after 14 days of exposure to 2 ***g*** through centrifugation, adult rats show profound deficits in righting responses, swimming and balance (Fox *et al.*, 1992). Recovery of normal orientation during swimming requires 4–24 h at 1 ***g***, whereas the righting reflex does not return to normal for 5–7 days (Ross and Tomko, 1998).

Neurophysiological studies have further documented an immediate post-flight alteration in vestibular activity (Boyle *et al.*, 2001). Within the first 16 h post-space flight, utricular afferents in the oyster toadfish were hypersensitive to translational acceleration, but had no change in directional selectivity. The afferent sensitivity returned to baseline levels by approximately 30 h postflight. Since anamniotes such as the toadfish have only type II hair cells, the reported increase in ribbon synapse number (Ross, 1997) could be one explanation for the afferent hypersensitivity.

Central vestibular system

Several studies have addressed the impact of altered gravitational environments on cells of the central vestibular system. Most of these studies have examined the vestibular nuclei and/or vestibulo-cerebellum, and have utilized either a morphological approach, or a marker for cellular activity. Early studies of the cerebellar nodulus of adult rats exposed to 18 days of space flight revealed ultrastructural alterations in Purkinje cell dendrites and mossy fiber terminals (Krasnov and Dyachkova, 1986, 1990). In these studies, the major modification described in the Purkinje cells was a widening of the synaptic cleft at contacts with Purkinje cell dendrites. In mossy fiber terminals, structural alterations included densely packed synaptic vesicles, with unusual clustering of such vesicles near the presynaptic membranes of axodendritic synapses, increased electron density of pre- and postsynaptic membranes, and enlargement of the synaptic gap. Similar changes are observed in nodular mossy fiber terminals from rats flown on 5–7 day missions and sacrificed within 8 h of landing. Moreover, ultrastructural alterations in the primary somatosensory cortex of these flight animals included a profound decrement in the number of axo-dendritic synapses, and an increase in the number of axon terminals showing "light" degeneration, as well as signs of "superexcitation" including an increase in the number of axon terminals showing dark degeneration. In concert with this finding, increased numbers of synaptic contacts have been demonstrated in neonatal swordtail fish after 16 days in microgravity, specifically in the nucleus magnocellularis of the primary vestibular brainstem region, area octavolateralis (Ibsch *et al.*, 2000).

More recently, the ultrastructure of the otolith-recipient zones of the cerebellar nodulus has been analyzed in tissue from flight and cage-control rats sacrificed in microgravity after 24 h of space flight (Holstein *et al.*, 1999). Qualitative observations of this tissue indicate that several structural alterations occur in the neuropil, and in the Purkinje cell cytoplasm, of the nodulus of rats exposed to 24 h of space flight. These anatomical alterations are not apparent in the cage control animals. Most notably, the cisterns of smooth endoplasmic reticulum that are normally present throughout Purkinje cells are substantially enlarged and more complex in Purkinje cells of the otolith-recipient zones of the nodulus. The increased complexity of the cisterns results in the formation of long, stacked lamellar bodies that are observed throughout entire Purkinje cells, including the somata, dendrites, thorns, and axon terminals. In addition, occasional enormous mitochondria, >1 μm in cross-sectional diameter, are present in the Purkinje cell somata of flight animals. Ultrastructural indications of degeneration and synaptic reorganization are also observed in the molecular layer of the nodulus from the flight animals, but not cage controls. Since these morphologic changes are not apparent in control animals, they are not likely to be due to caging or tissue processing effects. The particular nature of the structural alterations, including the formation of lamellar bodies and the presence of degeneration, suggests that excitotoxity may play a role in the short-term neural response to space flight.

In the granular layer of the nodulus of rats raised in a 2 ***g*** environment for 60 days, 80% of the glomeruli showed altered synaptic morphology, including changes in the density of pre- and postsynaptic membranes, increased thickness of the postsynaptic density, enlargement of the synaptic cleft, increased packing density of synaptic vesicles, enlarged mitochondria, and an increase in the number of microtubules (Krasnov and Dyachkova, 1986; Krasnov, 1991). Two days after return to a 1 ***g*** environment, the ultrastructure of the nodulus resembles that of control animals. The synaptic vesicle packing density is decreased, and the number of microtubules is diminished, suggesting reversibility of the gravity-induced effects. Taken together with the microgravity findings, these results more generally indicate that the *central* vestibular system responds to major changes in the gravitational stimulus with similar morphological restructuring, regardless of the direction (hypo- or hyper-) of that change.

In light of the ultrastructural findings suggesting that excitotoxicity may be a factor in early neuronal responses to altered gravitational environments (Holstein *et al.*, 1999), the recent study of glutamate receptor expression is of particular interest. This study examined glutamate receptor mRNA expression using RT-PCR on cells in the vestibular nuclei and vestibulo-cerebellum of rats exposed to hypergravity for 2 h to 7 days (Uno *et al.*, 2002). The results indicate that mRNA expression of GluR2 (an AMPA glutamate receptor subunit) and NR1 (the obligatory subunit of the NMDA glutamate receptor) in the nodulus/uvula and NR1 expression in the medial vestibular nucleus increase after 2 h of stimulation. This expression gradually returns to baseline during the 7 days of hypergravity exposure. As noted above, mRNA expression of GluR2 receptors in vestibular ganglion cells is reduced after 7 days of stimulation. Neither the mGluR1 metabotropic receptor nor the δ_2 glutamate receptor in the flocculus and nodulus/uvula is affected by hypergravity exposure for 2 h to 7 days. It was suggested that the immediate (2 h) adaptation to hypergravity involves enhanced cerebellar inhibition of the vestibular nuclei mediated by Purkinje cell NR1 and GluR2 receptors, whereas longer term adaptation involves decreased transmission from vestibular hair cells to primary afferent neurites mediated by down-regulation of postsynaptic GluR2 receptors on the primary afferents. One interesting aspect of this study is that functional NMDA receptors are not normally present on cerebellar Purkinje cells in adult mammals. Conceivably, an altered gravitational environment presents a sufficient stimulus to trigger their expression and activity. However, the activity of enzymes involved in energy metabolism (lactate dehydrogenase and creatinine kinase) in giant Deiters' neurons of the lateral vestibular nucleus and cells of the cerebellar nodulus in rats are not appreciably affected by 22 days of space flight (Krasnov, 1975).

Space flight-related studies utilizing markers for vestibular neuronal activity have concentrated primarily on the immediate early gene *c-fos*. The gene encodes for Fos protein, which reaches peak values within 2–4 h of the effective stimulus and returns to baseline within 6–8 h. Fos-related antigen (FRA) proteins, which are generated by multiple genes, are induced shortly after stimulation, but persist for days ("acute" FRAs such as FRA-1, -2, FosB, ΔFosB) or weeks ("chronic" FRAs, e.g., modified forms of ΔFosB). Fos protein is often utilized as a neural activity marker, since it can be identified with higher spatial resolution than metabolic indicators. Although the basal expression level of *c-fos* expression is low in the brain, increases in Fos-like immunoreactivity have been reported in vestibular structures following galvanic stimulation (Kaufman and Perachio, 1994; Mashburn *et al.*, 1997) and centripetal acceleration (Kaufman *et al.*, 1992), as well as space flight.

Results from space flight experiments on adult rat brain tissue indicate a trend toward increased numbers of Fos-immunopositive cells in the vestibular brainstem (particularly the medial and descending vestibular nuclei, MVN and DVN, respectively) 24 h postlaunch, and a statistically significant increase in the number of immunostained cells 24 h after return from a 17-day mission (Pompeiano *et al.*, 2002). The number of Fos-immunostained vestibular cells were equivalent in flight animals and ground controls by 13 days postlaunch and at 13 days postlanding. The pattern of FRA protein immunolabeling was qualitatively similar to that of Fos, except at 1 day after landing, when FRA-immunolabeled cells were observed throughout the entire DVN, but only in the caudal MVN while Fos staining was reported throughout the entire MVN. In addition, Fos- and FRA-like immunoreactivity in the vestibular portions of the inferior olivary complex of rats was unchanged 24 h postlaunch (d'Ascanio *et al.*, 2003). However, while Fos immunolabeling remained unchanged 24 h postlanding in these regions of the inferior olive, increased FRA-immunostaining was reported at that time. The authors attribute these findings to the mixed sensory signals derived from rapid fluctuations in *g*-force during launch. In contrast, the flight rats sacrificed in microgravity 1 or 13 days postlaunch had fewer Fos- and FRA-immunolabeled efferent vestibular neurons than ground controls (Balaban *et al.*, 2002), although no differences were observed between flight and control rats during the postlanding re-adaptation period. The results are interpreted as indicative of general physiological and morphological changes in the cells. Interestingly, Fos- and FRA-like immunoreactivity in autonomic regions such as area postrema and nucleus tractus solitarius of these flight rats were equivalent to control tissue during exposure to microgravity, but were significantly increased 24 h after landing (Pompeiano *et al.*, 2004).

Upregulation of genomic activity has also been demonstrated in several brainstem nuclei related to otolith pathways, particularly the dorsomedial cell column of the inferior olivary complex as well as MVN, DVN and the y-group, in head-fixed rodents exposed to 2 ***g*** centripetal acceleration (Kaufman *et al.*, 1992). This upregulation occurs following one-axis stimulation restricted to the plane of the saccule (Mashburn *et al.*, 1997) or the utricle (Kaufman *et al.*, 1991). Immunohistochemistry for fos protein and for FRA proteins has also been performed on vestibular tissue from 60-day-old rats exposed to 2 ***g*** or 4 ***g*** centrifugation in the plane of the saccule (hypergravity exposure), and in 60-day-old rats born and housed at 2 ***g***, then exposed for 90 min to 1 ***g*** (hypogravity condition) (Duflo *et al.*, 2000). This study found enhanced Fos-related labeling of the vestibular brainstem, particularly MVN, DVN, nucleus of Roller, the y-group and the inferior olivary nucleus, only in the hypergravity condition. Although the β subnucleus of the inferior olive was not immunostained in this study, nor in Mashburn *et al.*'s study (1997), it did display Fos-like immunoreactivity in the experiments of Kaufman and colleagues (Kaufman *et al.*, 1992, 1993), suggesting that utricular rather than saccular inputs activate β-subnucleus Fos expression.

Summary

It is clear that cellular and subcellular constituents of the vestibular pathways are modified in response to altered gravitational stimuli. In the vestibular periphery, opposite structural effects appear to result from hypo- and hypergravity stimulation. In the central vestibular system, neural circuits may well exhibit apparently identical structural changes in response to diverse hypo- and hyper-gravity stimuli, reflecting the more dynamic, highly regulated interactions of the central pathways. More research will be needed to resolve the inconsistencies in the current published literature.

SECTION 2: STUDIES OF THE EFFECTS OF MICROGRAVITY ON VESTIBULAR AND OCULOMOTOR FUNCTION IN THE RUSSIAN COSMOS PROJECT

Abstract

Experiments were performed while monkeys flew in space in the "Cosmos/Bion" Missions to determine the effect of microgravity on the oculomotor and vestibular systems. Eye-head coordination during gaze shifts to lateral targets (gaze fixation reaction, GFR) and multiunit activity in the medial vestibular nuclei (MVN) and cerebellar flocculus were studied in rhesus monkeys in the Bion 6 (Cosmos 1514) through Bion 11 projects. In the first few days of space flight, gaze displacement onto lateral targets became hypermetric, and the amplitude of head movements decreased. This was compensated for by increases in the gain of the angular vestibulo-ocular reflex (aVOR) that could last for the duration of the missions. Associated with this, there were increases in neuronal activity in MVN and flocculus. Sensitivities of the same populations of MVN neurons to linear acceleration, in general, increased gradually over the first 5–7 days in microgravity and then normalized over the course of the flight. These data indicate that the gain of the aVOR is increased during active lateral gaze fixations in space flight, and show that the underlying neural activity is appropriate to produce these changes.

Introduction

The primate research program "Bion" on the biosatellite "Cosmos" was planned in the 1970s to investigate vestibular dysfunction in space, with the aim that the outcome would benefit humans traveling in space. At that time the Space Adaptation Syndrome (SAS) was observed in 30–40% of cosmonauts, but there was no possibility of obtaining direct measurements of parameters related to vestibular, proprioceptive, motor, or other dysfunction during piloted flight, and most of the information on vestibulo-oculomotor dysfunction related to microgravity was obtained during postflight testing (Uganov, 1974). Moreover, changes observed after landing could be related not only to the effects of weightlessness but also to the factors that cosmonauts experience during landing (see Section 1). To obtain direct measurements of the vestibular dysfunction in microgravity, the space capsule "Vostok," which was originally designed for single-occupancy Cosmonaut flight and which had been used for six orbital flights, was adapted to handle two primate capsules (Gazenko and Ilyin, 1987). In this review we will refer to these experiments as Cosmos or Bion flights interchangeably.

There were three main vestibular studies in the Cosmos project. One was to study eye–head coordination and activity in the vestibular nerve, the medial vestibular nuclei (MVN) and the cerebellar flocculus associated with angular head movements in the horizontal plane during gaze shifts performed by the head and eyes to lateral targets, the gaze fixation reaction (GFR). Another project studied the sensitivity of central vestibular neurons to linear head displacements along the body axis. The third set of experiments studied the activity of leg flexor and extensor muscles during foot movements at different times of adaptation to microgravity. In the present review we will only cover the first two projects since the last project has been described in detail elsewhere (Edgerton *et al.*, 2000).

The results presented here are based on information from many sources (Shipov *et al.*, 1986; Sirota *et al.*, 1987, 1988a, 1989a,b, 1990a,b,c, 1991b,c; Kozlovskaya *et al.*, 1989, 1991, 1994; Yakushin *et al.*, 1989, 1990, 1992. However, the amount of data presented in these publications is limited, and the majority of quantitative data was taken from "Final Reports" that were submitted by investigators to the Institute of Biomedical Problem officials

Table 1

Flight numbers, times of launch, and monkeys in Cosmos flights

Cosmos flight number and date	Taking off time	Project number launch date	Monkey name	Flight duration (days)
1514	7:00 GMT	BION 6	Abrek	5
12.14.83	10:00 Moscow		Bion	
1667	3:21 GMT	BION 7	Vernyi	7
07.10.85	6:21 Moscow		Gordyi	
1887	12:43 GMT	BION 8	Drema	13
09.29.87	15:43 Moscow		Yerosha	
2044	6:28 GMT	BION 9	Jakonia #782	14
09.15.89	9:28 Moscow		Zabiaka #2483	
2229	13:40 GMT	BION 10	Ivasha #6151	12
12.29.92	16:40 Moscow		Krosha #7906	
*	13:50 GMT	BION 11	Lalik #484	15
12.24.96	16:50 Moscow		Multik #357	

*At this point, the Russian Space Agency stopped assigning flight numbers to the "Cosmos" Missions.

at the end of each project (Sirota *et al.*, 1984, 1986, 1988b, 1991a; Badakva *et al.*, 1993).

Methods

Male monkeys (*Macaca mulatta*) of 3–5 kg were used in these studies. Their names were assigned alphabetically; the first letter of the monkey's name corresponded to the sequential letters of Cyrillic alphabet (see Table 1). Although two primates traveled in each flight, information on vestibulo-oculomotor coordination and unit activity was not always available from both of them. Thus, although twelve animals took part in six Bion flights, valid data were obtained from only seven animals during four space flights of different durations. These results form the basis for the present report.

Eye–head coordination test

The gaze fixation reaction (GFR), which is a shift of gaze onto lateral targets using both head and eye movements, is a structural unit of daily operant behavior, and the dynamic characteristics of the GFR are similar in men and monkeys (Bizzi *et al.*, 1971, 1972). Small gaze shifts are usually performed first by an eye saccade and later by head movements. During gaze shifts larger than 20°, eye saccades are always accompanied by head movements (Tomlinson and Bahra, 1986; Phillips *et al.*, 1995). The test was structured so that the monkey first directed its gaze toward a fixation light in front in the primary position. The target was then repositioned laterally. In response, the animal first made an eye saccade toward the target at its new location. Generally, about

20–40 ms after the beginning of the eye saccade the head started to move toward the target. Since the eye saccade was much faster than the head movement, the eye was directed toward the target first. Thus, the head was still in motion at the end of the saccade, and the head performed only about 20–30% of the total motion.

Experimental conditions had an effect on GFR parameters: if the appearance of the lateral target was predictable, the head could start an anticipatory movement first, while the animal was still fixating the centrally located target. In other cases the head could delay the eye saccade significantly (Dichgans *et al.*, 1973; Grigaryan *et al.*, 1986). Thus, when the gaze jumped from the central position to a visual target located 40° laterally, the total gaze displacement was 40°. However, the contribution of the eyes and the head varied from trial to trial (Phillips *et al.*, 1995). To keep the gaze stationary on the new target, the eyes counter-rotated to compensate for the head motion (Bizzi *et al.*, 1972). As demonstrated in monkeys with a bilateral loss of vestibular function, the counter-rotation of the eyes after the saccade could also be anticipatory (Dichgans *et al.*, 1973). In normal animals, however, the counter-rotation of the eyes was due to activation of the angular vestibulo-ocular reflex (aVOR) (Dichgans *et al.*, 1973). Visual feedback after the saccade onto a newly located target also affected the ocular counter-rotation and provided the source for the adaptive modification of the aVOR gain. The major focus of the research was to determine the gain (eye velocity/head velocity) of the aVOR during these active gaze shifts onto target.

Although gaze shifts from one point to another are based on visual information, visual correction can occur only after the gaze has shifted to a new position. Moreover, since visual recognition has some delay, there was no visual feedback within about 80 ms immediately after the eye saccade. Therefore, the counter-rotation of the eyes reflected the current state of the aVOR gain. Since parameters of the GFR, such as amplitude and velocity of the eye, head and gaze movements, as well as counter-rotation of the eyes after a saccade, are determined by and reflect changes in sensitivity of the vestibular and proprioceptive sensors or brainstem structures, the GFR was considered a suitable reaction to study the effects of microgravity on vestibulo-oculomotor coordination.

Experimental paradigm utilized in Cosmos experiments

Gaze fixation reaction

Monkeys were trained to look at a central target and to position their heads straight ahead in the horizontal plane for 0.8 s. Head position was used as a feedback signal to trigger the next step. Targets located ±40° laterally from the center were presented for 1 s. The lateral targets were of two configurations. When the stimulus, a "C", was presented, the animals were required to press on the lever located in front, beneath the target plate. If the animal pressed the lever within 1 s from the time the lateral conditional target appeared, it got a reward of 0.3 ml of juice. If the animal did not press the lever within the required time, or if it pressed the lever in response to the stimulus, which was an "E", no juice was given. Additionally, the presentation of the next central target was delayed by 7–10 s. The appearance of targets on the left and right as well as the presentation of stimuli were randomized. Each program presentation was comprised of a set of 256 stimuli with 2/3 of them being conditional. This test was performed twice a day. In general, this test was accomplished within 20–25 m but, regardless of the animals' performance, only the first 20 min of the morning session were stored on tape for analysis (Sirota *et al.*, 1984, 1986, 1988b, 1991a).

MVN unit responses to linear acceleration

The response of MVN neurons to vertical linear acceleration was studied in each of the Cosmos 1667 to Cosmos 2229 flights. For this, the primate chair was elevated 45 mm slowly and then dropped suddenly to its original position with the aid of a spring mechanism. The total motion of the chair was 45 mm in all flights, but there was variation in the stimulus between flights. Most flight stimuli were comprised of a slow chair elevation over 8 s and a fast drop down to the original position. Peak acceleration for the upward motion was 0.14×10^{-4} ***g***. This is close to the threshold for detection of linear acceleration by the vestibular system, which is $\approx 0.1 \times 10^{-3}$ ***g*** for humans (Guedry, 1974). Thus, any responses observed during the elevation phase could have been due to random fluctuation or other factors. The stimuli parameters for the motion down in the drop varied on Earth and in Space. On Earth, the chair moved down over 0.6 s, and the downward pull of the spring was aided by the pull of gravity, to provide a downward acceleration of 25.4×10^{-3} ***g***. In Space the chair moved over 0.9 s, to provide a downward acceleration of 11.0×10^{-3} ***g***.

Surgical procedure

Two surgical approaches were used in the Cosmos experiments to implant the head holders. In the traditional method, used in Cosmos 1514, two bolts were implanted on the skull to fixate the head mount. A new technique that was minimally invasive was developed for later flights (Sirota *et al.*, 1988a), and is described in full detail elsewhere (Yakushin *et al.*, 2000b). EOG electrodes were implanted bitemporally to record the horizontal component of eye movements as well as above and below the left eye to record vertical eye movements.

Identification of the brain structures

Neurons recorded in the vestibular nuclei had various types of activation but most of the isolated neurons were modulated in phase with head velocity and were activated by eye position or eye velocity (Miles, 1974; Fuchs and Kimm, 1975; Keller and Daniels, 1975; Chubb *et al.*, 1984). Units recorded in the flocculus were similar to floccular units in previous studies (Lisberger and Fuchs, 1978a,b). Some of the recorded floccular cells had complex spikes that confirmed electrode locations near Purkinje cells in the cerebellar cortex. Attempts were also made to record units in the vestibular nerves in the Bion 8–11 projects, and some preliminary data are available (Kozlovskaya *et al.*, 1989, 1991; Correia 1998). Since there was no way to confirm that the recordings were taken from the same population of fibers on different flight days, these observations are omitted from this review.

Recorded signals and calibration

Head position in the horizontal plane was recorded with a special device placed on the center of the head-mount in the Bion 6–10 flights. The device was based on a compass principle. DC magnets were placed on either side of the primate chair at the level of the top of the head, where the device was mounted. The head position sensor was calibrated before flight. The horizontal EOG was calibrated using two assumptions based on the preflight performance: first, it was assumed that the final gaze position was equal to the angular position of the lateral target; second, under normal, preflight conditions, it was assumed that the counter-rotation of the eyes was fully compensatory during the GFR, and, therefore, that gaze (eye + head) position was stable after the eye saccade until the end of the head movement. Two channels capable of recording neural activity at frequencies ranging from 0.2 to 5 kHz were used in the Cosmos 1514. Two more channels with a frequency range from 0.2 to 10 kHz were used for unit recording in Cosmos 1667 flights, while four channels of this frequency band were used in later flights.

Electrode placements into MVN and the flocculus were based on the physiological responses of the identified units. The microelectrodes were introduced through metal guide-tubes implanted in the skull through 1 mm holes drilled through the scalp and skull in stereotaxic coordinates. Coordinates to reach MVN were P2–P4, lateral 2 mm. The flocculus was approached at the level of lateral vestibular nuclei, and the microelectrodes were tilted laterally so that they would not enter the brainstem after penetrating the flocculus. Three to four guide tubes targeted each structure. The microelectrodes were made of 80 μ tungsten wires covered with epoxy.

Data analysis

The methods of data processing varied with improving technology. Records obtained in the Cosmos 1514 flight were printed on a chart recorder. Since the amplitude and velocity of many parameters had relative calibrations, it was impossible to determine delays and durations of various GFR parameters precisely. Therefore, the results from the Cosmos 1514 flight are expressed as a percent relative to the preflight values. In all other flights, analog tapes recorded during flight were digitized with eight-bit resolution with a Motorola 6800 general-purpose mainframe computer. The channel that contained the marker of central and peripheral target presentation was digitized at 5 kHz. When the presentation of a lateral target was detected, analog signals, including the horizontal EOG, head position, lever-press and marker channel were digitized at the same frequency. Data were averaged over eight data points and stored in the computer at a 1.6 ms sampling rate for future analysis. Thus, although the data were stored at 625 Hz, each data point represented an average value of eight sequential data points and, therefore, the noise level due to the digitizing process was reduced. The following parameters of GFR were analyzed: latency, duration, amplitude and peak velocity of the eye, and head and gaze movements. There were no consistent changes in latencies and they will not be considered further.

Results: Gaze fixation on lateral targets; gaze fixation reaction (GFR)

Hypermetria of gaze

A typical saccadic gaze shift to the lateral target in preflight testing before the Cosmos 1514 flight is shown in Fig. 1A. A head movement in the same direction accompanied the eye movement. In the first inflight recording on the second day of space flight, the amplitude of the saccades was larger than before the flight, and the amplitude of head movement was approximately the same. Therefore, the gaze shift was hypermetric (Fig. 1B). Presumably, since the reward was dependent on accurate fixation of the target within one second, the animal overcame the hypermetric gaze shift after the initial period, as in Fig. 1F. Overall, the amplitude of saccades during gaze shifts performed with a single saccade in this monkey increased during flight. The increase was $11 \pm 3\%$ on day 2 and gradually increased during the flight, reaching a maximum of $42 \pm 7\%$ on day 5. Concurrently, the amplitude of head movement was smaller in flight, decreasing by 34% on flight day 2, and the head movements were still 27% lower than before flight on day 3. The head movements then normalized, and were only about 8% smaller on flight days 4 and 5. This animal performed the lateral gaze shift with one saccade before flight, using multiple saccades only 4% of time (Fig. 1C). The multisaccadic gaze shifts increased to 36% on the second day of space flight. The number of gaze shifts made with multiple

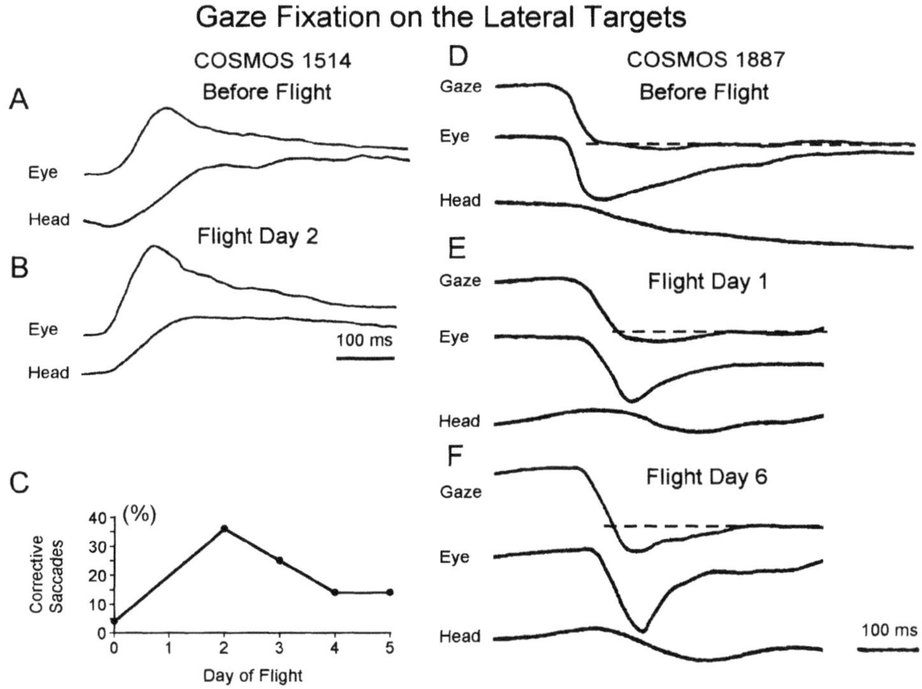

Fig. 1. A–C, Gaze fixation reaction onto a target located 40° to the right, performed by monkey Abrek before (A) and on the second day (22 h) of space flight (B). The data from each day had the same gain, and therefore the data could be compared. C, Percent of gain fixation reaction performed with corrective saccade before and during flight. D–F, Gaze fixation onto lateral targets performed by monkey Drema before (D) and during the first (E) and sixth (6) day of flight. The traces in D were reversed to facilitate comparison with E and F. Adapted from (Sirota *et al.*, 1984; Shipov *et al.*, 1986; Kozlovskaya *et al.*, 1989).

saccades was highest on day 2 and then decreased to 14% on flight days 4 and 5 (Fig. 1C).

Parameters of the GFR were similar in the monkey studied during the Cosmos 1887 flight. The gaze amplitude, which was $40.2 \pm 6.9°$ before flight, became hypermetric in space, going from $46.0 \pm 8.5°$ on the first day, to $58.3 \pm 7.0°$ on the 5th day. The gaze amplitude was slightly smaller (51–52°) when tested on days 8 and 10. Examples of these gaze shifts are shown in Fig. 1D–F. Before flight, gaze was stable on the lateral target over the entire period of head movement (Fig. 1D, dashed line). In flight, gaze was hypermetric, mostly due to the increased amplitude of the saccades, but gaze came back onto the target after the initial overshoots (Fig. 1E, F, dashed lines).

In Cosmos 2044, the gaze movements to the 40° lateral targets before flight were $38.0 \pm 0.4°$ and $37.0 \pm 0.6°$ in the two monkeys (Fig. 2). On day 2 in space, the gaze amplitudes were $47.0 \pm 1.3°$ in the monkey Jakonia (Fig. 2A) and 48.0 ± 1.9 in monkey Zabiaka (Fig. 2B). The gaze amplitudes normalized, however, within the next several days in both animals. Detailed analysis

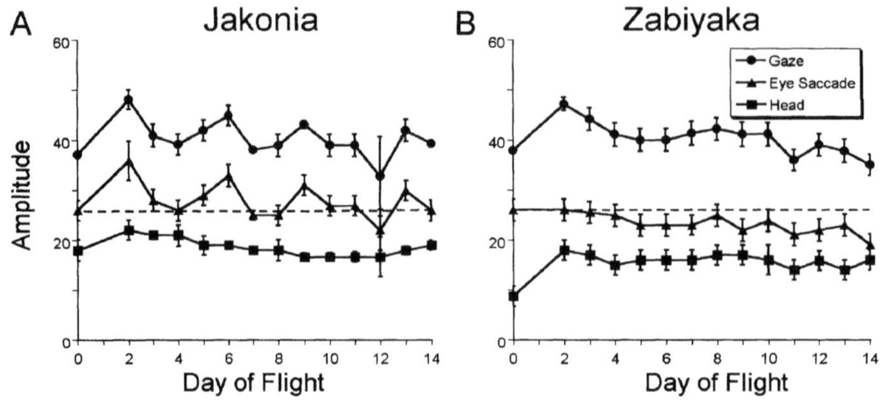

Fig. 2. Parameters of gaze fixation on the lateral targets performed by monkeys Jakonia (A) and Zabiyaka (B) during flight of Cosmos 2044.

revealed substantial differences in the source of observed changes. Although both animals had hypermetric gaze on day 2, the gaze overshot the target due to larger saccades in Jakonia, while the overshoot was due to larger head movements in Zabiaka.

In the Cosmos 1667 and 2229 flights, most gaze shifts were associated with multiple saccades. The amplitude of the associated head movements was decreased (Gazenko and Ilyin, 1987; Sirota *et al.*, 1987). Multiple saccades were not characterized quantitatively, but the gain of the compensatory aVOR was increased (see below). Thus, similar to the other flights, the lateral gaze displacements during the Cosmos 1667 and 2229 flights were also hypermetric.

In summary, gaze shifts to lateral targets in each of the six animals became hypermetric in the first few days of flight. This was observed whether the monkeys made the gaze shift in single or multiple saccades. In five of the six monkeys, the gaze overshoot was mainly due to an increase in the amplitude of the saccades, while in one animal it was due to an increase in the amplitude of the head movements (Fig. 2B). Thus, for successful visual fixation of target during head movements, the hypermetria had to be compensated for either by a corrective saccade or by a change in the gain of the aVOR (Fig. 1F).

Increase in the gain of the angular VOR (aVOR)

The gain of the aVOR was defined as the ratio of eye velocity to head velocity during counter-rotation of the eyes after the animals had made a saccade onto target. In Cosmos 2044, the instantaneous aVOR gain was studied as a function of time, starting from the end of the eye saccade over a 128 ms time period. When the eyes were stationary just after the saccade, the aVOR gain was zero. As counter-rotation began, the gain gradually rose. Under normal conditions the aVOR gain is 1.0, so the eye velocity is equal to head velocity over the counter-rotation period.

Nine superimposed gain curves obtained before flight in monkey Zabiaka are shown in Fig. 3A. The aVOR gain increased to unity within the first 30 ms after the saccade and then remained stable while the head was in motion. In most cases, the head movements ceased after 80 ms (Fig. 3E). Consequently, the aVOR gains were calculated as average values over the time interval from 32–64 ms after the initial saccade. Average instantaneous gain curves based on 30 responses before flight (Fig. 3B) were similar to those in Fig. 3A.

In flight there were substantial changes in aVOR gains. As shown in Fig. 3C, the gain of the aVOR increased to about 1.5 on the first recording of day 2 (Fig. 3C) and remained at this level until day 8 (Fig. 3D), when the increase became even larger (≈ 2.0). The gain was still about 1.5 on day 14 (Fig. 3E). Changes in the gains of the aVOR were similar during gaze shifts in either direction and were combined to obtain the average gain changes for this monkey (Fig. 3F, filled circles). The aVOR gain was also increased when the other animal of this flight, Jakonia, was tested for the first time in space on day 3 (Fig. 3F, filled squares). The gain increase was smaller subsequently, but was present over the entire flight.

The average gain values obtained from the instantaneous gain curves were compared to the aVOR gains obtained from taking the ratio of eye and head velocities at arbitrary points during the counter-rotation. The average aVOR gains were the same when measured with either method (Sirota *et al.*, 1991b). Thus, it was possible to compare the aVOR gain measurements in the different Cosmos projects. Individual gain curves for each of the monkeys in this report are shown in Fig. 3F. There were individual differences, but as a group, there was a substantial increase in the aVOR gains that persisted over the entire flight (Fig. 3G).

In summary, the general conclusion is that gaze became hypermetric upon entry into space on each of the Cosmos flights and that the positional errors generated by this gaze overshoot were compensated for by an increased gain of the aVOR. Although both of these changes could have occurred independently, since they occurred in parallel, it is more likely that one was primary, while the other was an adaptive response to the primary change.

The angular acceleration that activates the semicircular canals is the same on Earth and in space. Accordingly, the gain of the passive aVOR induced by steps of velocity and by voluntary sinusoidal head oscillation in darkness were not affected in microgravity (Benson *et al.*, 1986; Cohen *et al.*, 1992; Clarke *et al.*, 2000). If the hypermetric gaze was primarily due to exposure to microgravity, then the alteration in aVOR gain was a simple compensation for the positional error, driven by the visual feedback. However, there are reasons to question this explanation. First, the source of the hypermetric gaze varied among animals. In one, it was due to an increase in the amplitude of head movements (Fig. 2B), while in the others, it was due to the increase in the

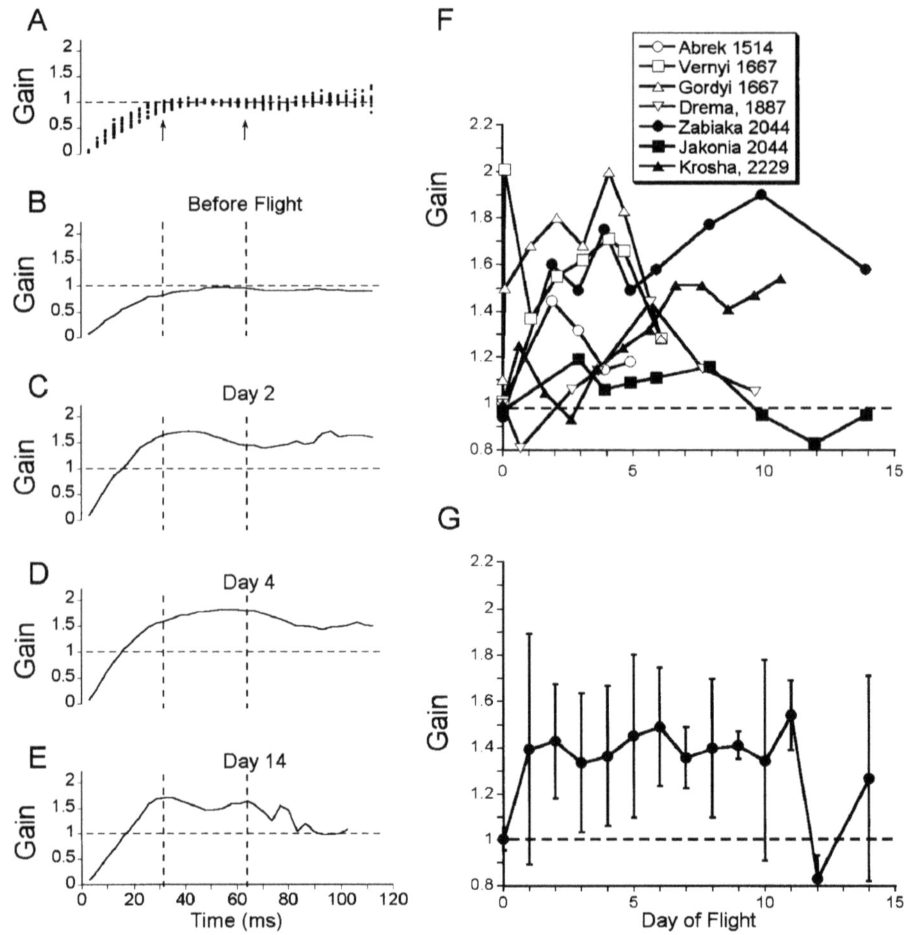

Fig. 3. Gain of the aVOR measured during eye counter-rotation during the gaze fixation task. A, Nine superimposed instantaneous gain curves obtained in monkey Zabiyaka (Cosmos 2044) before flight. The arrows in A and the vertical dashed lines in B–E indicate the regions where average values were measured. F, Individual aVOR gains for each animal tested during space flight. G, Average (± 1 SD) gains over space flight for the entire series, based on the individual responses shown in F.

amplitude of the saccades. Of interest is that similar increases in gaze amplitude have been observed in both human and monkeys after water immersion (Barmin *et al.*, 1983; Kreidich *et al.*, 1983; Badakva *et al.*, 2003). In monkeys the hypermetric gaze was due to an increase in saccade amplitude (Badakva *et al.*, 2003), while in humans the amplitude of the head movements increased (Barmin *et al.*, 1983; Kreidich *et al.*, 1983). There was no change in the acceleration of gravity in these experiments, yet the changes were similar to those during space flight.

An alternate explanation is that the increase in the gain of the aVOR gain was the primary event. Similar changes in aVOR gain were also observed after water immersion (Barmin *et al.*, 1983; Kreidich *et al.*, 1983; Badakva *et al.*, 2003), and as in the Cosmos 1667 flight, the first gain changes were observed after only two hours of immersion. Since changes in gravity were not a factor in these experiments, changes in proprioceptive inputs could have contributed to or been responsible for producing the gain changes. If the aVOR gain changes were the primary even, then parameters of the GFR should be different if the eye saccade was not accompanied by head motion. Kozlovskaya and colleagues demonstrated in cosmonauts tested before and after space flight that there are two types of the adaptive changes in GFR. One group had changes in the amplitude of the saccades and in aVOR gains, similar to those observed in the monkeys. The second group was comprised of spacecraft commanders, e.g., professional pilots. They performed gaze shifts in two stages: first by a saccade onto target and then by a head movement, which followed the saccade. In the pilots, changes in their aVOR gains were similar to those described above. Since the amplitude of the saccades during the gaze shift was accurate within $\pm 3°$, there was no visual/vestibular mismatch that could have driven the changes in the aVOR gains (Kozlovskaya *et al.*, 1985).

An alternate mechanism could also have been responsible for the changes in the gain of the aVOR. It was recently demonstrated that adaptive changes of the aVOR gain are a function of head position with regard to gravity (Tiliket *et al.*, 1993; Yakushin *et al.*, 2000a, 2003a,b, 2005). Thus, "normal" aVOR gains during active head movements may only exist under the level of gravity and/or proprioception in which the aVOR gains were adapted, and a change in gain of the aVOR due to insertion into microgravity could have been the primary change that produced the changes that were observed in microgravity.

If there were changes in the gain of the aVOR, there should be associated changes in the cellular activity in the vestibular nuclei and/or flocculus, which produced the adaptive changes in aVOR gain ((Lisberger, 1994; Lisberger *et al.*, 1994a,b), see Ito (1984) for review). Unit activity was recorded in the vestibular nuclei and flocculus during each space flight and is considered in the next section.

Neural activity in the vestibular nuclei and flocculus during space flight

Activity of a single lateral canal-related neuron in the right MVN of monkey Vernyi (COSMOS 1667) during sinusoidal head/body oscillation before flight is shown in Fig. 4A. The discharge rate increased in phase with ipsilateral head velocity (Fig. 4B). Therefore, this cell was a type I unit.[1] When this unit was well

[1]The vertical canals contribute to the neural response to head movement in the yaw plane. Therefore, vertical canal-related units would also be modulated with this stimulus, and it is possible that some of the activated neurons were actually more closely related to the vertical canals. Activity of the unit shown in Fig. 4 was modulated during horizontal but not vertical eye movements, and it is likely that this unit was a lateral canal-related neuron.

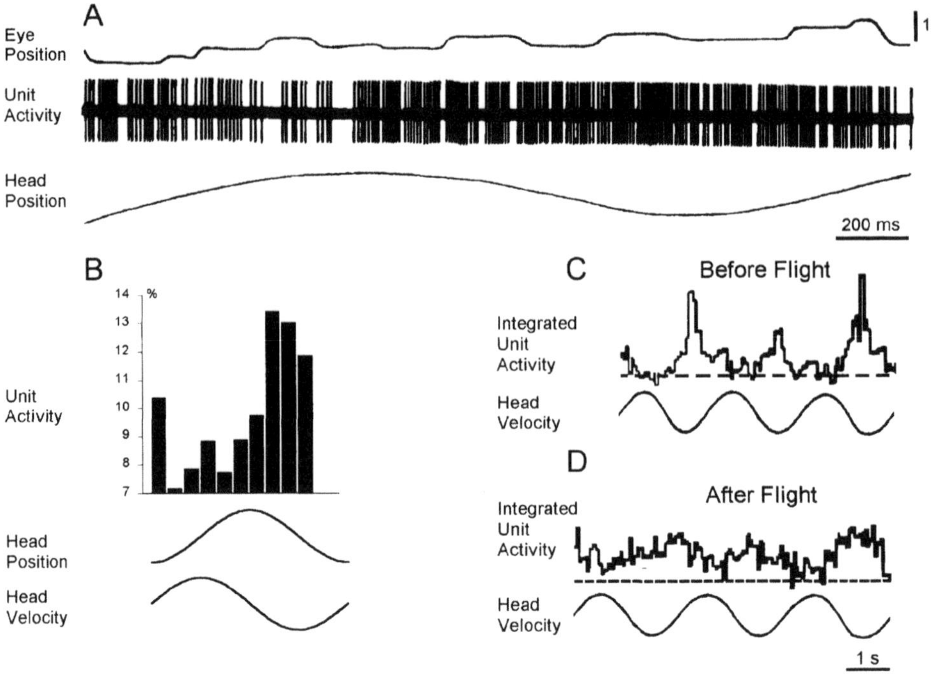

Fig. 4. Activity of the lateral canal-related type I neuron recorded the right medial vestibular nuclei in the monkey Vernyi (Cosmos 1667). A, Single unit recording at the site where one of the MVN electrodes was installed. Head position down is rotation to the right (ipsilateral). B, Poststimulus histogram for the unit shown in A. Traces below show head position and head velocity during sinusoidal rotation. C–D, Integrated unit activity recorded from the same electrode before (C) and after (D) seven days of the space flight. Note that the modulation in unit activity increased peaking just before the ipsilateral (rightward) head velocity had reached maximum. The head oscillation shown in A–D was driven manually with approximate frequencies 0.45 Hz (A, B), 0.5 Hz (C) and 0.4 Hz (D).

isolated from the background, the electrode was fixed at this location. Over time, this electrode recorded several neurons simultaneously. The multiunit activity recorded before and after the flight increased with ipsilateral rotation and had a similar phase relationship to head velocity (Fig. 4D) as the neuron in Fig. 4A. The electrodes over the various Cosmos flights were implanted in MVN in the region of type I units.

Multiunit activity recorded from three of four of the implanted electrodes in the two animals in Cosmos 1667 was of good quality. An example of a recording during a gaze shift to the right during flight is shown in Fig. 5A. Single neurons, shown as standard pulses, could be separated from the background (Fig. 5B). The quality of separation was similar for each flight day, although it was not certain that the same neurons were analyzed throughout the flight. Before flight, the peak activation occurred approximately at the time when ipsilateral head velocity reached a maximum (Fig. 5C, Before). The unit activity did not

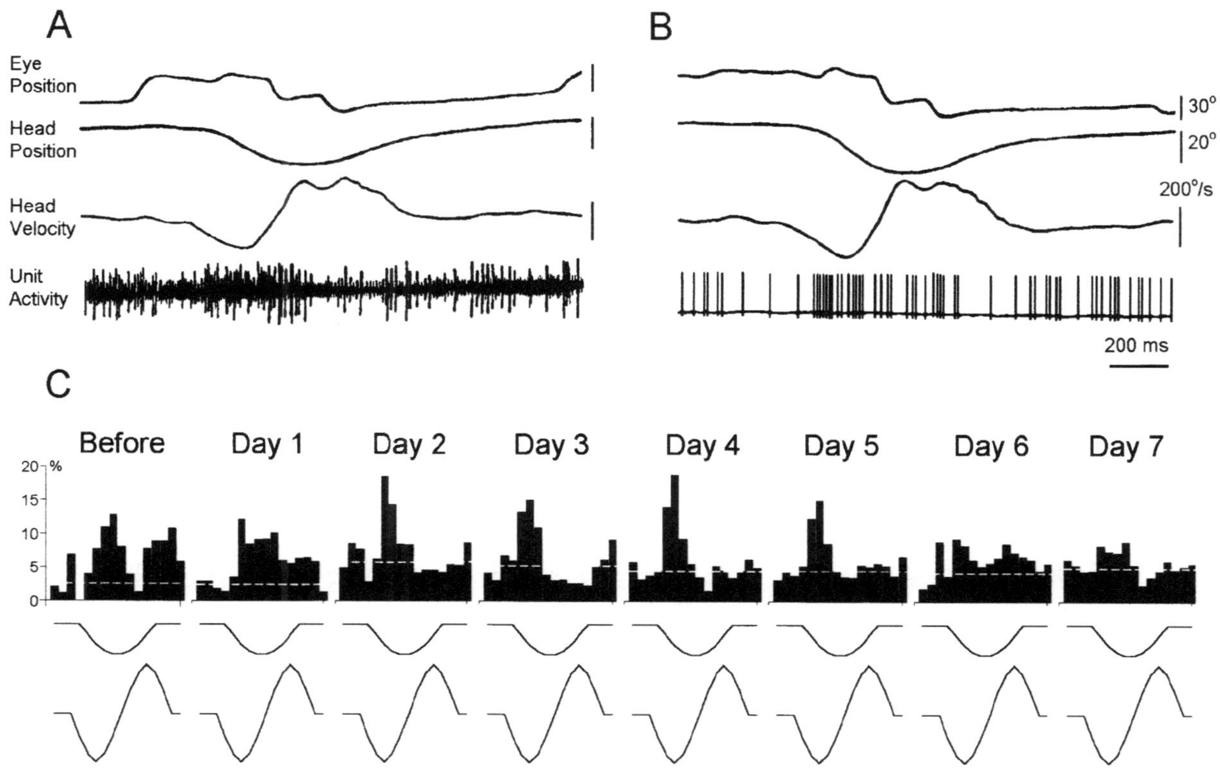

Fig. 5. A, Sample of multiunit activity recorded in the right vestibular nuclei in monkey Vernyi (Cosmos 1667) during the gaze fixation reaction onto the target on the right. The recording was obtained from the same electrode at the same location as in Fig. 4. B, Activity of the largest neuron from a sample similar to that shown in A and converted to standard pulses. C, Poststimulus histograms of this unit recorded at different times in flight. Thirty head movements of similar profile and amplitude were identified for each day. Head motion to the right and back to the center was divided into eight bins. An additional eight bins were used to characterize the data from before and after head motion. The idealized head position and head velocity are shown under each histogram.

decrease relative to the resting discharge rate (white dashed line) in association with contralateral head velocity. The profile of activation by ipsilateral head velocity was the same over all flight days. The amplitude of peak activation before flight (27 ± 4 imp s^{-1}) was comparable to the peak activation recorded during the second hour of space flight (23 ± 6 imp s^{-1}, day 1). This activation became larger on flight days 2–4 (50 ± 8 imp s^{-1}, day 2) and then decreased to normal on days 5–6 (25 ± 3 and 22 ± 3 imp s^{-1}, respectively). Thus, the activation was twice the normal value after 26 h in space (day 2). The changes were substantial over the next 2 days and then normalized. Data from this neuron are summarized in Fig. 6F (filled symbols). There was no significant change in the activity recorded during gaze shifts to the right after 2 h of space flight (day 1), but the activity gradually increased to a maximum on day 4, and returned back to normal for the rest of the flight.

The profile of the integrated multiunit activity was similar when multiunit activity recorded by the same electrode was integrated (Fig. 6A–C). That is, activation before flight (Fig. 6A) was similar to that on day 7 (Fig. 6C), and the increase in unit activity was larger on day 2 (Fig. 6B). The integrated activity from the multiunit recordings over the entire flight is shown in Fig. 6D (open symbols). Activation on flight day 1 after 2 h was the same as the preflight activity. This activity significantly increased on flight day 2 and then returned to normal by days 5–6 (Fig. 6F). The difference between the single and integrated multiunit recordings was presumably due to variation in response of the neurons in the vicinity of the recording electrode.

The multiunit activity in MVN was also studied in four other animals. The two animals in Cosmos 1887 had two electrodes implanted in MVN (Fig. 6D) and the two animals in Cosmos 2044 flight (Fig. 6E) each had one electrode in MVN. The sensitivities recorded in Cosmos 1887 flight were maximal on day 3 for all four electrodes and then decreased to normal in three of the four electrodes (Fig. 6E, filled circles). Changes in the sensitivities of the MVN neurons in the monkeys of Cosmos 2044 flight were more robust. The sensitivities increased significantly on day 3 and stayed high until day 8 when they began to normalize. In the animal Jakonia, the sensitivity stayed above the normal even on day 13 (Fig. 6F, filled symbols).

Multiunit activity was recorded in the flocculus from one electrode in each of the same four animals. In all cases the change in the sensitivity was maximal on the first day of flight (Fig. 6G, H). In two animals, the activity normalized by the middle of the flight (filled symbols), while in other two, it stayed above the norm until the last day of flight (Fig. 6G, H, open symbols).

A summary of all multineuronal recordings from the Cosmos flights is shown in Fig. 6I. Overall the sensitivity of the neural activity in both the vestibular nuclei and flocculus was maximal in the earlier days of flight, with the rise in activity occurring earlier in the flocculus than the MVN neurons. There was a striking similarity between the rise in activity of the flocculus (Fig. 6I, dashed line) and MVN multineuronal activity (Fig. 6I, solid line) and the

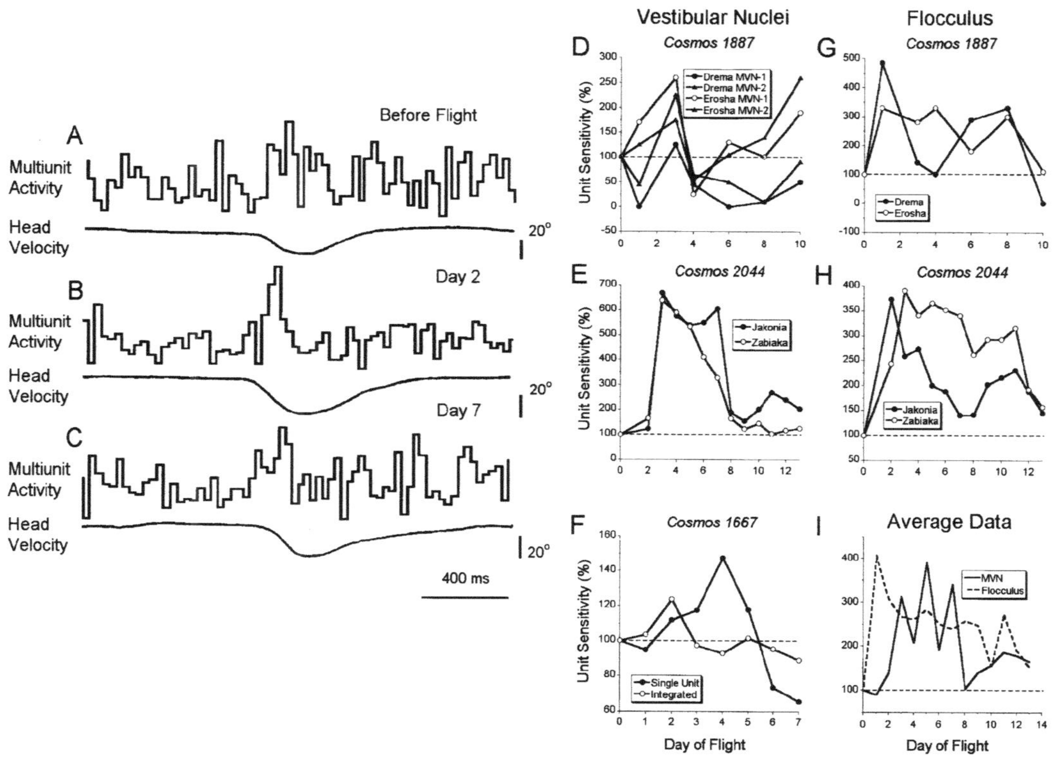

Fig. 6. A–C, Integrated (30 ms) multiunit activity from the right MVN recorded during angular head movement, when animal Vernyi performed gaze fixation onto the right lateral target, before (A), and on the second (B) and seventh (C) day of space flight (Cosmos 1667). D–F, Peak integrated MVN activity recorded on different days of space flight in monkeys Drema and Erosha (D, Cosmos 1887), Jakonia and Zabiaka (E, Cosmos 2044) and Vernyi (F, Cosmos 1667). G–H, Changes in flocculus activation during angular head rotation in the Cosmos 1887 (G) and Cosmos 2044 (H) flights. MVN-1 and MVN-2 (D) represent recordings from different sides of the brain. The location of the electrodes in the left or right MVN of the flocculus for Cosmos 1887 and Cosmos 2044 flights was not identified in the reports (Sirota *et al.*, 1988b, 1991b; Kozlovskaya *et al.*, 1989). I, Average activity in MVN (solid line) and the flocculus (dashed line) for animals tested in the COSMOS flights. D–I are normalized to the preflight date.

increases in aVOR sensitivity shown in Fig. 3G. The flocculus activity tended to rise earlier and persist longer than the MVN activity, but both were increased over the two weeks of flight. Thus, the changes in neural activity in MVN and flocculus during space flight mirrored the changes in aVOR sensitivity during the gaze fixation reaction. Presumably, the flocculus units were involved in the learning that changed the gain of the aVOR, while the increase in unit sensitivity in MVN to head rotation reflected the actual changes in gain in the compensatory processes.

Response of MVN units to linear acceleration in space

The first inflight recordings of central vestibular neurons during otolith stimulation in the Cosmos 1667 project utilized the same population of neurons in the right MVN of monkey Vernyi that was studied during angular rotation (Figs. 4, 5, 6A–C, F). A sample multiunit activity during otolith testing on day 1 is shown in Fig. 7A, and associated single unit activity in Fig. 7B. The stimulus cycle was divided into four phases for analysis. The spontaneous discharge rate (Phase 1) was obtained from a 1 s time period before each elevation of the chair. Unit activity during the elevation (Phase 2) and rapid drop (Drop, Phase 3) were taken as average values.[2] Additionally, activity was also utilized over a 1 s period immediately after the drop had terminated (Stop, Phase 4). Data were first expressed in imp s^{-1} and then normalized to activity in Phase 1 of each day. The bottom traces (Chair) in Figs. 7A, B, show the time intervals for each phase.

The resulting histograms for each flight day are shown in Fig. 7C. The first significant changes in the drop phase occurred on the first day after 2 hours of space flight. This activity normalized on days 2 and 3, but then increased again, reaching a much higher level (Fig. 7C, D). Sensitivity of the activity for the Drop in Phase 3 for each day is shown in Fig. 7D. The first significant changes were observed in the first day of flight. Activity was near normal on flight days 2 and 3 and then increased again on days 4–5. The direction of linear acceleration was opposite during the Drop and the Stop phases, but there was no difference between the two responses on any flight day. Since the head was not fixed during the Stop, it is not known whether the head continued to pitch at the time of the Stop, which could explain this apparent disparity. Therefore, the activity associated with the Stop Phase will not be considered further in the analysis.

Integrated multiunit activity from the second animal in the same flight (Gordyi) is shown in Fig. 8A–C. Before the flight, there was a brief increase in activity during the Drop (Fig. 8A). On the second day of flight there was a

[2]The acceleration in the preflight testing was higher during the rapid drop than inflight because gravity added a downward pull to that provided by the spring mechanism (see Methods). As described in Methods, the linear acceleration was only significantly above threshold for detection during the rapid drop in Phase 3 and the stop-reaction in Phase 4.

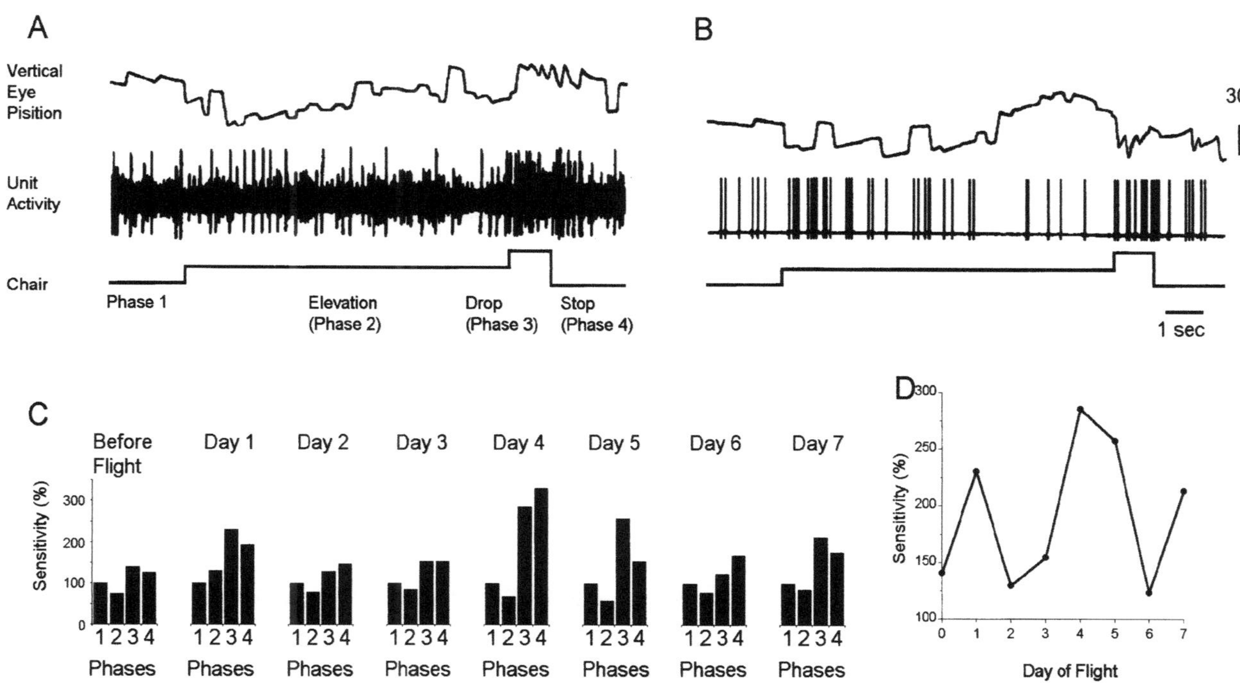

Fig. 7. A, Multiunit recording in the right MVN of monkey Vernyi (COSMOS 1667) during linear motion along the long body axis. Phase-1, Resting activity before stimulation; Phase-2, Slow elevation of the chair up; Phase-3, Fast movement down (Drop); Phase-4, Activity immediately after the Drop. B, Activity of a single unit discriminated from the multiunit record. The vertical eye position record indicates that there was up-beating nystagmus during Phase-3. C, Activity of the same unit in different days of flight normalized to the activity in Phase 1 on that day. D, Activity of the unit in Phase-3 on different days of space flight.

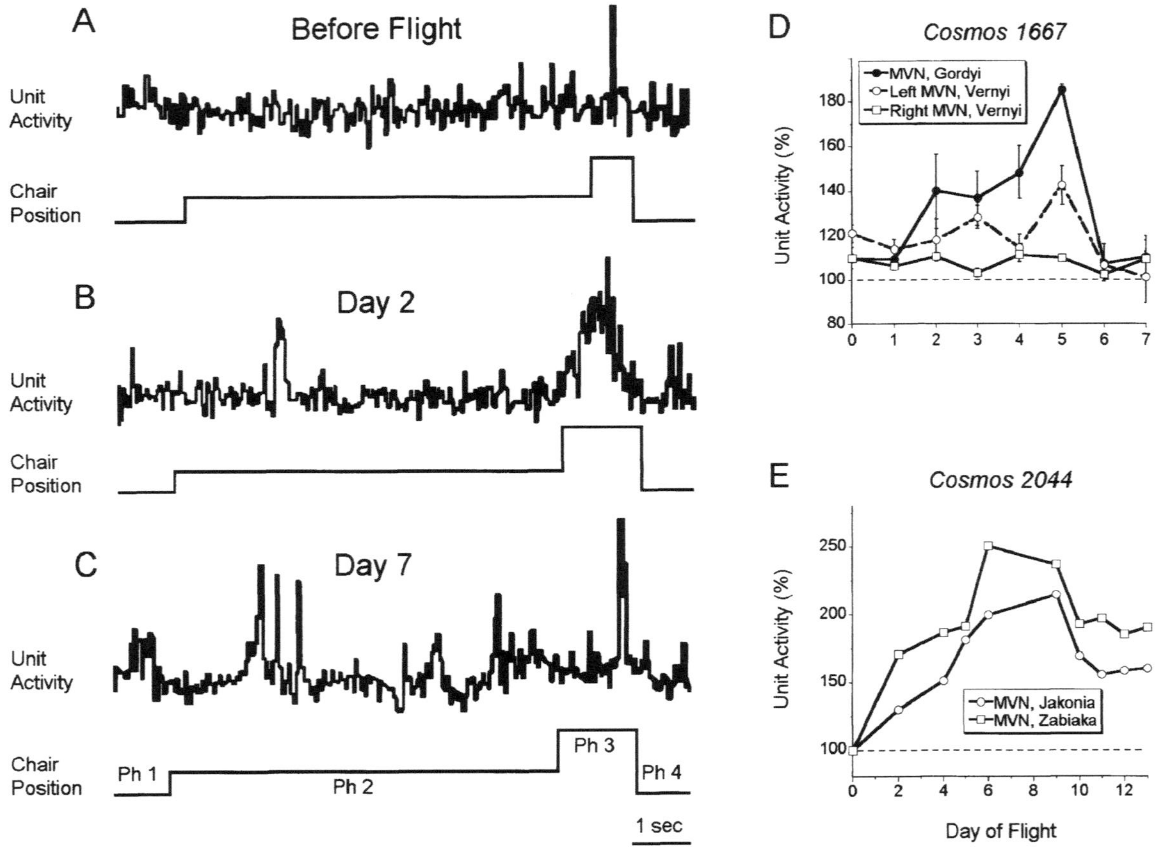

Fig. 8. A–C, Integrated (30 ms) multiunit activity recorded in monkey Gordyi during linear stimulation along the longitudinal body axis recorded before (A) on the second (B) and seventh (C) day of flight. D–E, Average multiunit activity in MVN in Phase-3 (see Fig. 7) on different days of the Cosmos 1667 (D) and Cosmos 2044 (E) flights. Multiunit activity in Phase-3 normalized to the activity observed before stimulation in Cosmos 1667 (D) and to the level of activation in Phase-3 recorded before flight in Cosmos 2044 experiment (E). Note that multiunit activity recorded in the right MVN of monkey Vernyi was not modulated in this test, but the single unit that was selected from this record (Fig. 7) was well modulated by the linear acceleration.

significant increase in activity during the Drop (Fig. 8B). The activity increased in response to the acceleration phase and then decreased to normal during the chair deceleration. The activity during the entire period on day 7 was more irregular than the other days (Fig. 8C), and the response to the Drop was again brief. Integrated activity recorded from both animals in the Cosmos 1667 flight is shown in Fig. 8D. The sensitivity of these MVN units was not changed on day 1, but was increased in two of the three electrodes on day 2. The activity increased further up to day 5, and then normalized.

Changes in the sensitivities of the MVN neurons in the monkeys Jakonia (Fig. 8E, open circles) and Zabiaka (Fig. 8E, open squares) from Cosmos 2044 were similar to those shown in Fig. 8D. The sensitivities gradually increased reaching maximal values on day 6 and then declined, although they were still above the baseline on the last day of flight (day 13). Thus, changes in the sensitivity of the MVN neurons to vertical linear acceleration mediated by the otoliths were similar in animals from the Cosmos 1667 and Cosmos 2044 flights. The sensitivity to linear acceleration gradually increased in space, reaching a maximum at days 5–6 and then decreasing toward normal values. The peak increases in sensitivity for the MVN neurons in Fig. 6D, E in response to angular acceleration during the gaze fixation reaction were between 300 and 600%, whereas the increases in sensitivity during linear acceleration were smaller, within 200–250%. Thus, the general pattern of response was similar to that obtained from angular acceleration, but the magnitude of the response to linear acceleration was smaller. This could represent a difference in response to angular and linear accelerations or could reflect a difference in the magnitude of convergence of the otolith and canal signals onto these neurons.

The responses to linear acceleration produced by sinusoidal oscillation along the body vertical axis in the Bion 11 Project were very similar to those observed in the Bion 9 (Cosmos 2044) flight. The multiunit activity of the neurons recorded in MVN in each animal was in phase with upward acceleration in monkey #484 and with downward acceleration in monkey #357 (Badakva *et al.*, 2000). In both cases, the sensitivity to linear acceleration increased only slightly on the first recording on Day 2, and reached a maximum on Days 4–6. Sensitivities then decreased but were still above normal up to the last day of flight.

Discussion

The data from the Cosmos flights show that the sensitivity of central vestibular neurons in MVN to both angular head rotation and vertical linear motion was affected by microgravity. In general, the sensitivity of these neurons increased in the earlier days in Space and then gradually normalized. In some instances changes were similar for several populations of neurons recorded on both sides of the brainstem from the same animal (Fig. 6D, E). In other cases, although average activity of the recorded neurons had similar changes in sensitivity during adaptation to microgravity (Fig. 6F, open symbols), the changes in sensitivity of a single unit could differ from the changes observed in the remaining population of units. A striking finding was that the observed changes in neuronal sensitivity were qualitatively similar to the changes observed in the aVOR gains and occurred over the same time course. Changes in sensitivity of the floccular neurons to the angular rotation were more uniform than the changes in MVN activity, presumably, because the flocculus was responsible for producing these adaptive changes.

Changes in sensitivity of the neuronal populations in the vestibular nuclei to the otolith stimulation were also relatively uniform and had approximately the same time course as the sensitivity to angular acceleration. The sensitivity to linear acceleration gradually increased, reaching a maximum by the end of the first week in space (Fig. 7D and Fig. 8D, E). This was different from the changes in sensitivity of the same neuronal population to angular rotation, which increased earlier in flight and was normalized by the time the otolith sensitivity had reached a maximum (Fig. 6I). This was somewhat surprising, since it would be expected that microgravity would affect otolith and not canal sensitivity. A majority of direct projections from otolith afferents go to the lateral and descending vestibular nuclei (Büettner-Ennever, 1999), but there are canal-sensitive neurons with some otolith sensitivity in MVN (Markham and Curthoys, 1972), which could serve a different purpose than maintaining balance or supporting the linear VOR. Otolith information is not necessary to induce changes in the angular VOR gain (Crane and Demer, 1999). As recently shown, however, the head position with respect to gravity in which the aVOR gain was adapted is expressed subsequently in changes in aVOR gain in every head position (Yakushin *et al.*, 2000a, 2003a,b, 2005). Therefore, the otolith signals could serve as a gravitational context for adaptation of the aVOR gain in specific head positions, and the changes in otolith sensitivity could be a reflection of changes in this gravitational context.

SECTION 3: NASA–RUSSIAN MONKEY EXPERIMENTS ON THE ANGULAR AND LINEAR VESTIBULO-OCULAR REFLEX

Abstract

The angular and linear vestibulo-ocular reflexes (aVOR and lVOR) of four rhesus monkeys were recorded before and after the 1988 and 1992–1993 Cosmos Space Flights 2044 and 2229 (Table 1). Two animals flew in each mission for approximately two weeks. Eye movements, induced by rotation with steps of velocity about a vertical axis, by constant velocity rotation about axes tilted from the vertical (off-vertical axis rotation, OVAR), and by horizontal and vertical translation were recorded binocularly with scleral search coils in two- and three-dimensions. Single unit recordings were also taken from semicircular canal afferents before and after flight.

Compensatory eye movements produced by the angular VOR (aVOR). Gains of semicircular canal-induced horizontal and vertical aVOR were unaffected in both flights, although the gain of the roll aVOR was diminished. Up/down asymmetries of vertical nystagmus present before flight were reduced for seven days after flight.

Activity of primary lateral canal afferents after space flight. The mean gain for nine different horizontal canal afferents, tested on the first postflight day of Cosmos 2044 with steps of velocity and sinusoidal rotation, was nearly twice that of 20 horizontal canal afferents similarly tested during preflight and postflight control studies. Adaptation of the afferent response to passive yaw rotation on the first postflight day was also greater. After the Cosmos 2229, however, afferent gains were reduced.

Spatial orientation of the aVOR. Spatial orientation of the aVOR was altered in two of the four monkeys after flight. In one, the time constants of postrotatory nystagmus, which had been shortened by head tilts with regard to gravity before flight ("tilt dumping"), was unaffected by the same head tilts after flight. In another animal, eye velocity, which tended to align with gravity before flight, moved closer toward a body axis after flight. This shift of orientation had disappeared by seven days after landing.

Compensatory eye movements produced by the linear VOR (lVOR). The gain of the high frequency compensatory lVOR was reduced for naso-occipital linear acceleration in one monkey, but maintained in a second monkey. Gain changes in the first animal lasted for 17 days after landing.

Orienting eye movements produced by the lVOR. The gain of the low frequency lVOR was tested using OVAR. Ocular counter-rolling (OCR) was reduced by about 70% during both the dynamic tilts experienced during OVAR and in response to static tilts. Similarly, modulation in vergence, in response to low frequency, naso-occipital linear acceleration during OVAR was reduced by over 50%. These changes in orienting eye reflexes persisted for 11 days after recovery.

Orientation of eye velocity induced by velocity storage. Steady state yaw axis horizontal eye velocities induced by OVAR were unaffected by space flight. The orientation of optokinetic after nystagmus (OKAN) and of vestibular nystagmus was altered, moving closer to a body than a spatial axis when tested shortly after landing in one animal.

Conclusion. There were both short and long term changes in otolith-ocular reflexes after adaptation to microgravity in both monkeys in the Cosmos 2229 flight, although horizontal and vertical semicircular canal-induced responses of the angular VOR to rotation were largely unaffected. The roll aVOR gain was also reduced. All of the reductions were greater in one animal (7906) across all tests. A comparison with data from astronauts suggests that maintenance of gain of both compensatory and orienting otolith ocular reflexes may depend on continuous exposure to linear acceleration during flight. Presumably, in future long duration space flights, this could be provided by centrifugation.

Introduction

The vestibular system is composed of two subsystems. One, comprised of the semicircular canals, senses angular acceleration of the head and generates compensatory eye movements that stabilize gaze during head and body movement over the angular vestibulo-ocular reflex (aVOR) (see Raphan and Cohen, 2002; Cohen and Gizzi, 2003; Cohen and Raphan, 2004 for review). Since the natural angular motion that excites the aVOR is essentially turning on a head- or body-centric axis, the horizontal aVOR would not be expected to change dramatically in a different gravitational environment. The second component of the vestibular system, however, the otolith organs, comprised of the saccules and utricles, sense both head orientation and head linear translation. The altered acceleration profiles experienced in microgravity should induce very different activation profiles in otolith afferents that could initiate morphological and behavioral adaptation of the vestibulo-ocular and vestibulo-spinal reflexes that depend on gravity. Thus, it would be expected that there might be changes in ocular counter-roll (OCR), vergence and spatial orientation after adaptation to microgravity but that the angular VOR would be less affected.

The linear vestibulo-ocular reflex (lVOR) can be further separated into high and low frequency components, using approximately 0.3 Hz as the divide between the two. The compensatory reflex provides ocular compensation against high frequency head translations (Schwarz *et al.*, 1989; Hess and Dieringer, 1991; Paige and Tomko, 1991a,b; Schwarz and Miles, 1991; Raphan and Cohen, 2002), and is used to maintain fixation on near targets during translation (Paige and Tomko, 1991a; Schwarz and Miles, 1991; Maruta *et al.*, 2001) or in response to centripetal acceleration generated by turning corners (Imai *et al.*, 2001). Vergence in response to high frequency linear acceleration along the naso-occipital axis is also compensatory, and supports fixation of near targets when moving forward (Paige, 1991; Dai *et al.*, 1996). There are also low frequency orienting otolith-ocular responses that tend to maintain the position of the retina in relation to the spatial vertical (Cohen *et al.*, 2001). These include horizontal, vertical and torsional shifts of the eyes, OCR, and sustained vergence in response to head tilts with regard to gravity (see Dai *et al.*, 1996 for review). Finally, angular eye velocity induced through activation of a central vestibular system known as "velocity storage" by the visual, vestibular and/or somatosensory systems, tends to orient to gravity or to the GIA when the GIA is tilted with respect to the head (Dai *et al.*, 1991, 1992).

In the pre- and postflight experiments described in this section, scleral search coil measurements of eye movements in response to controlled vestibular and visual stimulation were utilized in pre- and postflight experiments. The purpose was to demonstrate the effects of adaptation to microgravity on return to the 1 *g* environment of Earth. Experiments were also done on semicircular canal afferents to determine if the exposure to microgravity had significantly changed afferent activity after adaptation to microgravity.

Methods

Nineteen juvenile rhesus monkeys (*Macaca mulatta*) were candidates for the two Cosmos Biosatellite Flights. Of these, four animals were chosen for flight; the others served as controls. Monkeys 782 and 2483 flew in Cosmos Flight 2044. They were launched on 9/15/89 and recovered on 9/29/89 (Table 1). Animals 6151 and 7906 flew in Cosmos Flight 2229. These animals were launched on 12/29/92 and recovered on 1/10/93. Testing extended for 5 and 11 days postflight, respectively. In space, the monkeys sat in a capsule, 60 cm in diameter, in fur-lined chairs. The compartments that held the primate chairs were separated from each other, but the animals were within sight of one another. Their trunks were restrained, but their heads, arms and legs were free. The monkeys performed behavioral testing in space, moving their head and eyes toward lateral visual targets (see Section 2). The status of the animals was monitored by downlinked video, along with continuous recording of a wide range of other control signals from the capsule, including temperature, humidity, and food and water intake. Monkeys not chosen to fly were housed in comparable quarters on ground and served as controls. The experiments conformed to the Principles of Laboratory Animal Care (NIH Publication 85-23, Revised 1985), and were approved by the appropriate Institutional Animal Care and Use Committees.

Pre- and postflight testing

During experiments, the monkeys sat in a primate chair with their heads fixed to a plastic frame that held a square field coil, 25.4 cm on a side. The same primate chair and coil box was used in studies of high frequency linear oscillation. Yaw (horizontal), pitch (vertical) and roll (torsional) eye movements were recorded through two search coils, which were attached to the front and the top of the left eye. A frontal plane coil was also implanted on the right eye of the Cosmos 2229 monkeys. Voltages associated with eye position and with the position and velocity of the various axes were recorded through analog filters with a bandwidth of DC to 40 Hz. Eye positions and velocities were calibrated by assuming that the animals accurately tracked visual surround movement during rotation in light at 30°/s. Movements to the right and up caused upward trace deflections. Roll velocities were assumed to have a gain of 0.6. Eye movements were not recorded in the studies of afferent activity in the vestibular nerve.

Two multi-axis vestibular stimulators and three vestibular/oculomotor laboratories were transported from the United States to the Institute of Biomedical Problems in Moscow for these experiments. The apparatus and

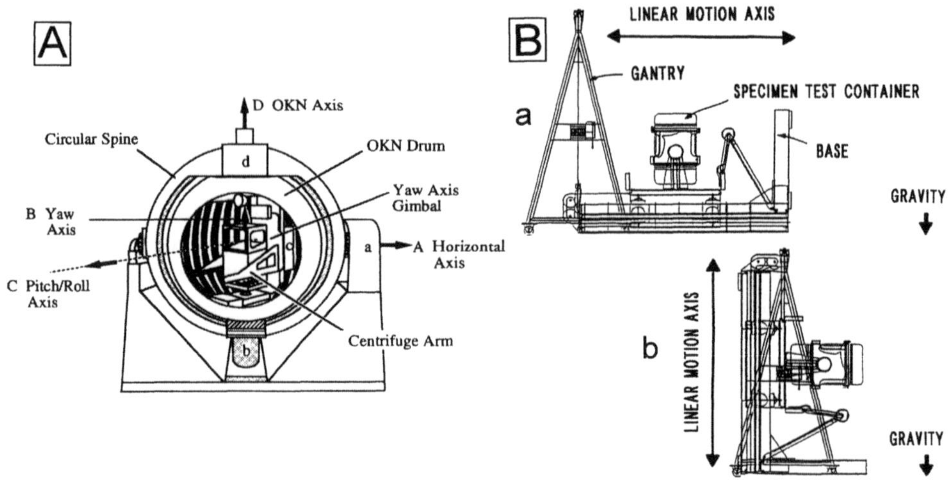

Fig. 9. A, Four axis vestibular and optokinetic stimulator for testing the aVOR, and for inducing OKN and OKAN and OVAR. Also shown are the primate chair and coil field apparatus used to activate the scleral search coils when recording eye position and eye movement. The same coil box and primate chair was also utilized in the linear oscillation apparatus. B, Portable linear sled for high frequency linear oscillation. A gimbaled, light-tight Specimen Test Container sat on a carriage supported by air bearings and "floated" on ceramic rails. The linear track was situated on a gantry that could be repositioned to give translation along earth-horizontal (Ba), oblique (Bb), or earth-vertical (Bc) axes. Stimuli were given in darkness, in a subject-stationary lighted visual surround (VSLVOR), or while viewing the visual surround that could move relative to the subject (VLVOR).

experiments are fully described in previous publications and reports (Cohen *et al.*, 1992; Correia *et al.*, 1992; Tomko *et al.*, 1993; Dai *et al.*, 1994, 1996). In brief, the apparatus shown in Fig. 9A was used to study the angular VOR (aVOR), optokinetic nystagmus (OKN), optokinetic after-nystagmus (OKAN), and the response to off-vertical axis rotation (OVAR). The response to angular acceleration was given by rotating the animals positioned in separate tests so that their yaw, pitch and roll axes were aligned with the spatial vertical. Optokinetic nystagmus (OKN) was induced by rotating the light-tight OKN shell around the animal's yaw axis with the yaw axis upright or in tilted positions. Off-vertical axis rotation (OVAR) was given by rotating the animals around a tilted yaw axis. The response to static tilts was tested by incrementally rotating the circular spine around the horizontal axis.

For tests of high frequency linear acceleration, the animals were oscillated on a linear sled, specially constructed at the NASA-Ames Research Center (Fig. 9B). Animals sat in a Specimen Test Container and moved along a linear track in Earth-horizontal (Fig. 9Ba), oblique (9Bb) or dorsoventral (9Bc) directions. In Ba, the animals were upright and in Bc they were either prone or supine. Motion was delivered along the interaural, naso-occipital, and dorsoventral axes as well as along intermediate oblique axes. lVORs were studied during 1.0 and 5.0 Hz Earth-horizontal head motion in darkness and while viewing a head-fixed (VSLVOR) or an earth-fixed (VLVOR) visual scene.

Afferent neural responses from the horizontal semicircular canals were tested by rotating the animals sinusoidally or at a constant velocity about a spatial vertical axis (not shown). To test potential sensitivity of the canal units to tilt before and after flight, the animals were also pitched along the naso-occipital axis.

Results

Compensatory eye movements from the angular VOR (aVOR), horizontal aVOR

For studying the gain of the aVOR, monkeys were tested with short (5 s) steps of constant velocity rotation from 30°/s to 180°/s in 30°/s intervals in darkness (Fig. 10A). For studying the time constant of the aVOR, monkeys were rotated at 60° /s in darkness to produce per-rotatory nystagmus. When the nystagmus died away, the animals were stopped to produce postrotatory nystagmus (Fig. 10A, B). The animals were subsequently rotated in the opposite direction and stopped. The slow phase eye velocities of the two per- and two postrotatory nystagmus profiles were fitted by single exponentials to extract the central vestibular time constants, and were averaged to generate a mean time constant. The gains (eye velocity/head velocity) of the response to short steps measured the status of the high frequency aVOR, and the time constants gave the state of the central velocity storage mechanism, which provides the low frequency characteristics of the aVOR and counters postrotatory nystagmus (see Raphan and Cohen, 2002 for review).

The induced horizontal aVOR slow phase velocities were close to stimulus velocities up to 120°/s for control monkeys, and fell slightly for stimulus velocities of 150°/s and 180°/s (Fig. 10B, C). Linear, least square regressions were used to characterize each set of pre- and postflight data. Slopes, intercepts and correlation coefficients were similar for the pre- and postflight data for both 782 (Fig. 10A) and for 2483 (Fig. 10B), and the data were not significantly different from the group means of the control pool. During the Cosmos 2229 Flight, horizontal aVOR gains were pooled for rotation to the right and left, since there was no difference between them. The preflight and postflight gains were the same (Dai *et al.*, 1994). Similarly, horizontal aVOR time constants were similar before and after flight. The mean aVOR time constant of 782 on three days of testing, about two months before flight was 22.6 ± 7.9 s. Pooled postflight day 1 and day 4 time constants were 22.6 ± 9.2 s and 18.8 ± 2.1 s. These means did not significantly differ from each other. Though not significant, the reduction in time constant on postflight day 4 was likely due to habituation due to repeated testing. These findings confirm findings in humans in a NASA–ESA Spacelab Flight that indicated that 14 days of exposure to microgravity had not affected the aVOR gain (Benson and Vieville,

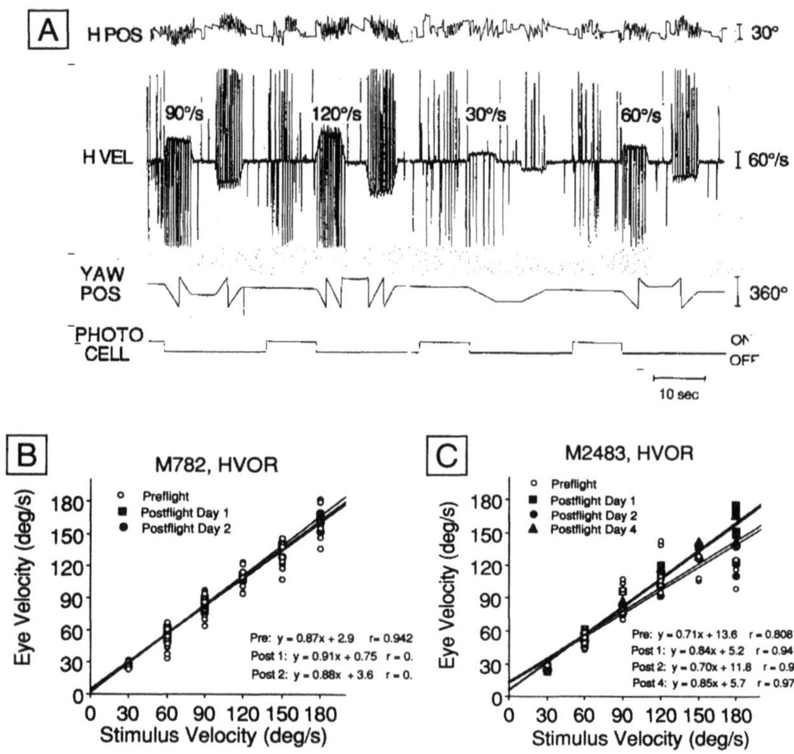

Fig. 10. A, Horizontal eye position (H POS) and eye velocity (H VEL) in response to 5 s steps of velocity (rotation about a vertical axis) from monkey 2483 after recovery in the Cosmos 2044 flight. Eye positions and eye velocities to the right are up in this and subsequent figures. The changes in yaw position of the animal, which were recorded by a potentiometer that reset every 360°, are shown in the third (YAW POS) trace. Downward movement of the trace is rotation to the left. The rotations were in darkness. The surround was lighted (PHOTO CELL) between velocity steps to extinguish any nystagmus that remained after the rotations. B, C, Horizontal slow phase velocities induced by steps of velocity as in A. The induced velocities were close to the velocity of rotation up to 180°/s and were not different before and after flight.

1986). Similarly, the time constant of the horizontal aVOR was unaffected in the NASA IML Shuttle Flight in 1992 (Oman and Balkwill, 1993).

Vertical and ROLL aVOR

Both flight and control monkeys in the Cosmos 2229 mission had an asymmetry of the vertical aVOR before flight when they were rotated on their side in pitch around a vertical axis (Fig. 11A). This stimulus only activates the vertical canals in contrast to pitch about a horizontal interaural axis while upright, which excites both the vertical canals and otolith organs. The mean gain for upward slow phases of vertical eye velocity before flight was 0.96 ± 0.03, and the mean

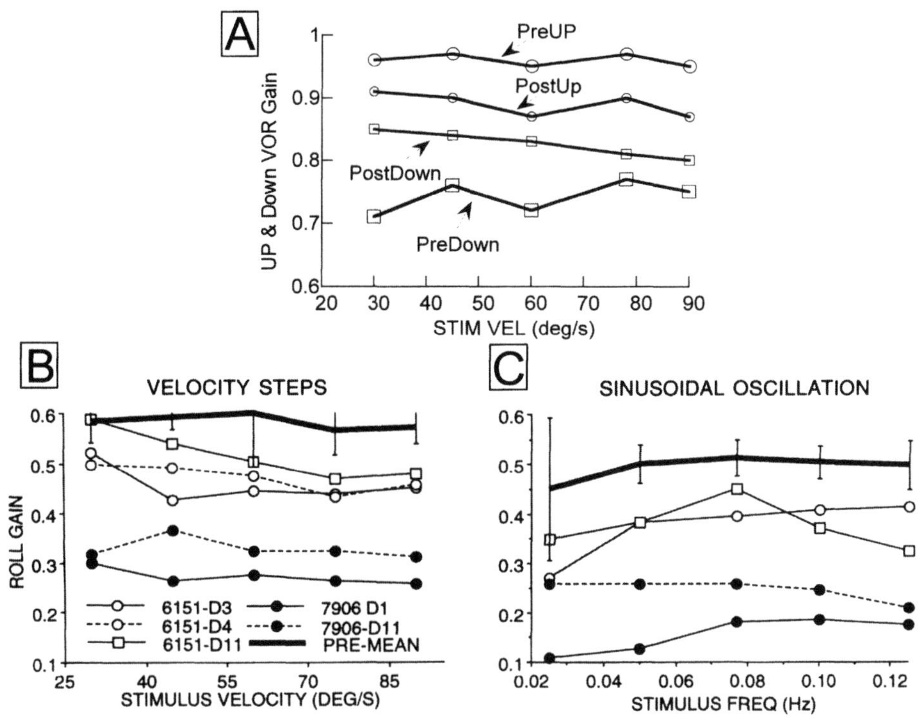

Fig. 11. A, Averaged gains of the vertical VOR before and after Cosmos flight 2229 for monkeys 6151 and 7906. The animals were rotated about a vertical axis with the left ear down in steps of velocity from 30 to 90°/s in 15°/s increments (accelerations >200°/s^2). Postflight data were taken one day after flight for 7906 and 3 days after flight for 6151. Before flight, the upward VOR (PreUP) gain was higher, and the gain dropped after flight (PostUp). The downward VOR had a lower gain before flight (PreDown) that was increased after flight (PostDown). Accordingly, the rather striking preflight Up/Down asymmetry of the vertical VOR was decreased after space flight. B, C, Gains of the roll VOR induced by velocity steps between 30 and 90°/s (B), and by sinusoidal oscillation at frequencies between 0.02–0.13 Hz (C). Animals were rotated about a vertical axis in the prone position. Means for preflight testing (±2 SD) are shown by the heavy solid lines (PRE-MEAN). There was a decrease in roll VOR gains for both 6151 (open symbols) and 7906 (filled symbols) after flight. The decrease was greater for 7906 (50%; filled symbols) than for 6151; open symbols).

downward gain was 0.75 ± 0.04 over a velocity range of 30 to 90°/s (Dai *et al.*, 1994). In the first postflight test, there was a 7.3% decrease in the gain of the upward aVOR to 0.90 ± 0.03, and a 7% increase in the gain of the downward aVOR to 0.82 ± 0.03 (Fig. 3A). These differences were small but statistically significant. Thus, the increase in the gain of the downward aVOR was at the expense of a decrease in gain for the upward aVOR.

By 7 days after landing, the upward aVOR gains had begun to return to preflight levels, and there was no significant difference between pre- and postflight values in both monkeys. The downward aVOR gain dropped to its original level in 6151, but not in 7906. There was no difference in the pitch aVOR gain before and after flight if gains for upward slow phases and downward slow phases were averaged. The pitch aVOR was also measured by rotating the animals sinusoidally at 0.1 Hz, peak velocity 60°/s, about an interaural horizontal axis. This stimulus activates both the vertical canals and the otolith organs. The vertical gain was slightly reduced after landing (6% for 7906 and 9% for 6151), and there was a reduction in the offset of mean velocity in the upward direction that had been present before flight. By the 11th day, the vertical gains had returned so that they were the same or slightly less than the preflight levels for both monkeys. Thus, there was a slight reduction in the gains of the vertical aVOR with otolith activation measured with sinusoids 1–3 days after landing, which had largely returned to the preflight level by the 11th day after recovery. Since the overall vertical gains that were the averages of the up- and down-responses, measured with steps, were unaffected by spaceflight, it is likely that the difference noted in the sinusoidal analysis was due to a change in the otolith contribution to the reflex.

The postflight upward and downward time constants in 782 were 21.3 ± 2.9 s and 10.1 ± 2.4 s, respectively, in response to rotation about a vertical axis while lying on their sides. This is consistent with the vertical time constants in other normal monkeys (Matsuo *et al.*, 1979; Matsuo and Cohen, 1984). Thus, the postflight change in vertical spontaneous nystagmus was not reflected in a change in the dominant time constants of the vertical aVOR.

The roll aVOR was measured in two experimental paradigms: Velocity steps were given with the monkeys in a prone position by rotating them about a vertical naso-occipital axis (Fig. 11B), and the monkeys were sinusoidally oscillated about a horizontal axis while upright (Fig. 11C). Both induce torsional nystagmus, but the first paradigm elicits pure activation of the vertical canals, whereas the second activates both the vertical canals and the otolith organs. The gain of the roll aVOR was reduced in both modes of stimulation, on average by 50% in 7906 and by 15% in 6151 (Fig. 11B, C). This implies that there had been adaptation in central pathways for torsion for activity originating in both the vertical canals and otolith organs during flight, as in humans during 1992 MIR mission (Clarke *et al.*, 1993, 2000).

Spontaneous vertical nystagmus

There was upward spontaneous nystagmus of about 5°/s in both monkeys before the second Cosmos flight, which is common in normal rhesus monkeys. This spontaneous nystagmus was reduced when the animals were first tested, three days after reentry.

High frequency linear vestibulo-ocular reflex

During interaural translation (5 Hz, 0.5 ***g***) postflight, 7906 had an approximate 2/3 reduction in the slope of the function relating lVOR sensitivity to vergence

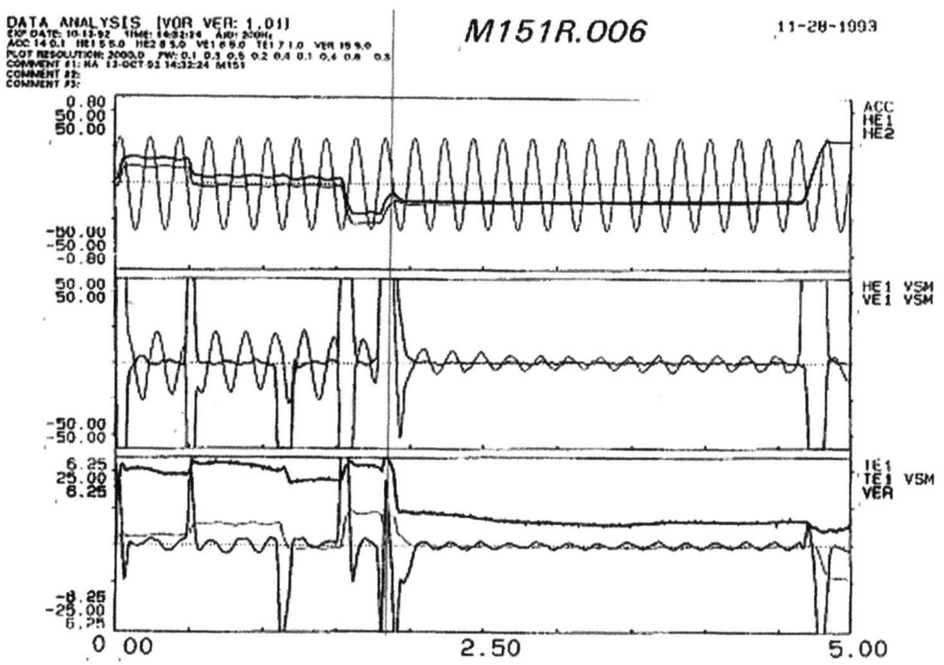

Fig. 12. Five-second sample of 5 Hz, 0.5 *g* peak inter-aural oscillation (ACC) in monkey 6151. In the first traces, ACC is the oscillation of the sled, and HE 1 & HE 2 are right and left horizontal eye position, respectively. In the center traces, HE 1 VSM and VE 1 VSM are right eye horizontal and vertical eye velocity. In the third trace, TE 1 and TE 1 VSM are torsional eye position and eye velocity and VER is vergence position. Note that when vergence was larger, at the left side of the third trace, the oscillations in horizontal eye velocity (large sinusoidal oscillations, middle traces) were larger than when vergence decreased toward the right side of the recording. Vergence had no effect on vertical eye velocity (middle traces). There was a small effect of vergence on torsional eye velocity (third traces).

after flight that had not recovered by 16 h after recovery (R + 16 h; Figs. 12, 13A). Under the same conditions, 6151 had almost identical responses pre- and postflight (not shown). During dorsoventral translation (5 Hz, 0.5 g) 7906 had from 35–60% reduction in the slope of the function relating the vertical lVOR sensitivity to vergence that had not recovered by R + 16 h (Fig. 13B). Under the same conditions, 6151 had responses immediately postflight that were almost identical to the preflight values. Pre- and postflight responses during naso-occipital motion were similar to one another for 6151, but the responses of 7906 were smaller and more variable postflight. This was also the case for motion along oblique axes between naso-occipital and dorsoventral.

Low frequency lVOR – off-vertical axis rotation (OVAR)

In preflight testing, steady state eye velocity during OVAR increased as a function of stimulus velocity and of tilt angle for the eight control monkeys,

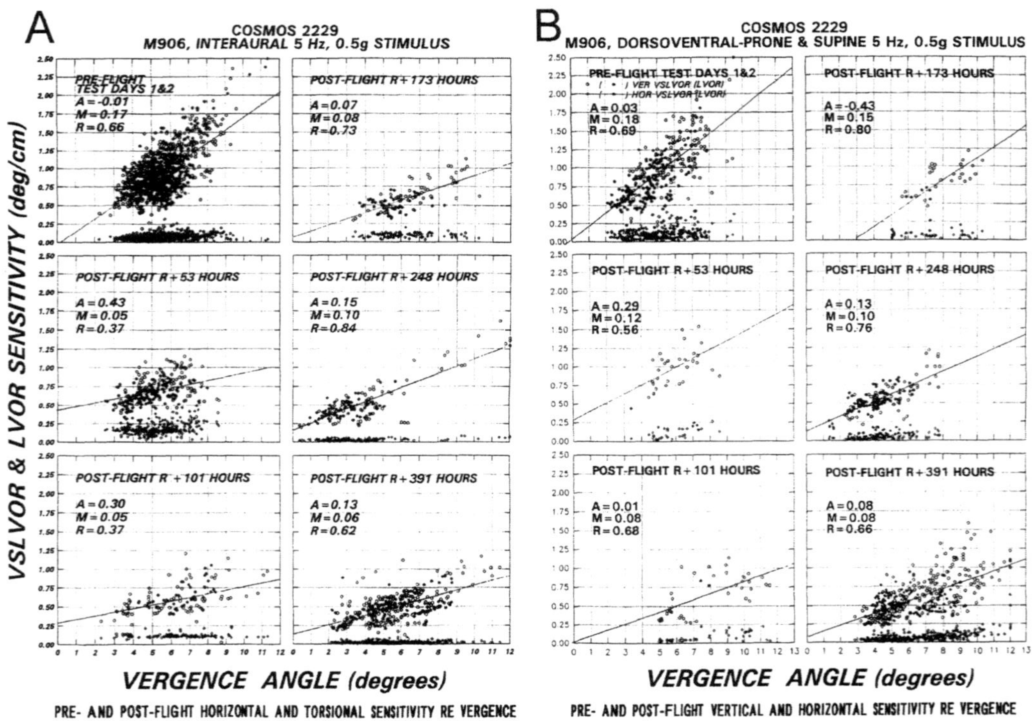

Fig. 13. A, Changes in sensitivity (degrees of eye movement/cm of oscillation of the linear sled) in horizontal (top clouds) and torsional eye velocity (bottom clouds) after 2229 flight in monkey 7906. VSL VOR represents trials in a stationary lighted surround, while LVOR were trials in darkness. In each graph, A is the intercept, M, the slope and R, the correlation coefficient of the linear fit. The increases in horizontal eye velocity remained linear after flight, but there were significant decreases in the sensitivity (slope) of the responses. Roll responses were relatively small and did not depend on vergence. B, Similar sensitivities for vertical (top clouds) and horizontal eye velocities (bottom clouds) to dorsoventral oscillation with the animal prone. The linear relationship between vergence and vertical eye velocity sensitivity was similar to that for horizontal eye velocity and vergence during interaural acceleration (A), and there was a similar decline in the sensitivity of this relationship after flight that remained for 16 days after landing. Horizontal sensitivity to dorsoventral acceleration was small in all instances.

saturating at about 45°/s (Fig. 14A). Steady state horizontal eye velocities induced by OVAR were the same after flight in the monkeys from both the 2044 flight and from the 2229 flight (Fig. 14B), and the phases of the modulations in horizontal slow phase eye velocity were also approximately the same before and after flight. There was an approximate doubling of the amplitude of modulation of horizontal slow phase velocity after the first flight, but not after the second flight. Thus, the animals appeared to be able to sense gravity through the otolith system normally after space-flight and to generate the same steady state level of horizontal slow phase eye velocity after, as before, flight. This implies that the neural mechanism for generating horizontal eye movements had not been substantially altered by adaptation to microgravity.

Ocular counter-rolling

In microgravity, there is no otolith-induced compensatory torsion of the eyes (ocular counter-rolling, OCR) in response to sustained head on body tilt (Clarke *et al.*, 1993, 2000). OCR, elicited by linear acceleration along the interaural axis and measured by an after-image method, was reduced in two cosmonauts for up to 14 days after landing (Yakovleva *et al.*, 1982), and recovered only at the next test point of 36 days. There was also anti-compensatory torsion ("paradoxical counter-rolling") in the direction of head tilt in some subjects after long-duration missions (Clarke *et al.*, 2000). In the Spacelab-1 Mission, OCR was measured in four subjects using a photographic technique (Young *et al.*, 1981). Expressed as a gain ratio, OCR in humans was reduced to one side by 28 to 56% one day after landing. OCR, measured by an after-image technique was reduced by 57% for five days in one astronaut after the 1992 Russian–German MIR Space Mission (Hofstetter-Degen *et al.*, 1993). The extensive literature is reviewed in Dai *et al.*, 1994; Clarke *et al.*, 2000; Moore *et al.*, 2001. Thus, there has been fragmentary evidence to indicate that there may be a reduction in the magnitude of OCR. On the other hand, there was no change in OCR in astronauts who flew in the 1998 Neurolab Mission (STS-90) (Moore *et al.*, 2001). The results, however, were quite different for the monkeys that flew on Cosmos 2229.

Dynamic OCR was assessed using OVAR. The OCR elicited by OVAR before and after space flight is shown in Fig. 14A (3rd trace) and Fig. 14B (1st trace). In both instances, OCR was induced by the response to a projection of the gravity vector that rotated relative to the head along the coronal plane. The maximum torsion was induced during OVAR when the gravity vector was aligned with the interaural axis. Responses of monkeys 6151 and 7906 before and after flight for all angles of tilt of the axis of rotation are summarized in Figs. 14C, D. After flight, the mean magnitude of OCR in the flight monkeys was $2.2 \pm 0.7°$ for tilt angles between 60 and 90°. This can be contrasted to the $6.3 \pm 0.7°$ of ocular torsion when the axis of rotation was tilted 60 to 90° in the control monkeys. The mean reduction in dynamic OCR after flight was about

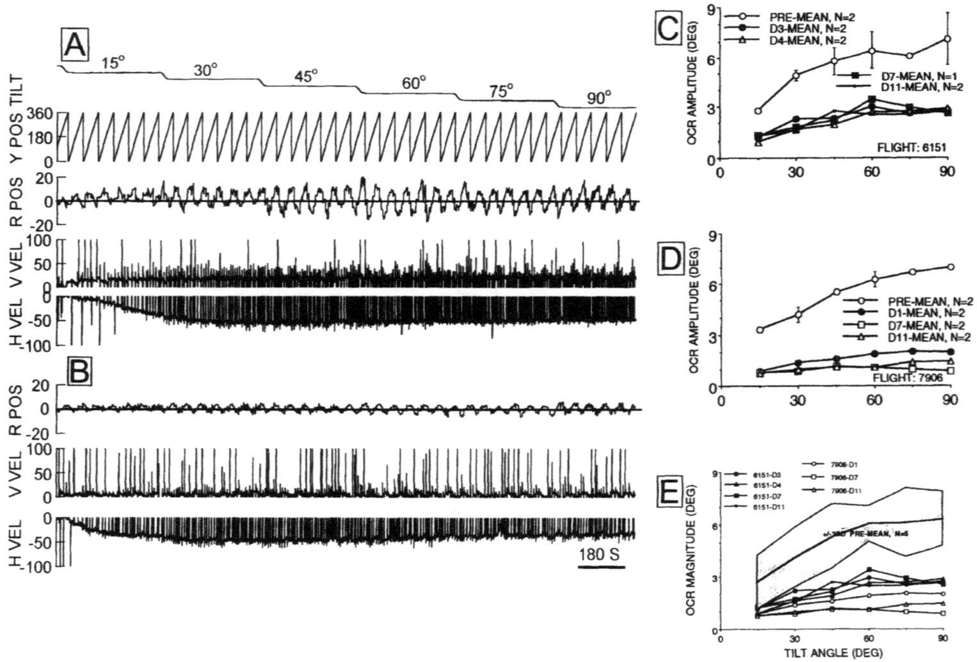

Fig. 14. A, B, Nystagmus induced by off-vertical axis rotation (OVAR) in darkness at 60°/s about tilted axes before (A) and after (B) COSMOS 2229 space flight for monkey 6151. In both A and B the axis of rotation (A, top trace) was tilted from 0-90° in the dark in 15° increments inducing OVAR nystagmus. The position of the animal about the yaw axis, recorded with a potentiometer that reset each 360°, is shown in the second trace (YAW POS). Upward spontaneous nystagmus was reduced after flight (VER VEL). The amplitude of the horizontal slow phase velocity (HOR VEL) was unchanged after flight, but the modulation in slow phase velocity before flight was increased. There was a prominent modulation in roll eye position (ROLL POS) before flight in response to the lateral tilts during OVAR (A, 3rd trace). After flight the modulation in roll position was decreased (B, 1st trace). C, D, Ocular Counter-Roll (OCR) induced during OVAR by tilts of the axis of rotation from 15 to 90° in 6151 (C) and 7906 (D), before (PRE-MEAN) and after flight (days 1, 7, 11). E, Comparison of postflight OCR in the two flight monkeys and six control monkeys and in the two flight monkeys. As above, PRE-MEAN is the mean of the tests before flight; D1–D11 indicate the tests done on recovery days 1 to 11. The OCR of 6151 and 7906 fell more than 2 D from the preflight means.

70%. The postflight OCR of the two flight monkeys fell more than two standard deviations below their own preflight means (Fig. 14C, D) and below that of five monkeys from the control group (Fig. 14E). Static OCR was $6.0 \pm 0.9^\circ$ for tilts of 90° in five monkeys tested before flight and in three control monkeys tested in the postflight period.

After space flight, the magnitude of static OCR was reduced in the two flight monkeys to a mean of $1.8 \pm 0.7^\circ$, a reduction of about 70%. The differences between preflight and postflight OCR, elicited by OVAR, which can also be seen in Fig. 14A, B (ROLL POS, bottom traces), remained over the 11 days of postflight testing and were significant in both flight animals ($p \ll 0.001$). In contrast, there was no change in the OCR of three control monkeys tested before and after flight. These data indicate that there had been long-lasting suppression of OCR after adaptation to microgravity.

Vergence

Vergence has been elicited by naso-occipital linear acceleration on a sled in both frontal and lateral eyed mammals, including humans (Paige, 1991; Paige *et al.*, 1998). Vergence increases when monkeys or humans are linearly accelerated forward at sinusoidal frequencies between 1 and 5 Hz, presumably, serving to help fixate near targets. At low frequencies, the gain of vergence and of other components of the lVOR drop significantly, but considerable vergence would still be expected in response to large linear accelerations during orienting responses. During off-vertical axis rotation (OVAR), the linear acceleration of gravity acts along all head axes in a sinusoidal fashion. For OVAR stimulus velocities of 60°/s with the axis of rotation tilted 90°, there would be sinusoidal linear acceleration of $\pm 1\ g$ at a frequency of 0.17 Hz. Along the naso-occipital axis, the maximal linear acceleration would be forward when the head is nose-up, similar to that at the onset of forward movement on a linear sled, and backward in the nose-down position. During OVAR, there was not only a continuous and systematic modulation in horizontal, vertical and torsional eye position, but the eyes also converged during each cycle of rotation (Dai *et al.*, 1996). Peak vergence (Fig. 15A, LT-RT POS, third trace) developed when the animal was in the nose-up position. Concurrently, the eyes moved down (VERT POS, fourth trace). When the animal was nose-down, vergence was minimal, and the eyes were elevated. The modulation of vergence and vertical movements of the eyes was 90° out of phase with modulations in torsional eye position, which were maximal in side-down positions.

Mean values of the peak-to-peak amplitude of modulation in vergence as a function of OVAR tilt angle for 10 monkeys and their associated standard deviations are shown in Fig. 15C. They were well approximated by a sinusoid (dashed line), indicating that there was a linear relationship between the amount of OCR and the angle of tilt. When the axis of rotation was tilted 90°, the maximal amplitude of modulation in vergence in these monkeys was 7.8°, and

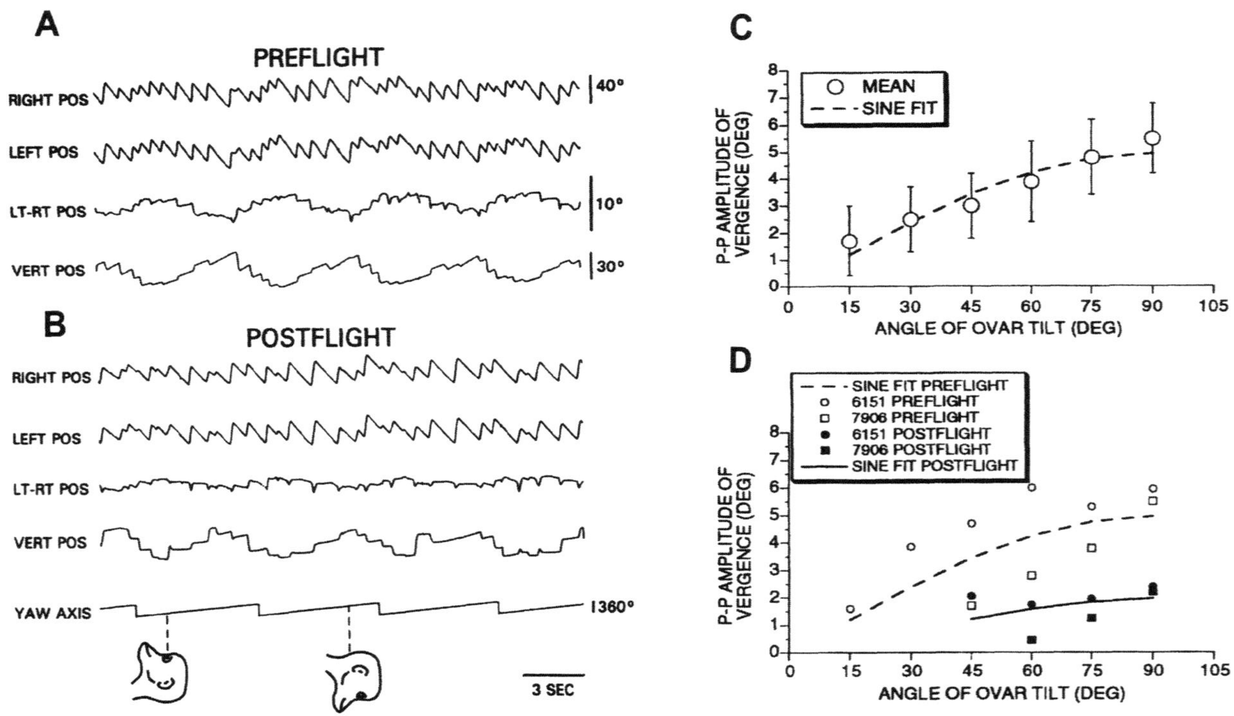

Fig. 15. A, B Comparison of modulation in vergence induced by OVAR in flight monkey 7906 in preflight and postflight tests from the Cosmos 2229 flight. From top to bottom the traces are right and left eye horizontal position (RIGHT POS, LEFT POS), vergence (LT-RT POS) and vertical eye position (VERT POS). Rotation positions about the animal's yaw axis is shown under the 4th trace in B and was the same for A. Modulation of vergence was reduced after space flight, but the phase of peak vergence relative to the nose-up position was unaltered. Although there was less vertical nystagmus after flight, the magnitude of modulation of vertical eye position was essentially the same after as before flight. C, Summary of the amplitude of ocular vergence modulation before space flight. Shown are means ±1 SD and a sinusoidal fit (dashed line) of the data of 10 monkeys, including 6151 and 7906 before flight. D, preflight vergence (open symbols) and postflight vergence (solid symbols) for 6151 (circles) and 7906 (squares). The sine fit of the preflight data (dashed line) can be compared to the sine fit of the postflight data (solid line). There was a significant decrease in the postflight vergence for both monkeys.

the minimum was 3.2°. The mean modulation of the peak-to-peak amplitude of vergence for all 10 animals at a 90° OVAR tilt angle was $5.5 \pm 1.3°$. Because the peak GIA occurred in the nose-up position, phases of peak vergence were determined relative to a nose-up position during OVAR. All animals had peak vergence when their heads were rotated close to the nose-up position. The mean phase was $-0.9 \pm 26.6°$. This supports the contention that the naso-occipital acceleration was the factor responsible for the vergence.

After space flight, the modulation in vergence was strikingly reduced in the two flight monkeys. Eye position data from the preflight and first postflight test for 7906 (Fig. 15B), demonstrates the reduction in the amplitude of vergence (LT-RTPOS). The magnitude of vertical eye modulation, which accompanies vergence during naso-occipital linear acceleration (VERT POS), was virtually the same before and after flight. Spontaneous vertical nystagmus was reduced during OVAR after space flight, and the beats of nystagmus were replaced by saccades (Fig. 15B, 4th trace). There was also a reduction in the nystagmus frequency in 7906 that was not found in 6151 or in the previous monkeys after recovery.

Results of postflight testing are summarized in Fig. 15D. As before flight, the data were fit by a sinusoid (Fig. 15D, solid line). Vergence fell at every tilt angle at which a modulation was measurable after flight, with peak vergence being reduced by over 50%. These reductions in modulation of vergence were significant. Test values throughout the postflight period were similar to those on postflight day 1. On postflight day 11, the peak modulation in vergence for 7906 was 1.8° with a phase of 16° with respect to the nose-up position for an OVAR tilt angle of 90°. Under the same condition, the vergence modulation for 6151 was 2.2° with a phase of 6.3°. By contrast, there was neither reduction in vergence modulation nor a significant phase change in two control monkeys that were available for testing in the postflight period.

Spatial orientation of velocity storage

Eye velocity during per- and postrotatory nystagmus, optokinetic nystagmus (OKN) and optokinetic after-nystagmus (OKAN) and the nystagmus induced by centrifugation tends to align with the direction of gravitoinertial acceleration (GIA) (Dai *et al.*, 1991; Wearne *et al.*, 1999; Moore *et al.*, 2004). A similar tendency has also been demonstrated during rapid reorientation of the head during postrotatory nystagmus and OKAN (Dai *et al.*, 1992; Fetter *et al.*, 1996). The process of alignment has three components: (1) The horizontal time constant is reduced according to the projection of the GIA onto the head yaw axis; (2) The vertical and/or roll time constants are increased according to the projection of the GIA onto the head vertical or roll axes; and (3) Vertical and/or torsional cross-coupled components appear, which are not aligned with the angular stimulus vector when the stimulus is given about the yaw axis. Optimally, the amount of axis shift can be assessed during OKAN, recorded in three dimensions with the head in tilted positions, or from vestibular nystagmus during centrifugation when the gravito-inertial acceleration vector is tilted with regard to the head and body, but there are also manifestations of the orientation of eye velocity to the GIA during OKN (Moore *et al.*, 2004). Partial testing can be used to characterize the spatial orientation of the aVOR, specifically, the rapid reduction in yaw eye velocity during postrotatory nystagmus (tilt-dumping) (Waespe *et al.*, 1985). On Earth, this results in a shorter horizontal aVOR time constant than in upright position. If the horizontal aVOR time constant were not to respond to head tilt with regard to gravity after space flight, it would indicate a reduction in sensitivity to spatial orientation of the aVOR.

Evidence for a change in spatial orientation of velocity storage came from two animals that were tested shortly after return from microgravity, one from Cosmos 2044 and the second from Cosmos 2229. Monkey 782 lost the ability to shorten its horizontal time constant when tilted with regard to gravity after the Cosmos 2044 flight (Fig. 15B). During postrotatory nystagmus, the animal was tilted 50° at the upward arrow. Time constants measured from the onset of tilt or at a comparable time 10 s after the onset of the postrotatory nystagmus were 6.7 s after tilt as against 28 s without tilt (Fig. 16A). After flight, there was no effect of tilt on postrotatory nystagmus in this monkey (Fig. 16B), and the time constant of each response, was the same as when it was upright in darkness during perrotatory nystagmus (19.3 ± 1.8 s ($n=6$) vs 22.1 ± 8.9 s ($n=6$). Findings were the same 2 and 4 days after landing in this monkey. Control time constants were 17.6 s ($\pm$ 2.3, $n=10$) vs tilt time constants of 16.6 s (± 3.0, $n=7$) on these days. This indicates that the loss of the ability to dump stored slow phase eye velocity persisted for several days after landing in this monkey. This deficit was not present in the second monkey from the same flight, however.

More complete testing for spatial orientation was performed after the Cosmos 2229 flight. Trajectories of eye velocity during OKAN in a normal monkey, upright and tilted with respect to gravity are shown in Fig. 16A–D. The angle of the eye velocity vector should be close to gravity in a phase plane plot at the end of OKAN when eye velocity approaches zero (Dai *et al.*, 1991). Before flight, the axis of eye rotation of 7906 for upward coupling from horizontal OKAN at a 90° tilt angle approached the spatial vertical at an angle of 5° (Fig. 17A). Immediately after the flight at R + 22 hours, the eye velocity vector was shifted 28° away from gravity toward a body orientation (Fig. 17B). Seven days later, the orientation of eye velocity had returned closer to gravity (Fig. 17C, 7°). Vectors calculated for OKAN with the animals tilted 30, 60 and 90° showed a similar pattern. That is, the spatial orientation of velocity storage towards gravity was altered after space flight and recovered after 7 days, suggesting that monkey 7906 had shifted the orientation of its eye velocity toward the body in Space. A similar shift in spatial orientation of eye velocity was not found in monkey 6151, but this animal was not tested until the third day, 72 h after recovery, by which time there may have been recovery of this function.

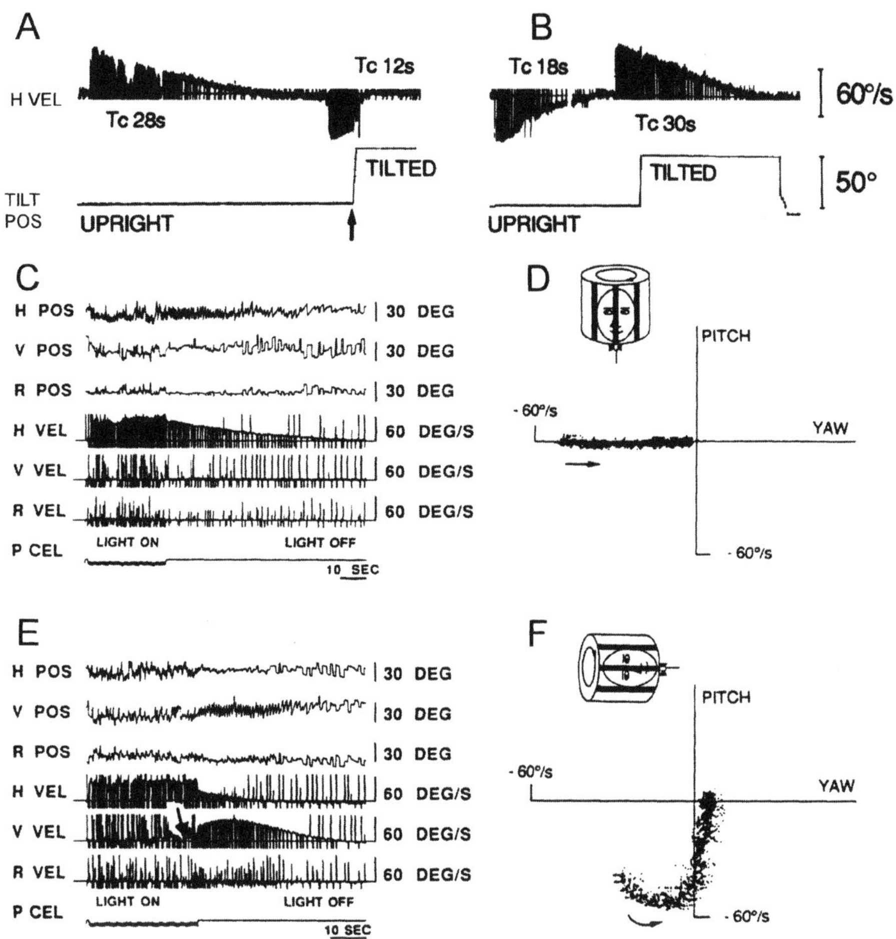

Fig. 16. A, B, Slow phase velocity of per- and postrotatory nystagmus induced by steps of velocity in darkness in monkey 782 before (A) and after (B) the Cosmos 2044 flight. In each set, the animal was rotated in darkness about a vertical axis, inducing per- and postrotatory nystagmus. After the eye velocity had declined to zero, the animal was stopped and 10 s later it was tilted 50° (TILT POS). Before flight (A), this maneuver caused a prompt decline in slow phase velocity. After flight (B), the tilt did not shorten the time course of decline in slow phase velocity. Per- and postrotatory time constants are shown below each response. C–F, Optokinetic nystagmus (OKN), optokinetic after nystagmus (OKAN) and trajectory of axis of eye velocity in horizontal and vertical dimensions on the right. C, D, Horizontal OKN and OKAN were induced with a normal monkey was upright. After the lights were extinguished, OKAN was horizontal and decayed to zero with its characteristic time course. D, The points in the phase plane plots start at the onset of OKAN, which are farthest from the origin, and proceed beat by beat toward the origin where the velocity was again zero. With the animal upright, the eye velocity vector was aligned with both the animal's yaw axis and the spatial vertical. E, F, When the animal was tilted on its side and given the same stimulus as in C, the time constant of the horizontal component of OKAN was shorter, and the OKAN had a cross-coupled vertical component, which rose and decayed. F, In the phase plane plot, the trajectory of OKAN was initially along the animal's yaw axis, but then curved toward the animal's pitch axis. This brought the eye velocity vector close to the spatial vertical at the end of OKAN.

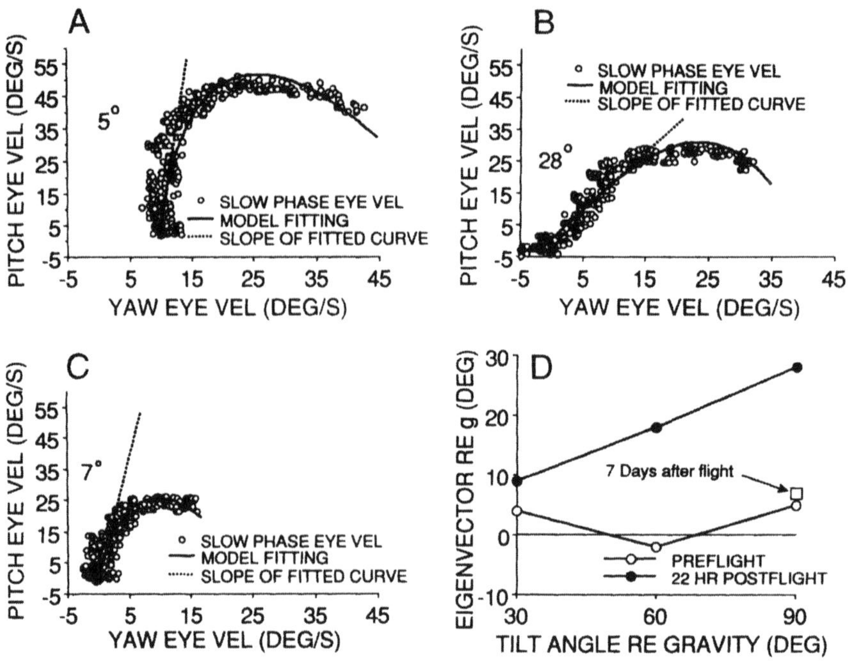

Fig. 17. A–C, Trajectory (phase plane) plots of vertical (ordinate) and horizontal (abscissa) eye velocity during cross-coupling of horizontal OKAN, similar to that shown in Fig. 16 F. The animal (7906) was in a 90° tilted, side-down position, and OKN was induced by yaw axis rotation of the visual surround at 60°/s. OKAN slow phase velocities began in each graph on the right, and the velocities progressed toward zero to the left in a curved fashion. Each circle represents the velocity of one slow phase. The solid curved line represents the fit of the data using a modified Levenberg–Marquardt algorithm. The straight line is the trajectory of the last part of the decaying OKAN, showing the slope of the fitted curve as the data approached zero. A, Before flight, the yaw axis eye velocity vector had an angle of 5° from the vertical. B, Twenty-two hours after flight, the eye velocity vector had shifted toward the body axis and was now 28° from the vertical. C, Seven days later, the eye velocity vector had returned to close to its original position and now was deviated 7° from the vertical. D, Analysis of eye velocity vectors (elgenvectors) obtained during OKAN cross-coupling, before (open circles) and 22 h (filled circles) and 7 days (open square) after flight. Values on the ordinate are deviations from the spatial vertical, which is represented by the horizontal line through zero. The abscissa shows the angle of tilt of the axis of rotation during the OKN and OKAN. Animals were tested at 30, 60 and 90° of tilt. Before flight and 7 days after flight, the eye velocity vectors lay close to the spatial vertical. Twenty two hours after recovery, there was a linear increase in the angle of deviation of the eye velocity vectors, representing deviations of the vectors from the spatial vertical. The deviation was maximal (28°) at 90° of tilt. Adapted from Dai *et al.*, Exp. Brain Res. 102, 45–56, 1994, with permission.

Studies of lateral semicircular canal afferents

Single unit recordings were made from horizontal semicircular canal afferents in four rhesus monkeys within the first 48 h following space flight recovery and for 8–11 days thereafter. Monkeys were exposed to step and sinusoidal rotations about an Earth-vertical (yaw) axis. After Cosmos 2044, the mean gain of 29 afferents in 107 tests was higher for four days and then returned to preflight levels. Figure 18 shows typical units from one of the flight monkeys before and after flight during velocity steps (Fig. 18A), sinusoidal rotation at 0.2 Hz, peak velocity 50°/s (Fig. 18B), sum of sines (Fig. 18C), and tilt (Fig. 18D). Note that the preflight and postflight units were not the same. Therefore, differences in resting discharge were not directly comparable. With this caveat, the modulation in responses to step and sine angular accelerations were substantially larger after flight (Fig. 18A–C).

The canal afferents had no response to tilt either before or after flight (Fig. 8D). This experiment was repeated after the Cosmos 2229 flight. On postflight day 1, mean gains from 12 afferents were lower than the mean of 24 afferents before flight ($p < 0.001$), and this difference persisted for 11 days postflight. Correia *et al.* note: "It is not clear why postflight recordings made from afferents following one flight (COSMOS 2044) showed mean parameters indicating an *increase* in gain and neural adaptation while the mean postflight

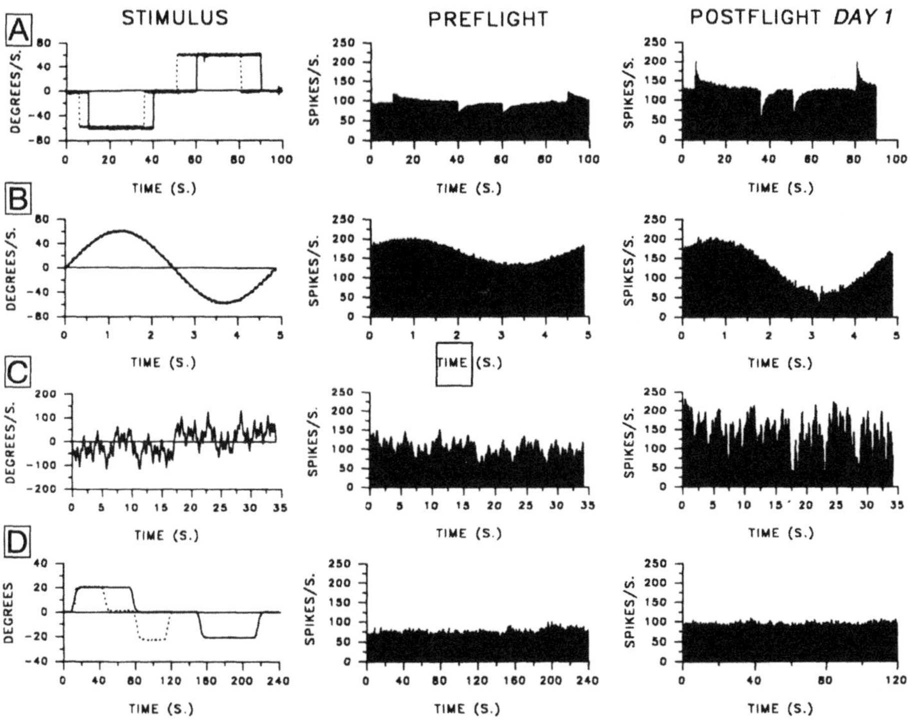

Fig. 18. Stimulus traces (left) and time histograms of typical responses of horizontal (lateral) semicircular canal afferents from Cosmos 2044, preflight and on postflight day 1. A, Rotation at 50°/s for 30 s. B, Sinusoidal rotation at 0.2 Hz, 50° peak velocity. C, Rotation with sum of sines. D, 20° tilt forward and back of the animal's yaw axis. Adapted from Correia *et al.*, J. Appl. Physiol., 1992, Aug. 73 (2 Suppl.): 1125–1265, with permission.

recordings made from a second flight (COSMOS 2229) show mean parameters indicating a *decrease* in gain and no change in neural adaptation. But for both flights, the horizontal semicircular canals were statistically significantly different from those of preflight and synchronous controls. These differences were present while concurrent tests of the horizonal VOR, using the same step paradigms, revealed no differences in slow phase eye velocity between control animals and those that flew in Space. These results suggest that compensation for the altered input from the vestibular sensory receptor (the semicircular canal) probably occurred within the CNS vestibular pathways."

Discussion

These results show that exposure to microgravity resulted in changes in the linear VOR, the angular VOR, and in the spatial orientation of slow phase velocity produced by the velocity storage component of the angular VOR. The gain of compensatory ocular movements produced by the high frequency sinusoidal oscillation was reduced in one monkey, and the gains of ocular counter-rolling (OCR) and vergence produced by low frequency, interaural and naso-occipital linear acceleration were also substantially lower in 2 monkeys after than before flight. Associated with this, there was a reversal of the asymmetry in vertical nystagmus, and a reduction in the gain of the roll aVOR. Finally, in the monkey that had the largest reduction in response to linear acceleration (7906), the orientation of slow phase eye velocity during OKAN and vestibular nystagmus shifted from a spatial toward a body axis. In contrast, high frequency, horizontal and vertical canal-related responses, generated by angular acceleration in a head coordinate frame of reference, were unaffected by microgravity in post-flight testing. Since the vertical canals had a normal gain after flight when tested with rotation in pitch, it is not likely that a change in the afferent input from the vertical canals was the cause for the reduction in the roll aVOR gain. Rather, there must have been changes in excitability in central pathways for torsion. The adaptive changes in otolith reflexes, in the roll aVOR and in the spatial orientation of eye velocity in the aVOR are likely to contribute to postural, locomotor and gaze instability on reentry after space flight.

The data in these experiments were taken from monkeys whose bodies were restrained. Therefore, these monkeys had only a very limited exposure to linear acceleration during flight. In Space, the otoliths are not activated by static lateral head tilt, and neck receptors do not contribute to OCR (Clarke *et al.*, 1993). Moreover, lateral translations, which are associated with OCR on Earth, mainly induce a horizontal component in Space (Clarke *et al.*, 2000). The relative absence of exposure to linear acceleration, therefore, could have been the cause for the reduction in the ocular/spatial reactions to linear acceleration upon return to Earth. Lateral translation when turning corners was also absent in Space, which might explain the 15–50% reduction in gain of the roll aVOR.

Of interest in this regard are recent results from the NASA Neurolab Mission (STS-90), which demonstrated that these otolith-ocular orienting responses were well maintained during space flight (Clément *et al.*, 2001; Moore *et al.*, 2001, 2004). It is possible that this difference in response to microgravity between the monkeys in the COSMOS flights and the astronauts in the Neurolab Mission was due to a species difference. More likely, however, is that the preservation of the otolith-ocular orienting and compensatory reflexes was due to the active movement and frequent centrifugation experienced by the astronauts in Space on STS-90. If this is correct, then a valuable lesson has come from the monkey experiments. Namely, that a program of active head movement with exposure to linear acceleration is an appropriate countermeasure for the reduction in the gain of OCR, vergence and the roll aVOR. Such a countermeasure could also help maintain the gain of vestibulo-sympathetic reflexes, which serve as a first line of defense in maintaining blood pressure upon standing (Kaufmann *et al.*, 2002). A drop in the gain of such reflexes could contribute to the orthostatic hypotension that commonly follows space flight. Thus, the data in these monkeys suggest that the gain of roll and vergence may be much more malleable than has commonly been assumed. Unfortunately, the only way to test this would be in Space, where there is a prolonged absence of gravity, and the opportunities for such experiments are few.

Our results also show that there is considerable variation between individuals exposed to essentially similar conditions in Space. Microgravity was associated with a striking reduction in the high frequency lVOR gain in 7906, whereas the compensatory lVOR was essentially normal in 6151, although 6151 was not tested until the third day after recovery. This mirrors similar variability in humans exposed to microgravity. On the other hand, there was a striking similarity in the increased sensitivity of the aVOR during active head turns, in the saccadic hypermetria during lateral gaze shirts, and in the increase in the sensitivity of units in MVN and flocculus to head turns in some of these same monkeys. Afferent activity from the lateral semicircular canals also varied considerably between Cosmos flights 2044 and 2229 (Correia *et al.*, 1992; Correia, 1998) in some of these same monkeys, although the passive aVOR gains were unaffected (Cohen *et al.*, 1992; Dai *et al.*, 1994). Some of these changes may have been a consequence of the behavioral state or the experimental conditions, interacting with exposure to microgravity, but there also seems to be an individual difference of response to essentially similar conditions.

Eye velocity during nystagmus due to excitation of the slow component of the aVOR tends to align with the spatial vertical when the head is tilted with regard to the GIA (Dai *et al.*, 1991, 1992). In both flights, this spatial orientation was affected in one of the two monkeys available for testing shortly after landing. The shift in orientation disappeared over a time course of several days after reentry into a gravitational environment (Cohen *et al.*,

1992; Dai *et al.*, 1994). The reduction in the direct orienting response of OCR and vergence persisted for a longer period than the indirect orienting response of OKAN cross-coupling, which is produced through a different mechanism, i.e., through velocity storage. An important aspect of re-adaptation upon the reentry is to reorient to gravity so that the body posture is maintained during motion. This may explain why the reorientation of velocity storage, as reflected in OKAN cross-coupling, occurred sooner than the recovery of other ocular orienting reactions produced by the lVOR.

In this chapter, we have summarized experiments that provide a consolidated body of new data that demonstrate changes that can be expected in the vestibulo-ocular reflex and the central vestibular system after space flight. The experiments in Sections 2 and 3 also show the unique value of the monkey as a model for studying effects of human vestibular adaptation in Space, wherein behavior and single unit recordings can be combined in a single animal that has many similarities to humans. It is hoped that the work described in this chapter will aid in understanding the complex phenomena that accompany adaptation to microgravity, especially in relation to long-duration space flight. There is a larger reason to promote the use of experimental animals in Space. There is no other way to understand the critical role of gravity on the formation and structure of the nervous system. Consequently, an essential component of future space-related experiments should be directed toward determining how afferent input from the otoliths and semicircular canals is altered and how central processing is changed in microgravity in experimental animals.

Acknowledgment

We thank Jun Maruta for assistance. This work was supported by NASA Grants NAG 2-573 (BC), NAG 2-703 (BC), NASA Contract NAS 9-19441(BC), NCC2-1173 (GRH), NAG 2-946 (GRH), NIH Grants: DC02451 (GRH), DC04996 (SY), DC05204, and EY01867.

References

Aizikov, G.S. and Markin, A.S. Posture, movement and balance in rats after Biosatellite flights. Space Biol. Aviaspace Med. 15(6), 33–38 (in Russian).

Aizikov, G.S., Markin, A.S. and Mokrousova, A.V. (1981) Specificity of the lift-reaction in white rats after prolonged spaceflight (effects of microgravity and hypergravity). Ann. Exp. Biol. Med. 12, 722–725 (in Russian).

Aizikov, G.S., Markin, A.S. and Shipov, A.A. (1992) Flipping-reaction in rats after the "Cosmos 936" spaceflight. Aviaspace Ecol. Med. 26(1), 62–64 (in Russian).

Anken, R., Kappel, T. and Rahmann, H. (1998) Morphometry of fish inner ear otoliths after development at 3g hypergravity. Acta Oto-Laryngol. 118, 534–539.

Apanasenko, Z.I. (1977) Vestibulo-tonic reflexes of rat's legs after the "Cosmos 605" spaceflight. Space Biol. Aviaspace Med. 11(4), 28–34 (in Russian).

Badakva, A.M., Miller, N.V. and Eron, J.N. (2003) Effects of head-out water immersion on eye-head coordination in rhesus monkey. J. Gravitat. Physiol. 10, P5–P8.

Badakva, A.M., Zalkind, D.V., Miller, N.V. and Riazansky, S.N. (2000) Head pitch movement and vestibular neuronal activity in response to otolith stimulation of monkeys in space. J. Gravit. Physiol. 7, S99–S105.

Badakva, A.M., Zalkind, D.V., Miller, N.V., Sirota, M.G., Beloozerova, I.N., Kozlovskaya, I.B. (1993) The Final Report on Cosmos 2229 Experiments. Chapter 4. Effects of microgravity on neuronal circuitry involved in forming of the vestibular and oculomotor responses: results of primate studies in BION projects. *In Institute of Biomedical Problems*, pp. 92–104, Russian Academy of Science, Moscow.

Balaban, E., Centini, C. and Pompeiano, O. (2002) Tonic gravity changes alter gene expression in the efferent vestibular nucleus. NeuroReport 13, 187–190.

Barmin, V.A., Kreidich, Y.V. and Kozlovskaya, I.B. (1983) Influence of optokinetic stimulation and immersion on eye-head coordination in man. Physiologist 26, S83–S85.

Bauerle, A., Anken, R., Hilbig, R. and Rahmann, H. (2004) Histology of the utricle in kinetotically swimmiing fish: a parabolic aircraft flight study. Acta Oto-Laryngol. 124, 19–22.

Beier, M., Anken, R. and Rahmann, H. (2002) Influence of hypergravity on fish inner ear otoliths: II. Incorporation of calcium and kinetotic behaviour. Adv. Space Res. 30, 727–731.

Benson, A.J., Kass, J.R. and Vogel, H. (1986) European vestibular experiments on the Spacelab-1 mission: 4. Thresholds of perception of whole-body linear oscillation. Exp. Brain Res. 64, 264–271.

Benson, A.J. and Vieville, T. (1986) European vestibular experiments on the Spacelab-1 mission: 6. Yaw axis vestibulo-ocular reflex. Exp. Brain Res. 64, 279–283.

Bizzi, E., Kalil, R.E. and Morasso, P. (1972) Two modes of active eye-head coordination in monkeys. Brain Research 40, 45–48.

Bizzi, E., Kalil, R.E. and Tagliasco, V. (1971) Eye-head coordination in monkeys: evidence for centrally patterned organization. Science 173, 452–454.

Boyle, R., Mensinger, A.F., Yoshida, K., Usui, S., Intravaia, A., Tricas, T. and Highstein, S.M. (2001) Neural readaptation to Earth's gravity following return from space. J. Neurophysiol. 86, 2118–2122.

Büettner-Ennever, J.A. (1999) A review of otolith pathways to brainstem and cerebellum. Ann. N. Y. Acad. Sci. 871, 51–64.

Chubb, C.M., Fuchs, A.F. and Scudder, C.A. (1984) Neuron activity in monkey vestibular nuclei during vertical vestibular stimulation and eye movements. J. Neurophysiol. 52, 724–742.

Clarke, A.H., Grigull, J., Mueller, R. and Scherer, H. (2000) The three-dimensional vestibulo-ocular reflex during prolonged microgravity. Exp. Brain Res. 134, 322–334.

Clarke, A.H., Teiwes, W. and Scherer, H. (1993) Vestibulo-oculomotor testing during the course of a space flight mission. Clinical Investigations 71, 740–748.

Clément, G., Moore, S.T., Raphan, T. and Cohen, B. (2001) Perception of tilt (somatogravic illusion) in response to sustained linear acceleration during space flight. Exp. Brain Res. 138, 410–418.

Cohen, B. and Gizzi, M. (2003) The physiology of the vestibulo-ocular reflex. Chapter 41. *In* Luxon, L., Martini, A. and Furman, J.D.S. (eds.). *Textbook of Audiological Medicine*, pp. 701–716. Oxford University Press, New York.

Cohen, B., Kozlovskaya, I., Raphan, T., Solomon, D., Helwig, D., Cohen, N., Sirota, M. and Yakushin, S. (1992) Vestibuloocular reflex of rhesus monkeys after space flight. J. Appl. Physiol. 73, 121S–131S.

Cohen, B., Maruta, J. and Raphan, T. (2001) Orientation of the eyes to gravito-inertial acceleration. N. Y. Acad. Sci. 942, 241–258.

Cohen, B. and Raphan, T. (2004) The physiology of the vestibulo-ocular reflex (VOR). Anatomy and physiology of the central and peripheral vestibular system. *In* Highstein, S., Popper, A. and Fay, R.R. (eds.). *Springer Handbook of Auditory Research.* Springer Verlag, New York.

Correia, M.J. (1998) Neuronal pasticity: adaptation and readaptation to the environment of space. Brain Res. Rev. 28, 61–65.

Correia, M.J., Perachio, A.A., Dickman, J.D., Kozlovskaya, I.B., Sirota, M.G., Yakushin, S.B. and Beloozerova, I.N. (1992) Changes in monkey horizontal semicircular canal afferents after space flight. J. Appl. Physiol. 73, 112S–120S.

Crane, B.T. and Demer, J.L. (1999) A linear canal-otolith interaction model to describe the human vestibulo-ocular reflex. Biol. Cybern 81, 109–118.

Dai, M., McGarvie, L., Kozlovskaya, I.B., Raphan, T. and Cohen, B. (1994) Effects of space flight on ocular counterrolling and spatial orientaion of the vestibular system. Exp. Brain Res. 102, 45–56.

Dai, M., Raphan, T. and Cohen, B. (1991) Spatial orientation of the vestibular system: dependence of optokinetic after nystagmus on gravity. J. Neurophysiol. 66, 1422–1438.

Dai, M., Raphan, T., Kozlovskaya, I. and Cohen, B. (1996) Modulation of vergence by off-vertical yaw axis rotation in the monkey: normal characteristics and effects of space flight. Exp. Brain Res. 111, 21–29.

Dai, M.J., Raphan, T. and Cohen, B. (1992) Characterization of yaw to roll cross-coupling in the three-dimensional structure of the velocity storage integrator. Ann N.Y. Acad. Sci. 656, 829–831.

d'Ascanio, P., Balaban, E., Pompeiano, M., Centini, C. and Pompeiano, O. (2003) Fos and FRA protein expression in rat precerebellar structures during the Neurolab Space Mission. Brain Res. Bull. 62, 203–221.

Dichgans, J., Bizzi, E., Morasso, P. and Tagliasco, V. (1973) Mechanisms underlying recovery of eye-head coordination following bilateral labyrinthectomy in monkeys. Exp. Brain Research 18, 548–562.

Duflo, S.G.D., Gestreau, C. and Lacour, M. (2000) Fos expression in the rat brain after exposure to gravito-inertial force changes. Brain Res. 861, 333–344.

Edgerton, V.R., Roy, R.R., Hodgson, J.A., Day, M.K., Weiss, J., Harkema, S.J., Dobkin, B., Garfinkel, A., Konigsberg, E. and Koslovskaya, I. (2000) How the science and engineering of space flight contribute to understanding the plasticity of spinal cord injury. Acta Astronautica 47, 51–62.

Einstein, A. (1911) Ueber den Einfluss der Schwerkraft auf die Ausbreitung des Lichtes. Annalen der Physik 35, 898–908.

Fetter, M., Heimberger, J., Black, R.A., Hermann, W., Sievering, F. and Dichgans, J. (1996) Otolith-semicircular canal interaction during postrotatory nystagmus in humans. Exp. Brain Res. 108, 463–473.

Fox, R.A., Daunton, N., Corcoran, M. and Wu, L. (1992) Time-course of behavioral adaptation to hypergravity. *In* Woollacott, M. and Horak, F. (eds.). *Posture and Gait: Control Mechanisms.* Vol. I, pp. 178–181. University of Oregon Books, Portland.

Fuchs, A.F. and Kimm, J. (1975) Unit activity in vestibular nucleus of the alert monkey during horizontal angular acceleration and eye movement. J. Neurophysiol. 38, 1140–1161.

Gazenko, O.G., Grigorian, R.A., Kitaev-Smik, L.A. and Klochkov, A.I. (1965) Increase of extensor muscular tonus in microgravity in cats with complete and partial cerebellum removal. Ann. Exp. Biol. Med. 7, 7–12 (in Russian).

Gazenko, O.G. and Ilyin, E.A. (1987) Physiological investigations of primates onboard biosatellites Cosmos-1514 and Cosmos-1667. The Physiologist 30, S31–S35.

Grigaryan, R.A., Gazenko, O.G., Kozlovskaya, I.B., Barmin, V.A. and Kreidich, Y.V. (1986) The vestibulo-cerebellar regulation of oculomotor reactions in microgravitational conditions. *In* Keller, S.L. and Zee, DS (eds.). *Adaptive Processes in Visual and Oculomotor Systems.* Vol. 57, pp. 121–127. Pergamon Press, New York.

Guedry, F.E. (1974) Psychophysics of vestibular sensation. *In* Kornhuber, H.H. (ed.). *Handbook of Sensory Physiology.* Vol. 6, pp. 3–154. Springer Verlag, Berlin.

Hess, B.J.M. and Dieringer, N. (1991) Spatial organization of linear vestibulo-ocular reflexes of the rat: Responses during horizontal and vertical linear acceleration. J. Neurophysiol. 66, 1805–1818.

Hofstetter-Degen, K., Weizig, J. and von Baumgarten, R. (1993) Oculovestibular interactions under microgravity. Clinical Investigations 10, 749–756.

Holstein, G.R., Kukielka, E. and Martinelli, G.P. (1999) Anatomical observations of the rat cerebellar nodulus after 24 hr of space flight. J. Grav. Physiol. 6, 47–50.

Ibsch, M., Anken, R.H. and Rahmann, H. (2000) Weightlessness during space flight results in enhanced synapse formation in a fish brain vestibular nucleus. Neurosci. Letts. 296, 13–16.

Imai, T., Moore, S.T., Raphan, T. and Cohen, B. (2001) Interaction of the body, head and eyes during walking and turning. Exp. Brain Res. 136, 1–18.

Ito, M. (1984) *Cerebellum and Neural Control*. Raven Press, New York.

Kaufman, G.D., Anderson, J.H. and Beitz, A.J. (1991) Activation of a specific vestibulo-olivary pathway following centripetal acceleration in the rat. Brain Res. 562, 311–317.

Kaufman, G.D., Anderson, J.H. and Beitz, A.J. (1992) Fos-defined activity in rat brainstem following centripetal acceleration. J. Neurosci. 12, 4489–4500.

Kaufman, G.D., Anderson, J.H. and Beitz, A.J. (1993) Otolith-brainstem connectivity: evidence for differential neural activation by vestibular hair cells based on quantification of Fos expression in unilateral labyrinthectomized rats. J. Neurophysiol. 70, 117–127.

Kaufman, G.D. and Perachio, A.A. (1994) Translabyrinth electrical stimulation for the induction of immediate-early genes in the gerbil brainstem. Brain Res. 646, 345–350.

Kaufmann, H., Biaggioni, I., Voustaniuk, A., Diedriich, A., Costa, F., Clark, R., Gizzi, M., Raphan, T. and Cohen, B. (2002) Vestibular control of sympathetic activity: An otolth-sympathetic reflex in humans. Exp. Brain Res. 143, 463–469.

Keller, E.L. and Daniels, P.D. (1975) Oculomotor related interaction of vestibular and visual stimulation in vestibular nucleus cells in alert monkey. Exp. Neurol. 46, 187–198.

Kitaev-Smik, L.A. (1963) Animal posture in microgravity. *In Aviation and Space Medicine*, pp. P247–250. Moscow (in Russian).

Kozlovskaya, I.B., Barmin, V.A., Kreidich, Y.V. and Repin, A.A. (1985) The effects of real and simulated microgravity on vestibulo-oculaomotor interaction. The Physiologist 28, 51–56.

Kozlovskaya, I.B., Ilyin, E.A., Sirota, M.G., Korolkov, V.I., Babayev, B.M., Beloozerova, I.N. and Yakushin, S.B. (1989) Studies of space adaptation syndrome in experiments on primates performed on board of Soviet biosatellite "Cosmos-1887". The Physiologist 32, 45–48.

Kozlovskaya, I.B., Sirota, M.B., Yakushin, S.B., Beloozerova, I.I., Babayev, B.M., Badakva, A.M., and Zalkind, D.V. (1994) The influence of weightlessness on parameters of the horizontal vestibulo-ocular reflex (HVOR) in monkeys. (In Russian). *In Space Biology and Aviacosmic Medicine. Report Abstracts of X Conference*, 7-10/VI, pp. 113–114. Slovo, Moscow.

Kozlovskaya, I.B., Sirota, M.G., Beloozerova, I.N., and Yakushin, S.B. (1991) Vestibulo-motor interaction in microgravity. *In International Meeting "Biosatellite Cosmos"*. p. 52. (In Russian) IMBP, Moscow, Leningrad 12–15 August.

Krasnov, I.B. (1975) Quantitative cyto- and histochemical studies of the Deiter's nucleus and nodular cortex of cerebellum in rats exposed to weightlessness. Aviat. Space Environ. Med. 46, 1119–1122.

Krasnov, I.B. (1987) Increase in the sensibility of otolith apparatus in weightlessness. Morphological evidence. Intercosmos Space Biol. Med. Grp. Mtng. 10, 127.

Krasnov, I.B. (1991) The otolith apparatus and cerebellar nodulus in rats developed under 2-G gravity. The Physiologist 34, S206–S207.

Krasnov, I.B. and Dyachkova, L.N. (1986) Ultrastructure of the cortex of the cerebellar nodulus in rats after a flight on the biosatellite KOS 1514. Kosm. Biol. Aviakosm. Med. 20, 45–48.

Krasnov, I.B. and Dyachkova, L.N. (1990) The effect of space flight on the ultrastructure of the rat cerebellar and hemisphere cortex. The Physiologist 33, 29–30.

Kreidich, Y.V., Repin, V.A., Barmin, V.A. and Kozlovskaya, I.B. (1983) Effect of immersion hypokinesia on man's eye and head movements during the gaze fixation reachion. (In Russian) Space Biol. and Aerospace Med. 5, 41–45.

Lim, D.J., Stith, J.A., Stockwell, C.W. and Oyama, J. (1974) Observations on sacculus of rats exposed to long-term hypergravity. Aerospace Med. 45, 705–710.

Lisberger, S.G. (1994) Neural basis for motor learning in the vestibuloocular reflex of primates. III. Computational and behavioral analysis of the sites of learning. J. Neurophysiol. 72, 974–998.

Lisberger, S.G. and Fuchs, A.F. (1978a) Role of primate flocculus during rapid behavioral modification of vestibuloocular reflex. I. Purkinje cell activity during visually guided horizontal smooth-pursuit eye movements and passive head rotation. J. Neurophysiol. 41, 733–763.

Lisberger, S.G. and Fuchs, A.F. (1978) Role of primate flocculus during rapid behavioral modification of vestibuloocular reflex. II. Mossy fiber firing patterns during horizontal head rotation and eye movement. J. Neurophysiol. 41, 764–777.

Lisberger, S.G., Pavelko, T.A., Bronte-Stewart, H.M. and Stone, L.S. (1994a) Neural basis for motor learning in the vestibuloocular reflex of primates. II. Changes in the responses of horizontal gaze velocity Purkinje cells in the cerebellar flocculus and ventral paraflocculus. J. Neurophysiol. 72, 954–973.

Lisberger, S.G., Pavelko, T.A. and Broussard, D.M. (1994b) Neural basis for motor learning in the vestibuloocular reflex of primates. I. Changes in the responses of brain stem neurons. J. Neurophysiol. 72, 928–953.

Lychakov, D.V., Bojajieva-Mikhailova, A., Christov, I., Pashchinin, A.N., Evdokimov, I.I. and Markov, A.A. (1988) Changes in the otolith apparatus of rats and fish after prolonged exposure to acceleration. (In Russian) Kosm. Biol. Aviakosm. Med. 22, 27–33.

Lychakov, D.V., Pashchinin, A.N., Bojajieva-Mikhailova, A. and Khristov, I. (1989) Study of receptor organs of the vestibular apparatus of Cosmos-1667 rats. (In Russian) Kosm. Biol. Aviakosm. Med. 23, 27–33.

Lychakov, D.V. and Lavrov, E.A. Study of the vestibular end organ structure and ionic composition of the body of young tadpoles (Xenopus Laevis) after spaceflight. Space Biol. Aviaspace Med. 19(3), 48–52.

Markham, C.H. and Curthoys, I.S. (1972) Convergence of labyrinthine influences on units in the vestibular nuclei of the cat. II. Electrical stimulation. Brain Res. 43, 383–396.

Maruta, J., Simpson, J.I., Raphan, T. and Cohen, B. (2001) Orienting otolith-ocular reflexes in the rabbit during static and dynamic tilts and off-vertical axis rotation. Vision Research 41, 3255–3270.

Mashburn, T.H., Kaufman, G.D., Purcell, I.M. and Perachio, A.A. (1997) Saccule contribution to immediate-early genes production in the gerbil brainstem with posterior galvanic or hypergravity stimulation. Brain Res. 761, 51–58.

Matsuo, V. and Cohen, B. (1984) Vertical optokinetic nystagmus and vestibular nystagmus in the monkey: Up-down asymmetry and effects of gravity. Exp. Brain Res. 53, 197–216.

Matsuo, V., Cohen, B., Raphan, T., de Jong, V. and Henn, V. (1979) Asymmetric velocity storage for upward and downward nystagmus. Brain Res. 176, 159–164.

Miles, F.A. (1974) Single unit firing patterns in the vestibular nuclei related to voluntary eye movements and passive body rotation in conscious monkeys. Brain Res. 71, 215–224.

Moore, S.T., Clement, G., Raphan, T. and Cohen, B. (2001) Ocular counter-rolling (OCR) induced by centrifugation during orbital space flight. Exp. Brain Res. 137, 323–335.

Moore, S.T., Cohen, B., Raphan, T., Berthoz, A. and Clement, G. (2004) Spatial orientation of optokinetic nystagmus and ocular pursuit during orbital space flight. Exp. Brain Res, In press.

Oman, C.M. and Balkwill, M.D. (1993) Horizontal angular VOR, nystagmus dumping, and sensation duration in Spacelab SLS-1 crewmembers. J. Vesti. Res. 3, 315–330.

Paige, G.D. (1991) Linear vestibulo-ocular reflex (LVOR) and modulation by vergence. Acta Otolaryngol. (Stockh.) 481 (Suppl.), 282–286.

Paige, G.D., Telford, L., Seidman, S.H. and Barnes, G.R. (1998) Human vestibuloocular reflex and its interactions with vision and fixation distance during linear and angular head movement. J. Neurophysiol. 80, 2391–2404.

Paige, G.D. and Tomko, D.L. (1991a) Eye movement responses to linear head motion in the squirrel monkey. I. Basic Characteristics. J. Neurophysiol. 65, 1170–1182.

Paige, G.D. and Tomko, D.L. (1991b) Eye movement responses to linear head motion in the squirrel monkey. II. Visual-vestibular interactions and kinematic considerations. J. Neurophysiol. 65, 1183–1196.

Pedrozo, H.A. and Wiederhold, M.L. (1994) Effects of hypergravity on statocyst development in embryonic *Aplysia californica*. Hearing Res. 79, 137–146.

Phillips, J.O., Ling, L., Fuchs, A.F., Siebold, C. and Plorde, J.J. (1995) Rapid horizontal gaze movement in the monkey. J. Neurophysiology 73, 1632–1652.

Pompeiano, O., d'Ascanio, P., Centini, C., Pompeiano, M. and Balaban, E. (2002) Gene expression in rat vestibular and reticular structures during and after space flight. Neurosci. 114, 135–155.

Pompeiano, O., d'Ascanio, P., Balaban, E., Centini, C. and Pompeiano, M. (2004) Gene expression in autonomic areas of the medulla and the central nucleus of the amygdala in rats during and after space flight. Neurosci. 124, 53–69.
Rahmann, H., Slenzka, K., Kortje, H. and Hilbig, R. (1992) Synaptic plasticity and gravity: ultrastructural, biochemical and physico-chemical fundamentals. Adv. Space Res. 12, 63–72.
Raphan, T. and Cohen, B. (2002) The vestibulo-ocular reflex (VOR) in three dimensions. Exp. Brain Res. 145, 1–27.
Razumeev, A.N. and Grigorian, R.A. (1976) *Cerebellum and Gravity*, P453, Nauka, Moscow (in Russian).
Ross, M.D. (1987) Implications of otoconial changes in microgravity. The Physiol. 30, 90–93.
Ross, M.D. (1993) Morphological changes in rat vestibular system following weightlessness. J. Vestib. Res. 3, 241–251.
Ross, M.D. (1994) A space flight study of synaptic plasticity in adult rat vestibular maculas. Acta Oto-Laryngol Suppl. 516, 3–14.
Ross, M.D. (1997) Morphological evidence for local microcircuits in rat vestibular maculae. J. Comp. Neurol. 379, 333–346.
Ross, M.D. (2000) Changes in ribbon synapses and rough endoplasmic reticulus of rat utricular macular hair cells in weightlessness. Acta Otolaryngol. 120, 490–499.
Ross, M.D., Donovan, K. and Chee, O. (1985) Otoconial morphology in space-flown rats. The Physiol. 28, 219–220.
Ross, M.D. and Tomko, D.L. (1998) Effect of gravity on vestibular neural development. Brain Res. Revs. 28, 44–51.
Schwarz, C., Busettini, C. and Miles, F.A. (1989) Ocular responses to linear motion are inversely proportional to viewing distance. Science 245, 1394–1396.
Schwarz, C. and Miles, F.A. (1991) Ocular responses to translation and their dependence on viewing distance. I. Motion of the observer. J. Neurophysiol. 66, 851–864.
Shipov, A.A. and Aizikov, G.S. (1992) Physical state, equilibrium function and vestibulo-motor reflexes in the rats after space flights on board biosatellites of Cosmos series. Physiologist Suppl., 35(1), S206–207.
Shipov, A.A. and Ovechkin, V.G. (1980) Rat semicircular canal function after the "Cosmos 936" spaceflight. Space Biol. Aviaspace Med. 14(2), 25–30 (in Russian).
Shipov, A.A., Sirota, M.G., Beloozerova, I.N., Babaev, B.M. and Kozlovskaya, I.B. (1986) Results of tests on the primate vestibulo-visualmotor reactions in biocosmos experiments. *In* Keller, S.L. and Zee, D.S. (eds.). *Adaptive Processes in Visual and Oculomotor Systems*, Vol. 57, pp. 129–132. Pergamon Press, New York.
Shipov, A.A. and Tabakova, L.A. (1982) Specificity of nystagmus in rats after the "Cosmos 1129" spaceflight. Space Biol. Aviaspace Med. 16(6), (in Russian).
Sirota, M.G., Babayev, B.M., Beloozerova, I.B., Nyrova, A.N., Yakushin, S.B. and Kozlovskaya, I.B. (1987) Characteristics of vestibular reactions to canal

and otolith stimulation at an early stage of exposure to microgravity. The Physiologist 30, S82–S84.

Sirota, M.G., Babayev, B.M., Beloozerova, I.N., Nyrova, A.N., Yakushin, S.B. and Kozlovskaya, I.B. (1988a) Neuronal activity of nucleus vestibularis during coordinated movement of eyes and head in microgravitation. The Physiologist 31, 8–9.

Sirota, M.G., Beloozerova, I.N., Babaev, B.M., Yakushin, S.B., Ivanov, A.M., Nyrova, A.N., Kozlovskaya, I.B. (1988b) The final report on Cosmos 1887 experiments. Chapter 7.2. The vestibular function and vestibulo-oculomotor interactions in microgravity. *In Institute of Biomedical Problems*, pp. 1–50. Ministry of Health of the USSR, Moscow.

Sirota, M.G., Beloozerova, I.N., Babaev, B.M., Kozlovskaya, I.B. (1984) The final report on Cosmos 1514 experiments. Chapter 2.4. Study of the vestibular function and vestibulo-oculomotor interaction in monkey. *In Institute of Biomedical Problems*, pp. 1–31. Ministry of Health of the USSR, Moscow.

Sirota, M.G., Beloozerova, I.N., Babaev, B.M., Yakushin, S.B., Ivanov, A.M., Kozlovskaya, I.B. (1986) The final report on Cosmos 1667 experiments. Chapter 7. The vestibular function and vestibulo-oculomotor interaction in microgravity. *In Institute of Biomedical Problems*, pp. 119–162. Ministry of Health of the USSR, Moscow.

Sirota, M.G., Beloozerova, I.N., Babaev, B.M., Yakushin, S.B., Kozlovskaya, I.B. (1989a) Effects of microgravity on vestibular function in monkey. Results of space flight studies. *In Eleventh Ann. Meeting IUPS Commission on Gravitational Physiology*, 2 p, Lyon, France, 25–27 Sept.

Sirota, M.G., Beloozerova, I.N., Yakushin, S.B., Kozlovskaya, I.B. (1989b) Eye-head coordination during adaptation to microgravity. (In Russian). *In Regulation of Sensor-Motor Functions*, pp. 140. Vinniza, USSR.

Sirota, M.G., Beloozerova, I.N., Yakushin, S.B., Babaev, B.M., Kozlovskaya, I.B. (1990a) Changes in neuronal activity of vestibular nuclei during active head movements in microgravity. *In XII International Meeting of Gravitational Physiology*, 63 p (In Russian). IMBP, Moscow, Leningrad, 14–19 October.

Sirota, M.G., Beloozerova, I.N., Yakushin, S.B., Babaev, B.M., Kozlovskaya, I.B. (1990b) Influence of microgravity to vestibular function of monkey. *In XII International Meeting of Gravitational Physiology*, pp. 63–64. (In Russian) IMBP, Moscow, Leningrad, 14–19 October.

Sirota, M.G., Beloozerova, I.N., Yakushin, S.B., Ivanov, A.M., Kozlovskaya, I.B. (1990c) Vestibulo-motor interaction of monkey in microgravity. Eye-head coordination test. *In XI Meeting in Space Biology and Medicine*, pp. 342–343. (In Russian) IMBP, Kaluga, Moscow.

Sirota, M,G., Beloozerova, I.N., Babaev, B.M., Yakushin, S.B., Kozlovskaya, I.B. (1991a) The final report on Cosmos 2044 experiments: Chapter 6. Effects of the space flight on primate's motor and vestibular functions. *In Institute of Biomedical Problems*, pp. 223–295. Ministry of Health of the USSR, Moscow.

Sirota, M.G., Beloozerova, I.N., Yakushin, S.B., Ivanov, A.M., Kozlovskaya, I.B. (1991b) Gain of vestibulo-ocular reflex of monkey in microgravity. Experiment on biosatellite "Cosmos-2044" boarad. (In Russian) *In International Meeting "Biosatellite Cosmos"*, 114 p, IMBP, Moscow, Leningrad, 12–15 August.

Sirota, M.G., Beloozerova, I.N., Yakushin, S.B., Ivanov, A.M., Kozlovskaya, I.B. (1991c) Kinematic of saccades and head movement in eye-head coordination test of monkey. Experiment on biosatellite "Cosmos-2044" board. (In Russian) *In International Meeting "Biosatellite Cosmos"*, 113 p, IMBP, Moscow, Leningrad, 12–15 August.

Tiliket, C., Shelhamer, M., Tan, H.S. and Zee, D.S. (1993) Adaptation of the vestibulo-ocular reflex with the head in different orientations and positions relative to the axis of body rotation. J. Vestib. Res. 3, 181–195.

Tomko, D., Kozlovskaya, I.B., Paige, G. and Badakva, A.M. (1993) *Adaptation to Micro-Gravity of Oculomotor Reflexes (AMOR): Otolith-Ocular Reflexes*, 25 p, National Aeronautics and Space Agency (NASA), Washington, D.C.

Tomlinson, D. and Bahra, P.S. (1986) Combined eye-head shifts in the primate I. Metrics. J. Neurophysiol. 56, 1542–1557.

Uganov, E.M. (1974) Vestibular testing of cosmonauts during space flight of "Voshod" spacecraft. *Microgravity*, Vol. 1 (In Russian), pp. 83–88. Moscow.

Uno, T., Horii, A., Umemoto, M., Hasegawa, T., Doi, K., Uno, A., Takemura, T. and Kubo, T. (2000) Effects of hypergravity on morphology and osteopontin expression in the rat otolith organs. J. Vestib. Res. 10, 283–289.

Uno, T., Horii, A., Uno, A., Fuse, Y., Fukushima, M., Doi, K. and Kubo, T. (2002) Quantitative changes in mRNA expression of glutamate receptors in the rat peripheral and central vestibular systems following hypergravity. J. Neurochem. 81, 1308–1317.

Vinnikov, J.A., Lychakov, D.V., Palmbach, L.R., Govardovsky, V.I., Adanina, V.O., Allachverdov, V.L. and Pogorelov, A.G. (1980) Study of the vestibular apparatus in the frog xenopus laevix and rats exposed to long-term weightlessness. Zh. Evol. Biochem. Phyziol. 16, 574–579.

Vinnikov, Y.A., Lychakov, D.V., Palmbax, L.R, Govardovsky, V.I., Adanin, V.O., Allaxverdov, B.L. and Pogorelov, A.G. (1980) Study of the vestibular apparatus of frogs (Xenopus Laevis) and rats in prolonged microgravity. J. Evolut. Biochem. Physiol. 26(6), 574–579 (in Russian).

von Baumgarten, R.J., Baldrighi, G. and Shillinger, G. (1972) Vestibular behavior of fish during diminished G-force and weightlessness. Aerospace Med. 43, 626–632.

von Baumgarten, R.J., Simmonds, R.C. and Boyd, J.F. (1975) Effects of prolonged weightlessness on the swimming pattern of fish aboard Skylab-3. Aviat. Space Environ. Med. 46, 902–906.

Waespe, W., Cohen, B. and Raphan, T. (1985) Dynamic modification of the vestibulo-ocular reflex by the nodulus and uvula. Science 228, 199–201.

Wearne, S., Raphan, T. and Cohen, B. (1999) Effects of tilt of the gravito-inertial acceleration vector on the angular vestibuloocular reflex during centrifugation. J. Neurophysiol. 81, 2175–2190.

Weiderhold, M.L., Harrison, J.L., Parker, K. and Nomura, H. (2000) Otoliths developed in microgravity. J. Grav. Physiol. 7, 39–42.
Weiderhold, M.L., Pedrozo, H.A., Harrison, J.L., Hejl, R. and Gao, W. (1997) Development of gravity-sensing organs in altered gravity conditions: Opposite conclusions from an amphibian and a molluscan preparation. J. Grav. Physiol. 4, 51–54.
Yakovleva, I.Y., Kornilova, L.N., Serix, G.D., Tarasov, I.K. and Alekseev, V.N. (1982) Results of vestibular function and spatial perception of the cosmonauts for the 1st and 2nd exploitation on station of salut 6. Space Biol (Russia) 1, 19–22.
Yakushin, S.B., Beloozerova, I.N., Sirota, M.G. and Kozlovskaya, I.B. (1989) Cerebellar and brain stem structures activity during semicircular canals stimulation in microgravity. *Regulation of sensomotor functions, (In Russian)*, 179 p, Vinniza, USSR.
Yakushin, S.B., Beloozerova, I.N., Sirota, M.G. and Kozlovskaya, I.B. (1990) Neuronal activity of brain stem vestibular structures and cerebellum of monkey in space flight. (In Russian). *In IX Meeting in Space Biology and Medicine*, Vol. 1, 359 p, Vinniza, Moscow.
Yakushin, S.B., Raphan, T. and Cohen, B. (2000a) Context-specific adaptation of the vertical vestibuloocular reflex with regard to gravity. J. Neurophysiol. 84, 3067–3071.
Yakushin, S.B., Reisine, H., Büttner-Ennever, J., Raphan, T. and Cohen, B. (2000b) Functions of the nucleus of the optic tract (NOT). I. Adaptation of the gain of the horizontal vestibulo-ocular reflex. Exp. Brain Res. 131, 416–432.
Yakushin, S.B., Raphan, T. and Cohen, B. (2003a) Gravity specific adaptation of the vertical angular vestibulo-ocular reflex; dependence on head orientation with regard to gravity. J. Neurophysiol. 89, 571–586.
Yakushin, S.B., Palla, A., Haslwanter, T., Bokisch, C.J., Straumann, D. (2003) Dependence of adaptation of the human vertical angular vestibulo-ocular reflex on gravity. Exp. Brain Res. 152, 137–142.
Yakushin, S.B., Xiang, Y., Raphan, T., Cohen, B. (2004) Spatial distribution of gravity dependent gain changes in the vestibulo-ocular reflex. J. Neurophysiol. Article in Press, DOI, 10.1152/jn. 01269.
Yakushin, S.B., Sirota, M.G., Beloozerova, I.N., Babayev, B.M., Kozlovskaya, I.B. (1992) Eye-head coordination of monkey (Macaca mulatta) in gaze fixation reaction during "Cosmos" biosatellites flight. *In XVIIth Barany Society Meeting,* Czechoslovakia, June 1–5.
Young, L.R., Lichtenberg, B.K., Arrott, A.P., Crites, T.A., Oman, C.M. and Edelman, E.R. (1981) Ocular torsion on earth and in weightlessness. Ann. N.Y. Acad. Sci. 374, 80–92.

Experimentation with Animal Models in Space
G. Sonnenfeld (editor)

DOI: 10.1016/S1569-2574(05)10006-9

Effect of Space Flight on Circadian Rhythms

Gianluca Tosini and Jacopo Aguzzi

Neuroscience Institute, Morehouse School of Medicine, Atlanta, GA, USA

Introduction

Daily rhythmicity is a ubiquitous feature of living systems. Generally, these rhythms are not just passive consequences of cyclic fluctuations in the environment, but, rather they originate within the organisms. They are generated and controlled by endogenous physiological oscillators. The fundamental adaptive function of an endogenously programmed rhythmicity is to provide an optimal and anticipatory temporal organization of physiological processes and behavior in relation to the environment. Endogenously controlled daily rhythms are called circadian rhythms, referring to the fact that they have a period of about a day. Under conditions of temporal isolation this period, in most cases, slightly deviates from exactly 24 h. Synchronization of endogenous circadian rhythms to the environment occurs through daily adjustment of the clock by external time cues or zeitgebers. The most important and reliable cue for synchronization is the daily light–dark cycle (Pittendrigh and Daan, 1976).

Properties of circadian rhythms

Since some of the terminology used in this review is specific to the circadian specialty, we will begin this chapter with an explanation of the most common terms used to describe circadian rhythms. A circadian clock is a part of an organism which is capable of generating a self-sustained rhythm with a period close to 24 h. A rhythm is characterized by three features: period, amplitude and phase (Fig. 1). The period is the length of time necessary to complete one full cycle, the amplitude is, roughly, the difference between the maximum and the minimum levels observed during a full cycle, and the phase refers to the temporal relationship between a specific identifiable point on the cycle and a point on a reference cycle.

When an organism is maintained under constant conditions its circadian rhythms persist with a period close to, but different from 24 h. The fact that rhythmicity persists under constant condition demonstrates that the rhythm is endogenous. The period of a rhythm under constant conditions is called its

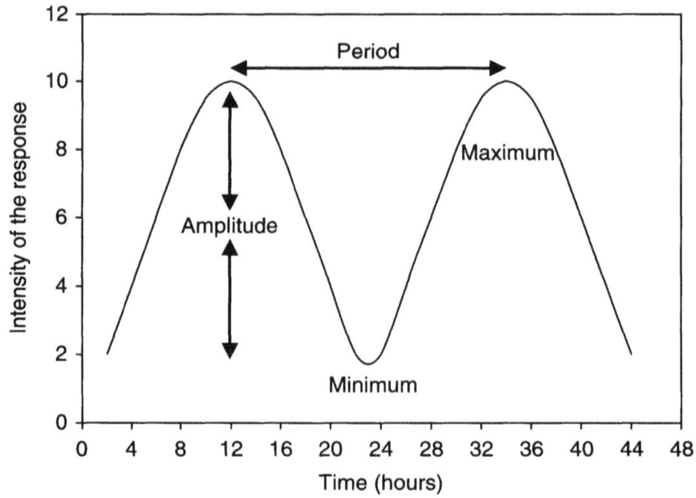

Fig. 1. Diagram illustrating the formal properties of a circadian rhythm. The period is the time required to complete one full cycle (24 h). The amplitude is the difference between the maximum and the minimum values. The phase is the relationship that the rhythm assumes with respect to a referencing time scale (i.e., the light cycle).

free-running period. The free-running period of a rhythm is directly related to the period of the biological clock which generates it.

To be of any use, circadian clocks must be synchronized with the 24 h rotation of the earth on its axis. The process by which a circadian rhythm is synchronized with the environmental day/night cycle is called entrainment. Many periodic factors present in the environment can entrain circadian rhythms, but the environmental light cycle is by far the most important. Most organisms can be entrained by temperature cycles and also by food and drink availability as well as social interaction. A circadian rhythm is entrained when its period matches that of the entraining stimulus (zeitgeber).

Molecular aspects of the circadian clock

The core of the molecular mechanisms responsible for the generation of the circadian oscillation is constituted by interacting positive and negative transcriptional and posttranscriptional feedback loops (see Reppert and Weaver, 2002 for a recent review and Fig. 2). The positive autoregulatory feedback loop begins with the heterodimer CLOCK/BMAL1 binding to the E-box (i.e., a short DNA sequence to which transcription factors of the bHLH family of proteins can bind) of the *Period* (*1-3*) genes promoter, initiating the transcription of these genes. This results in the rise in the levels of the PERIOD (1-3) proteins in the cytoplasm. CLOCK/BMAL1 also acts as a transcriptional activator of the two mammalian cryptochrome genes. Once the

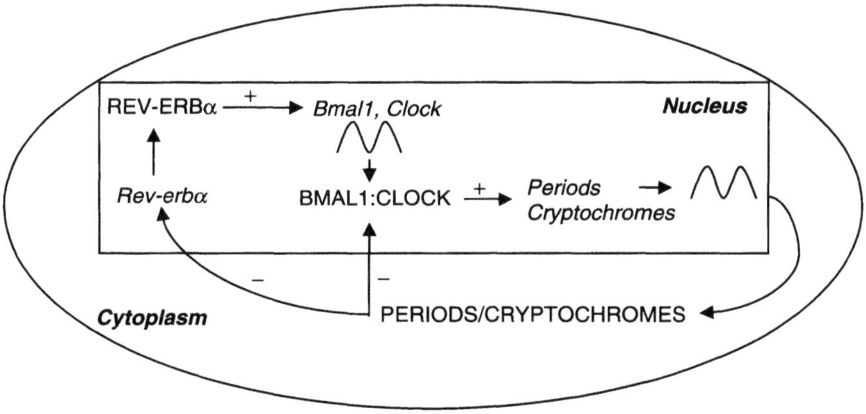

Fig. 2. Schematic and simplified representation of the regulation of genes believed to be involved in the circadian clock. In the nucleus, the genetic information encoded in these genes is converted in mRNA (in italics) and then is moved to the cytoplasm where it will be used to generate the protein (in capital). The BMAL1 : CLOCK complex positively (+) regulate the transcription of *Periods* and *Cryptochromes*. PERIODs and CRYPTOCHROMEs accumulate in the cytoplasm, re-enter the nucleus where they negatively (−) regulate their own transcription by acting on BMAL1 : CLOCK complex and REV-ERBα, which controls the transcription of *Bmal1* and *Clock* mRNA (see text for more details).

PERIOD proteins and the CRYPTOCHROME (1-2) proteins have reached determined levels, they form heterodimers that translocate to the nucleus. Once they have reached the nucleus, the heterodimers inhibit the transcription of their own genes through inactivation of the positive effectors (the CLOCK/BMAL1 complex). At this point, the level of the PERIOD and CRYPTOCHROME proteins in the cytoplasm begins to decrease, leading to a parallel decrease in the formation of the inhibiting heterodimers. Ultimately, the level of heterodimers will be insufficient to inhibit the transcription of the *period* and *cry* genes, which will once again fall under the positive control of the CLOCK/BMAL1 complex. A recent study has revealed that REV-ERBα an orphan nuclear receptor controls the cyclic transcription of *Bmal1* and *Clock* and its own transcription is activated by CLOCK/BMAL1 complex and repressed by PERIOD and CRYPTOCHROME proteins (Preitner *et al.*, 2002).

Circadian organization

In mammals, including human beings, the central pacemaker controlling rhythms is localized in the suprachiasmatic nuclei (SCN) of the hypothalamus (Klein *et al.*, 1991) (Fig. 3). This pacemaker or "clock" has been shown to be responsible for the temporal organization of a wide variety of functions, ranging from sleep and food intake to physiological measures such as body temperature, heart rate and hormone release. However many experimentally generated data suggested that mammals possess circadian oscillators outside

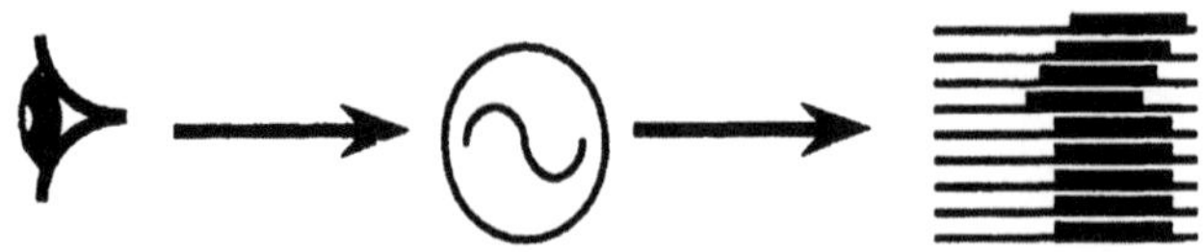

Fig. 3. Simplified model of the mammalian circadian system. Light is perceived by specialized photoreceptors located in the retina. The photic information is then conveyed to the central circadian pacemaker in the brain (SCN). The circadian pacemaker transmits this information to the rest of the organism.

the SCN. The most convincing evidence of this comes from studies that demonstrated the persistence of some circadian rhythms in animals with SCN lesions. One of the best examples of such a phenomenon is represented by the observation that the circadian rhythm in photoreceptor disc shedding persists in SCN-lesioned animals (review in: Tosini and Fukuhara, 2003). Additional experimental evidence supporting the idea that mammals have multiple circadian oscillators came after the cloning of the circadian clock genes. As soon as these genes were cloned it became evident that "clock" genes were not only expressed in the SCN, but their expression was widespread within the body, and in several tissues/organs the expression was rhythmic (Shearman *et al.*, 1997; Zylka *et al.*, 1998; Fukuhara *et al.*, 2000). Indeed, recent studies by Menaker and collaborators have demonstrated that many isolated tissues/organs of mammals can also express a circadian rhythm when cultured *in vitro* (Yamazaki *et al.*, 2000). This research team, using transgenic animals (rat and mouse) in which luciferase expression was placed under the control of the *Period 1* promoter, demonstrated that *in vitro* culture of several organs and tissues showed circadian rhythms in bioluminescence, although the rhythmicity in these tissues and organs tended to dampen out after few cycles, while the bioluminescence rhythms in the SCN continued to oscillate for several more days (Yamazaki *et al.*, 2000). This result suggests that there must be important differences between the "central" and the "peripheral" oscillators; and that the mechanism of the circadian clock located in the SCN, must somehow be different from the mechanism of the peripheral tissues. In addition, it was also shown that different tissues/organs respond in different ways to a phase shift, since the peripheral oscillators entrain to LD cycles much less promptly than the SCN. One of the most intriguing aspects of the mammalian circadian system remains the unveiling of how each part of the system communicates with the others and thus how the system is kept synchronized.

Experimental evidence suggests that both neural and humoral factors are involved in this process. A recent study has indicated that the SCN directly projects to several peripheral areas, suggesting an active role for neural connections in controlling the expression of circadian rhythms in the periphery (Ueyama *et al.*, 1999). At the same time, it must also be mentioned that restoration of the circadian rhythm in locomotor activity in SCN-lesioned animals has also been achieved following the transplantation of SCN which has been encapsulated in a semi-permeable polymeric capsule, preventing neural outgrowth but allowing the diffusion of humoral signals (Silver *et al.*, 1996). As mentioned earlier, the possibility of entraining animals to feeding regimens has been reported in several studies, and two recent papers have demonstrated that in animals subjected to a regimen of restricted feeding, the circadian oscillators in the liver can be entrained independently from the SCN and the LD cycle (Stokkan *et al.*, 2001). Abrupt changes in the feeding time lead to a gradual resetting of the rhythmic gene expression, thus, indicating that phase resetting is mediated by clock-dependent mechanisms. Interestingly, the food-induced re-synchronization proceeds faster in the liver than in the other peripheral tissues suggesting that as in the case of the resetting action of light on the SCN, different tissues (cells) may preferentially employ specific synchronizing signals.

As we have already mentioned, light is the most important synchronizing cue for the entrainment of circadian rhythm in mammal. In mammals, light information from the eye reaches the primary circadian pacemaker (SCN) via a distinct neural projection called the retino-hypothalamic tract (Moore and Lenn, 1972). This tract arises from a small population of retinal ganglion cells and forms a relatively small percentage of the fibers of the optic nerve. Until recently, it was believed that the rods and cones were responsible for collecting the light information necessary to the SCN for synchronization to light. However, experiments conducted in the early 1990s on mice with genetically determined impairments of the eye (Foster *et al.*, 1991) cast some doubts on this assumption. In a series of very elegant experiments Foster and collaborators investigated the effects of photoreceptor loss on circadian photic entrainment in mouse models that lacked all rods and cones and they discovered that the absence of rods and cones did not influence photoentrainment or pineal melatonin suppression (Freedman *et al.*, 1999; Lucas *et al.*, 1999). The finding that these responses are not removed by the absence of the rod and cone photoreceptors (Freedman *et al.*, 1999; Lucas *et al.*, 1999) suggests that a novel photoreceptor and photopigment must be present in the inner layers of the retina. Recent studies indicate that the most likely candidate for this is the mammalian homologue of *Xenopus* melanopsin (Provencio *et al.*, 1998). Mammalian melanopsin is expressed within the inner retina in a small number of ganglion cells and in an even smaller number of cells within the amacrine cell layer in the mouse retina. The function of melanopsin is unknown, but its homology to other opsins and its localization to the inner retina suggest melanopsin to be a good candidate for non-visual photoreception (Provencio *et al.*, 2000). More recently it has been demonstrated that a subpopulation of retinal ganglion cells are directly photosensitive (Berson *et al.*, 2002) and contains melanopsin. Finally, studies using transgenic animals (melanopsin knock-out) have demonstrated the involvement of this new photopigment in the mechanisms of the photic entrainment (Panda *et al.*, 2003).

Circadian rhythms in space

One the most important aspects of space flight is the absence of geophysical 24-h cycles, which, of course, affects overall temporal organization. In this section, we will review the current knowledge of the effect the space flight (or altered gravity) exerts on the circadian system.

The data available on the effect of space flight on the circadian rhythm of animals are limited. A few studies have investigated the effects of space flight on the circadian system of the insect. In the beetle *Trigonoscellis gigas* circadian rhythms of locomotor activity persisted in space albeit the period showed some alteration (increase or decrease) (Alpatov *et al.*, 1998; Hoban-Higgins *et al.*, 2003).

The data available in rodents are also very limited. The effects of space flight on the rat's circadian rhythms were monitored during the flight of Spacelab-3. During this flight the animal displayed a circadian rhythm in the body temperature that was similar to the rhythm observed in the same animals before the flight (Fuller *et al.*, 1989). However, although the animals were maintained in Light : Dark (LD) cycle, the body temperature rhythms showed a periodicity different from 24 h (mean 24.4 h) thus indicating that the animals were free-running and not entrained to the LD cycle present in the environment. Interestingly, it was observed that the period of the heart rate and drinking rhythm maintained a 24 h periodicity (Fuller *et al.*, 1989). Such a result suggested that space flight (including microgravity) may affect different rhythms (and/or circadian pacemaker) in different manners. Another study (Kwarecki *et al.*, 1980) reported that urinary calcium concentration rhythms in space cannot be resynchronized after a phase-shift of 180°.

Additional data on the circadian rhythm of locomotor activity in the rat were recorded during the Neurolab mission. The analysis of the data revealed that space flight increased the length of the free-running period compared to what was observed on the Earth (Fuller *et al.*, 1999). Finally, a study investigated the effect of microgravity on the development of circadian rhythmicity. Pregnant rats were flown aboard the Space Shuttle (STS-70) for gestational days 11–20, and then returned to the Earth where the pups were delivered. No significant differences in the rhythms of body temperature and activity were observed between the animals exposed to space flight during the embryonic development and the control (Hoban-Higgins *et al.*, 1999).

The first set of data considering the effect of space flight on the circadian rhythms of the pig-tailed macaque (*Macaca nemestrina*) was obtained during the space flight of Biosatellite III. The data collected during this mission suggested that the rhythm of body temperature persisted but it was free-running and did not entrain to the ambient lighting (Hahn *et al.*, 1971); in addition, a fragmentation of the sleep–wake cycles was also observed (Hanley and Adley, 1971; Hoshizaki *et al.*, 1971). Another set of data (body temperature) was collected on Rhesus monkey (*M. mulatta*) during the Cosmos 1514 mission.

In this study also it was observed that although the animals were exposed to LD cycle, the body temperature showed a non-temperature rhythm (Sulzman *et al.*, 1992).

More recently, four male monkeys (*M. mulatta*) were been flown on two Cosmos missions (2044 and 2229), while several physiological parameters (activity, heart rate and body temperature) were monitored. In all subjects, the body temperature rhythm was delayed with respect to that observed in ground controls, whereas no differences were observed in the activity and heart rate (Fuller *et al.*, 1996). A similar result was obtained in *M. mulatta* flown on Bio 6–10 (Sulzman *et al.*, 1992) and on Bio 11 (Alpatov *et al.*, 2000).

In conclusion, it appears that during the space flight the body temperature rhythm is delayed; such delay is probably due to a small increase of the period of the circadian pacemaker controlling the body temperature rhythm. However, it must be noted that the circadian period remains within the entrainment range, thus preventing the free-running of the circadian rhythm.

Since the possibility of conducting circadian experiment during space flights is very limited, and obtaining a microgravity environment on the earth is impossible; many studies have investigated the effects that exposure to an increased gravitational environment (hypergravity) via centrifugation produce on the circadian timing system. In the desert beetle (*Trigonoscelis giga*), exposure to a gravitational field of 2G induced a longer free-running period of the locomotor activity rhythm in DD, whereas in LL the free-running period was no different from that observed at 1G (Hoban-Higgins *et al.*, 2003). In rat, exposure to a 2G gravitational field disrupted the locomotor activity rhythms only during the first 7–10 days; also the amplitude of the circadian rhythm of body temperature was significantly affected (i.e., reduced) during the same period (Fuller, 1994; Fuller *et al.*, 1989, 1994). Another study reported that acute exposure to 2G (1 h) phase shifted the circadian rhythm of locomotor activity (Hoban-Higgins *et al.*, 1995) and, in addition, it has been shown that when the hypergravity is applied periodically (1 h out of every 24 h) the animals can entrain (Fuller *et al.*, 1992). Similar results were also obtained in the mouse (Murakami and Fuller, 2000).

An additional study examined the effect of the hypergravity on the function of the retinohypothalamic tract. Rats were exposed to either 2 days or 21 days of 2G via centrifugation and during the last hour of 2G exposure, one series of rats was exposed to a 1 h phase-shifting light. Surprisingly, the number of c-Fos positive after neurons within the SCN was significantly reduced in the rats exposed to 2 days of 2G. A recovery in the effect of light to induce c-Fos reactivity within SCN neurons occurred in the rats exposed to 21 days of 2G. These data suggest that exposure to 2G may suppress, at least temporarily, the responsiveness of the SCN to the phase-shifting effects of light mediated by the RHT (Murakami *et al.*, 1998).

Finally, a recent study using transgenic mice (C57BL/JEi-het) lacking macular otoconia has demonstrated that the vestibular system, and in particular the vestibular macular receptors, is responsible for the effects observed in the circadian system following exposure to increased gravitational (2G) forces (Fuller *et al.*, 2002).

Conclusions

From the data collected so far it is evident that gravity can influence the circadian timing system in several ways. Changes in the gravitational field can affect the phase, period and amplitude of several circadian rhythms. The effects of gravity are probably mediated by the vestibular system via a vestibulohypothalamic connection. However, further studies are needed to fully understand the effects of space flight on the regulation of the circadian system.

Acknowledgments

The laboratory of G. Tosini is supported by grants from National Institute of Neurological Disorders and Stroke and by NASA Cooperative Agreement NCC 9-58 with the National Space Biomedical Research Institute.

References

Alpatov, A.M., Hoban-Higgins, T.M., Fuller, C.A., Lazarev, A.O., Rietveld, W.J., Tschernyshev, V.B., Tumurova, E.G., Wassmer, G. and Zotov, V.A. (1998) Effects of microgravity on circadian rhythms in insects. Journal of Gravitational Physiology 5, 1–4.

Alpatov, A.M., Hoban-Higgins, T.M., Klimovitsky, V.Y., Tumurova, E.G. and Fuller, C.A. (2000) Circadian rhythms in Macaca mulatta monkeys during Bion 11 flight. Journal of Gravitational Physiology 7, S119–S123.

Berson, D.M., Dunn, F.A. and Takao, M. (2002) Phototransduction by retina ganglion cells that set the circadian clock. Science 295, 1070–1073.

Foster, R.G., Provencio, I., Hudson, D., Fiske, S., De Grip, W. and Menaker, M. (1991) Circadian photoreception in the retinally degenerate mouse (*rd*/*rd*). Journal Comparative Physiology [A] 169, 39–50.

Freedman, M.S., Lucas, R.J., Soni, B., Von Schantz, M., Munoz, M., David-Gray, Z. and Foster, R.G. (1999) Regulation of mammalian circadian behavior by non-rod, non-cone, ocular photoreceptors. Science 284, 502–504.

Fukuhara, C., Dirden, J.C. and Tosini, G. (2000) Circadian expression of Period 1, Period 2, and Arylalkylamine N-acetyltransferase mRNA in the rat pineal gland under different light conditions. Neuroscience Letters 286, 167–170.

Fuller, C.A., Murakami, D.M. and Sulzman, F.M. (1989) Gravitational biology and the mammalian circadian timing system. Advances in Space Research 9, 283–292.

Fuller, C.A., Murakami, D.M. and Pesce, V.D. (1992) Entrainment of circadian rhythms in the rat by daily one hour G pulses. The Physiologist 35, S63–S64.

Fuller, C.A. (1994) The effects of gravity on the circadian timing system. Journal of Gravitational Physiology 1, P1–P4.

Fuller, C.A., Hoban-Higgins, T.M., Klimovitsky, V.Y., Griffin, D.W. and Alpatov, A.M. (1996) Primate circadian rhythms during space flight: results from Cosmos 2044 and 2229. Journal of Applied Physiology 81, 188–193.

Fuller, C.A., Murakami, D.M., Hoban-Higgins, T.M., Fuller, P.M. and Wooley, D.E. (1999) Effects of spaceflight on the regulation of the rat circadian timing system. Society for Neuroscience Abstract 25, 869.

Fuller, P.M., Jones, T.A., Jones, S.M. and Fuller, C.A. (2002) Neurovestibular modulation of circadian and homeostatic regulation: vestibulohypothalamic connection? Proceeding National Academy of Sciences USA 99, 15723–15728.

Hahn, P.M., Hoshizaki, T. and Adey, R.W. (1971) Circadian rhythm of the Macaca nemestrina monkey in Biosatellite III. Aerospace Medicine 42, 295–304.

Hanley, J. and Adley, R.W. (1971) Sleep and wake states in the Biosatellite III monkey: visual and computer analysis of telemetred electroencephalographic data from earth orbital flight. Aerospace Medicine 43, 204–213.

Hoban-Higgins, T.M., Murakami, D.M., Tandon, T. and Fuller, C.A. (1995) Acute exposure to 2G phase shifts the rat circadian timing system. Journal of Gravitational Physiology 2, P58–P59.

Hoban-Higgins, T.M., Murakami, D.M., Tang, I.H., Fuller, P.M. and Fuller, C.A. (1999) Development of circadian rhythms in rat pups exposed to microgravity during gestation. Journal of Gravitational Physiology 6, 71–79.

Hoban-Higgins, T.M., Alpatov, A.M., Wassmer, G.T., Rietveld, W.J. and Fuller, C.A. (2003) Gravity and light effects on the circadian clock of a desert beetle, *Trigonoscelis gigas*. Journal of Insect Physiology 49, 671–675.

Hoshizaki, T., Durhan, R. and Adey, W.R. (1971) Sleep-wake pattern of a Macaca nemestrina monkey during nine days of weightlessness. Aerospace Medicine 42, 228–296.

Kwarecki, K., Debicc, H. and Kater, Z. (1980) Rhythms of electrolyte and hydroxyproline excretion in urine of rats after three weeks of weightlessness. *In* Hideg, J. and Gazenko, O. (eds.). *Advances in Physiological Sciences Gravitational Physiology*, Vol. 19, pp. 33–38.

Klein, D.C., Moore, R.Y. and Reppert, S.M. (1991) *Suprachiasmatic Nucleus. The Mind's Clock*, Oxford University Press, New York, NY.

Lucas, R.J., Freedman, M.S., Munoz, M., Garcia-Fernandez, J.M. and Foster, R.G. (1999) Regulation of the mammalian pineal by non-rod, non-cone, ocular photoreceptors. Science 284, 505–507.

Moore, R. and Lenn, N. (1972) A retinohypothalamic projection in the rat. Journal Comparative Neurology 146, 1–14.

Murakami, D.M. and Fuller, C.A. (2000) The effect of 2G on mouse circadian rhythms. Journal of Gravitational Physiology 7, 79–85.

Murakami, D.M., Tang, I-H. and Fuller, C.A. (1998) Chronic 2G exposure affects C-Fos reactivity to a light pulse within the suprachiasmatic nucleus. Journal Gravitional Physiology 5, 71–79.

Panda, S., Provencio, I., Tu, D.C., Pires, S.S., Rollag, M.D., Castrucci, A.M., Pletcher, M.T., Sato, T.K., Wiltshire, T., Andahazy, M., Kay, S.A., Van Gelder, R.N. and Hogenesch, J.B. (2003) Melanopsin is required for non-image-forming photic responses in blind mice. Science 301, 525–527.

Pittendrigh, C.S. and Daan, S.A. (1976) A functional analysis of circadian pacemakers in nocturnal rodents I. Stability and lability of spontaneous frequency. Journal Comparative Physiology 106, 233–252.

Provencio, I., Jiang, G., DeGrip, W.J., Hayes, W.P. and Rollag, M.D. (1998) Melanopsin: An opsin in melanophores, brain and eye. Proceeding National Academy of Sciences 95, 340–345.

Provencio, I., Rodriguez, I.R., Jiang, G., Hayes, W.P., Moreira, E.F. and Rollag, M.D. (2000) A novel human opsin in the inner retina. Journal of Neuroscience 20, 600–605.

Preitner, N., Damiola, F., Lopez-Molina, L., Zakany, J., Duboule, D., Albrecht, U. and Schibler, U. (2002) The orphan nuclear receptor REV-ERBalpha controls circadian transcription within the positive limb of the mammalian circadian oscillator. Cell 26, 251–260.

Reppert, S.M. and Weaver, D.R. (2002) Coordination of circadian timing in mammals. Nature 418, 935–941.

Shearman, L.P., Zylka, M.J., Weaver, D.R., Kolakowski, L.F., Jr. and Reppert, S.M. (1997) Two period homologs: circadian expression and photic regulation in the suprachiasmatic nuclei. Neuron 19, 1261–1269.

Silver, R., Le Sauter, J., Tresco, P.A. and Lehman, M.N. (1996) A diffusible coupling signal from the transplanted suprachiasmatic nucleus controlling circadian locomotor rhythms. Nature 382, 810–813.

Stokkan, K.A., Yamazaki, S., Tei, H., Sakaki, Y. and Menaker, M. (2001) Entrainment of the circadian clock in the liver by feeding. Science 291, 490–493.

Sulzman, F.M., Ferraro, J.S., Fuller, C.A., Moore-Ede, M.C., Klimovitsky, V., Magedov, V. and Alpatov, A.M. (1992) Thermoregulatory responses of rhesus monkeys during spaceflight. Physiology Behavior 51, 585–591.

Tosini, G. and Fukuhara, C. (2003) Photic and circadian regulation of retinal melatonin. Journal of Neuroendocrinology 15, 364–369.

Ueyama, T., Krout, K.E., Nguyen, X.V., Karpitskiy, V., Kollert, A., Mettenleiter, T.C. and Loewy, A.D. (1999) Suprachiasmatic nucleus: A central autonomic clock. Nature Neuroscience 2, 1051–1053.

Yamazaki, S., Numano, R., Abe, M., Hida, A., Takahashi, R., Ueda, M., Block, G.D., Sakaki, Y., Menaker, M. and Tei, H. (2000) Resetting central and peripheral circadian oscillators in transgenic rats. Science 288, 682–685.

Zylka, M.J., Shearman, L.P., Weaver, D.R. and Reppert, S.M. (1998) Three *period* homologs in mammals: differential light responses in the suprachiasmatic circadian clock and oscillating transcripts outside of brain. Neuron 20, 1–20.

Experimentation with Animal Models in Space
G. Sonnenfeld (editor)

DOI: 10.1016/S1569-2574(05)10007-0

Development as Adaptation: A Paradigm for Gravitational and Space Biology

Jeffrey R. Alberts[1] and April E. Ronca[2]

[1]*Department of Psychology, Indiana University, Bloomington, IN, USA*
[2]*Department of Obstetrics and Gynecology, Wake Forest University School of Medicine, Winston Salem, NC, USA*

Abstract

Adaptation is a central precept of biology; it provides a framework for identifying functional significance. We equate mammalian development with adaptation, by viewing the developmental sequence as a series of adaptations to a stereotyped sequence of habitats. In this way development is adaptation. The Norway rat is used as a mammalian model, and the sequence of habitats that is used to define its adaptive-developmental sequence is (a) the uterus, (b) the mother's body, (c) the huddle, and (d) the coterie of pups as they gain independence. Then, within this framework and in relation to each of the habitats, we consider problems of organismal responses to altered gravitational forces (micro-$\boldsymbol{g}$ to hyper-$\boldsymbol{g}$), especially those encountered during space flight and centrifugation. This approach enables a clearer identification of simple "effects" and active "responses" with respect to gravity. It focuses our attention on functional systems and brings to the fore the manner in which experience shapes somatic adaptation.

We argue that this basic developmental approach is not only central to basic issues in gravitational biology, but that it provides a natural tool for understanding the underlying processes that are vital to astronaut health and well-being during long duration flights that will involve adaptation to space flight conditions and eventual re-adaptation to Earth's gravity.

Introduction

All multicellular organisms develop, that is they undergo growth and differentiation. But growth and differentiation are abstractions. What grows when and, more importantly, what kinds of functions emerge is the actual development. To get from abstraction to the direct analysis of an organic process requires a framework. In the present chapter, we offer one such framework. Put simply: *Development is adaptation.*

Here is what we shall do: First we review briefly the concept of adaptation, recognizing that it has several distinct, but legitimate meanings, more than one of which is pertinent to our topic. This is an important preliminary step, because it is critically important that the uses of the term adaptation be recognized at each stage of the discussion.

Next, we present an overview of how development is adaptation. Here we focus on a single mammalian species, the Norway rat (*Rattus norvegicus*), though the perspective that we offer is applicable to other species. Once this perspective is defined in a preliminary way, we go into each of these ontogenetic habitats and discuss some examples of how the developing rat adapts to it. We then turn to related issues in gravitational biology, drawing on empirical data from space flight and centrifuge studies, as possible. For each of the developmental habitats we attempt to integrate gravity-related data and issues with the larger, adaptive-developmental framework.

Finally, we discuss how such developmental analyses are pertinent to gravitational and space biology in general, and to contemporary problems in space biomedicine in particular.

Meanings of adaptation

Adaptation is a central precept of biology. Etymologically, the word derives from *ad* (toward) + *aptus* (fit). It implies a condition in which there is a match or a fit of one feature with another. Listed below are some examples of different, legitimate uses of the terms "adapt" and "adaptation," based on a particularly lucid discussion by Pittendrigh (1958). Note that these are distinct, but interrelated uses:

(1) *The relation between the organism and its environment.* Here we say, for example, that rats are adapted to living in burrows. This usage implies a variety of specializations that enable rats to be fit for a particular habitat. Such relations are asymmetrical in the sense that organisms are adapted to environments and not vice versa.

(2) *A feature of the organism that serves some proximal end.* The organismal "features" denoted here can be anatomical, physiological, or behavioral and the "proximal end" or the function served might be the acquisition of food or a mate, hiding cryptically, attracting attention, or some other function that can be discerned. Thus, we might note that the intestines of infant mammals are histologically distinct. The cells and the enzymes in the intestinal lining are specially adapted for utilizing a high fat, liquid diet such as mother's milk.

(3) *The process of acquiring adaptation within the lifespan of an individual.* Rats that are exposed to cold conditions during development usually display greater capabilities to generate metabolic heat as adults. The development of insulative fur can be similarly influenced. Such adaptation is sometimes called "somatic adaptation" which is meant to indicate that there is somatic change without change in the genotype, whereas meanings 1 and 2, earlier, imply an historical change in genotype. The capability to adapt this way may itself be considered an adaptation, in the sense of meaning 2.

(4) *A historical process whereby adaptation is shaped, usually by natural selection.* If we were to note that some rodents adapted to life in arid desert environments we are recognizing processes that spanned many generations and there is implied a change in genotype.

These different uses are common and commonly confused. We will see later in this chapter, and in other chapters in the present volume, that somatic adaptation (meaning #3) is a common usage in the language of space biology where much attention is given to somatic changes that occzur when terrestrially born (not to mention terrestrially evolved) organisms are placed in a weightless or fractional-***g*** environment.

To make matters slightly more complicated, we must now consider how the term adaptation appears in developmental thinking.

Development as adaptation: A mammalian model

Mammals, as a class of vertebrate species, are defined largely in terms of their reproductive and developmental processes. Fertilization of eggs by sperm is internal. Gestation of the offspring occurs within the body of the female. The young are born live and free (as opposed to other species in which offspring are born still encased in a developmental envelope such as an egg or shell). Mammals are uniquely and universally equipped with mammary glands by which the post parturient females provide nutrition to their offspring with mother's milk. The young, in turn, are uniquely equipped to suckle. Provision of mother's milk is typically part of postnatal parental behavior, i.e., coordinated activities that increase the probability of the offspring survival after it leaves the female's body (Lott and Dale, 1973).

Features common to one group and absent in others are excellent exemplars of shared ancestry. This, coupled with specialized variants of shared general features, emphasize underlying historical processes, often but necessarily implying some changes in genotype, that lead to systematic changes in phylogeny. Developmental changes are usually the avenue of the alterations in phylogenetic patterns (e.g., Gould, 1977; Raff, 1996).

The broad and general considerations reviewed earlier, translate here into a general framework and a specific model. The general framework is that the mammalian reproductive and developmental processes create a distinct and invariant sequence of environments to which the developing mammal adapts. These environments are, in effect, definable and demanding habitats through which the young mammal moves in time. To meet the demands of each habitat requires suites of morphological, physiological and behavioral adaptations in the fetus, newborn, infant, and juvenile. As the propagule moves from one habitat to the next, it must change. In some cases these changes are gradual. In other instances they are dramatically transformative. Some commentators have even likened some of these transitions to metamorphosis, to emphasize their rapidity and drama (e.g., Yeh and Moog, 1974).

In this general framework, development can be characterized as a sequence of adaptive changes to a sequence of habitats. Unfortunately, further discussion of these ideas is beyond the bounds of the present chapter, but the interested reader may relate this framework to kindred constructs such as "ontogenetic adaptation" (Galef, 1981; Oppenheim, 1981; Alberts and Gubernick, 1985; Alberts and Cramer, 1988) and "ontogenetic niche" (Alberts and Cramer, 1988; West *et al.*, 1988).

A model and a model organism

The specific model to which we now turn will extend our discussion of the basic framework and enable consideration of data and topics in space and gravitational biology. Specifically, we shall use the domesticated Norway rat (*Rattus norvegicus*) as a "model organism." Conveniently, the Norway rat is a standard subject for many basic and biomedical studies and it has been used extensively in space flights and ground-based gravitational studies.

Figure 1 illustrates the basic model used here. It depicts four representative scenes in the ontogenetic process of *R. norvegicus*. Importantly, it indicates these scenes as a sequence. More specifically, each of the images in Fig. 1 (A–D) represent a habitat in which a developing rat lives or, in an active and adaptive sense, "makes its living." In this way, the rat has a job to do and it accomplishes its job with adaptations as a tool. In the process, the developing rat adapts to each habitat.

Umwelt and its roles in ontogeny

We are about to see that, in addition to the fixed sequence of habitats that mammals confront, there is also a fixed sequence to the onset of sensory function. In Norway rats, the onset of function of some sensory systems occurs prenatally, whereas other modalities begin to function at different points postnatally (Alberts, 1984). Thus, some systems are functional in the Uterine Habitat (Fig. 1A) while others are completely non-functional there. As the

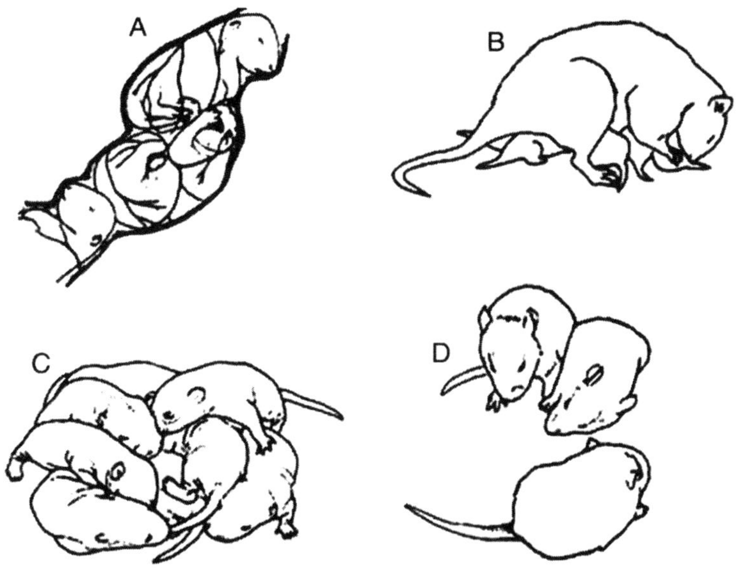

Fig. 1. Four habitats that characterize early adaptation in Norway rats. (A) The uterus as habitat. (B) The mother as habitat. (C) The huddle as habitat. (D) The coterie or social cohort.

pup enters subsequent habitats, additional sensory channels open to stimulation. In every case there is further development. These are critical points, especially within the kind of framework that we are using here.

There are two important dimensions to each habitat. One is the physical dimension, which can be described in terms of its measurable parameters. The other important dimension is the organism's *experience* of the habitat, as determined by the organism's sensory-perceptual systems, i.e., the functional effects of the stimuli that reach the organism via its sensoria.

In the scientific study of animal behavior, or ethology, the concept of *Umwelt* (Uexkull, 1909), which denotes the perceptual world of an animal, is an important topic. The *Umwelt* of different species varies dramatically, depending on the kinds of sense organs that they have. (Some bats discriminate objects by their echoes, other animals use magnetic fields to orient or navigate, while others "see" heat with nasal pit infrared detectors. Many animals perceive olfactory cues with greater acuity and in dimensions not available to humans.)

Umwelt is applicable to our developmental concerns here. The sensory capabilities of fetuses and other immature forms are often dramatically different (usually more limited) than in the adult form. Such differences are important in understanding behavior at each stage of development. In addition, contemporary behavioral neuroscience has established that experience [sensory function = experience (cf., Gottlieb, 1971)] plays important, determinative roles in subsequent neural and behavioral development. The celebrated studies of Hubel and Weisel, (1998) on visual system development emphasized the critical roles of visual stimulation in the development of the visual system, including formation of cell assemblies and synaptic architecture in central nuclei, as well as the development of visually guided behavior. The rule is that function and development are bidirectionally linked (function ← → development). More recently, there is accumulating new evidence of *inter*-sensory development, i.e., that stimulation and function in one modality affects the development of other sensory modalities (e.g., Kenny and Turkewitz, 1986; Lickliter, 1993; Reynolds and Lickliter, 2004a,b).

There is an invariant sequence of onset of function for the sensory systems that have been analyzed in comparative developmental studies (Alberts, 1984; Gottlieb, 1971). The sequence is: Tactile–Vestibular–Auditory–Visual. The chemical senses appear to be among the early onset systems [in rat, probably after vestibular onset and well before auditory and visual (cf., Alberts, 1984)]. Specifically, on the basis of anatomical, physiological, and behavioral evidence it appears that by embryonic day 16, the fetal rat reacts to punctate tactile stimulation to selected regions of the body (Narayanan *et al.*, 1971). Vestibular function is also apparent prenatally, as are the chemical senses (see Alberts, 1984). But, in every case, this is simply onset of function. The subsequent timing, rate, and extent of further development (e.g., threshold sensitivities, range of detection, topographic spread, etc.) appears determined, at least in part, by experience. Auditory and visual function in rat have postnatal onsets, around postnatal day 12 and 15, respectively.

To summarize briefly: in *R. norvegicus*, tactile, vestibular, and chemical sensing begins prenatally and continues after birth. Thus, we can ask whether and to what extent the fetus experiences the tactile, vestibular, and chemical features of the uterine habitat. Similarly, we must extend our questions into a rat's postnatal life. Auditory onset of function begins around postnatal day 12 (P 12). As the pups approach the time of egression from the nest, the eyelids unseal and onset of vision is initiated (about P 15). Although all sensory systems are operational by P 15, development continues thereafter.

For our present concern with gravitational questions, the early onset of vestibular function is profoundly important. Several implications should be noted. First, we emphasize that the onset of vestibular function is dramatically early in every vertebrate species that has been studied. Thus, all sensory systems that develop subsequently are, in theory, susceptible to influence by the previously initiated vestibular function. Perhaps this makes it less surprising that the vestibular system is integral to visual function, to proprioception, to spatial learning, and in other non-vestibular systems. But mysteries abound. Precious little is known about the development of vestibular function itself, and the prospect of finding that the development of other, perhaps *all* other sensory–perceptual systems may be influenced by vestibular development is truly exciting, if not sobering.

With these considerations in mind, we can now enter the rat world(s) and consider development as adaptation.

The uterus as habitat and the fetus adaptations

Figure 1A depicts the initial habitat in which all prenatal mammals reside. It is the mother's uterus and this figure shows a few fetal rats *in situ*. Norway rats, like most mammals, produces multiple offspring with each parity. In rat, there are usually about 10 fetuses, arranged in line in each of the two uterine horns. The figure shows just one portion of one horn. Each fetus resides individually in an amniotic sac and, in the model used here, this is its prenatal habitat.

Consider what a special and demanding habitat this is. Essentially, it is an aqueous habitat for the fetus (though, as a presumptive rat, this is an organism destined for terrestrial life in gaseous habitat). In the uterus, the fetus is physically connected to the uterine wall, tethered by the umbilical cord and placenta. The cord is a conduit for arterial and venous circulation and supports homeostatic functions including respiration, nutrition and waste removal. The uterine habitat changes considerably during the three weeks or so gestation, but again, this is beyond the scope of the present presentation and the interested reader can consult many sources for full discussions (e.g., Dawes, 1968; Liggins, 1982).

It is in the uterine habitat that the rat begins its behavioral life, that is, movement begins. These events are lawful, organized, rhythmic, and functional. Breathing movements begin prenatally, though it is lung fluid rather than air, that is moved through the trachea (Dawes, 1974; Oliver, 1981). Although unrelated to prenatal gas exchange functions, fetal respiratory movements contribute to the remarkable postnatal ability of the respiratory musculature to maintain vigorous, continuous, coordinated function after birth (Liggins, 1982; Dawes, 1968).

Trunk and limb movements also begin prenatally. Reductions in embryonic limb movements, even for a relatively brief period, can permanently block joint flexibility. Limb movements are part of the formative process of the musculoskeletal apparatus (Drachman and Sokolov, 1966; Moessinger, 1983). Similarly, manipulations that disrupt fetal swallowing and tongue movements have been directly associated with impairments in gastrointestinal development (Liggins, 1982).

The fetus has two jobs: It must "earn a living" (i.e., adapt) in an aqueous, restrictive, and uniquely structured habitat—a uterus, while busily becoming an entirely different type of organism, an orally feeding, air-breathing quadruped. The phenomena of prenatal movements provide multiple lessons. First, we can ask whether the movements are an aspect of adaptation to the uterus. Fetal swallowing contributes to the reduction in amniotic fluid volume, a mechanical necessity as the propagules grow within the confines of the dam's body. Movements appear to be part of successful (and adaptive) growth and differentiation in the uterine habitat. Interrupting the expression of such movements can compromise the formation of functional joints or disrupt the development of other systems. Thus, movements in utero contribute both to immediate and future development.

The uterus is a stimulating environment. The uterine habitat is contained in another body, that of the mother. Many fancy that such an encased habitat must be the epitome of quiet, serene, undisturbed existence. Not so.

There is now a sizeable body of evidence showing that uterus is bombarded by stimuli. Maternal behavior and maternal physiology are major contributors of the stimuli (e.g., Bradley and Mistretta, 1975; Decasper and Fifer, 1980; Hofer, 1981; Fifer and Moon, 1988; Ronca *et al.*, 1993). For example, we (Ronca *et al.*, 1993) have shown that during gestation, fetuses *in utero* are exposed to tactile and vestibular stimuli associated with the mother's behavior. Specifically, fetuses are exposed to accelerations as the dam ambulates, circles, and rears. When she grooms her abdomen or lies prone, the fetuses are compressed, and they are vibrated as she scratches with a hindlimb. These activities occur at high levels throughout pregnancy, exposing fetuses to *hundreds* of sensory inputs each day, many of which occur within just 24 h of birth (Ronca *et al.*, 1993).

During labor, fetuses are pitched and rotated by uterine contractions. They are exposed repeatedly to powerful compressions of the head and body until the newborn is finally squeezed through the birth canal. Upon delivery from the birth canal, the dam removes the birth membranes, enabling air to teach pups' nares for the first time. She licks and handles pups, removing amniotic fluid from the skin as she lifts and rotates them, providing extensive cutaneous stimulation and repeated angular accelerations. The postpartum thermal environment is much cooler than the intrauterine environment (21°C vs. 37.5°C), causing thermally fragile newborn pups to cool to room temperature within minutes of birth. They are soon re-warmed, however, when they are gathered in a nest and brooded over by the dam.

In an earlier section of the present chapter, we discussed the onset of sensory function in vertebrates, noting that in rat, tactile and vestibular sensitivity have prenatal onsets. A pertinent question, then, is whether the fetal rats' initial sensory capabilities are sufficient to detect the stimuli that impinge upon them *in utero*? We asked this question by first measuring the stimuli that reach a rat fetus *in utero* (Ronca *et al.*, 1993). We then created simulations of these stimuli and presented them at realistic levels to fetal rats that had been externalized from the mother's body, while maintaining the umbilical connections. We measured stimulus detection by the fetuses via changes in heart rate and behavior that had been documented as valid and reliable measures (Ronca and Alberts, 1990). From the tests with simulations, we learned that fetuses do indeed experience the vestibular, tactile, chemosensory and thermal stimuli associated with life in the uterus and parturition (see Alberts and Ronca, 1996; Ronca and Alberts, 1994). Such experiences are involved in fetal learning, arousal, and birth-associated reorganizations that are vital to perinatal adaptation.

Demonstrations of responses to vestibular stimuli in fetal rats are exciting data for space biology, for it is currently more feasible to fly pregnant rats than to sustain a lactating rat and her suckling offspring in the weightless environment of orbital space flight. With this in mind, we turn to reflect on past flight experiments, which have provided some enticing findings.

Gravid without gravity: Consequences of space flight during gestation

To date, there have been three occasions in which prenatal development of rats has been challenged with space flight conditions. The first such flight experiment was on Cosmos 1514, launched in December 1983. This was followed by two Small Payload experiments flown on the Space Shuttle (STS-66 and -70) in 1993 and 1995, respectively. Figure 2 illustrates some of the mission parameters (for additional detail, see Ronca, 2003). Postflight studies included analyses of the mothers as well as the offspring. In each of these investigations, some of the fetuses were studied soon after they returned to Earth and others were left to gestate and then were studied as newborns and postnatally. In the present section we shall discuss the studies of the fetal development and the gestating dams.

Fetuses from Cosmos 1514

A major goal of the Cosmos 1514 mission was to test the ability for mammalian pregnancy and prenatal development to proceed in microgravity (Alberts *et al.*, 1985, Serova, 1993). The unmanned satellite carried ten, time-mated pregnant rats into low Earth orbit. The fetuses inside these Flight rats were the first mammalian specimens to undergo a portion of their gestation in the absence of Earth-normal gravity. All of the ten pregnant Flight dams returned alive and still pregnant.

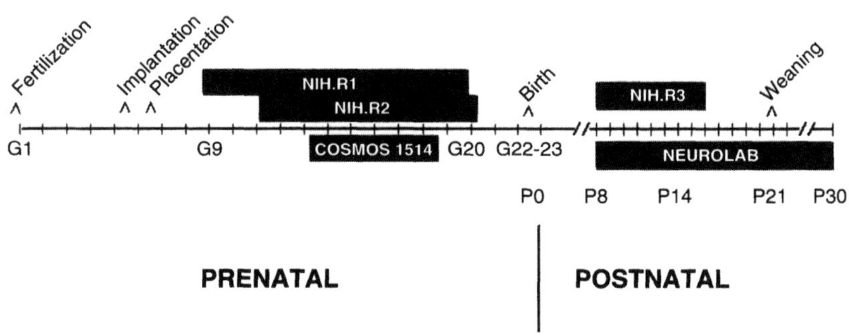

Fig. 2. Overview of spaceflight missions that carried prenatal and young postnatal rats in relation to developmental milestones (G = gestation day; P = postnatal day).

Five of the ten Flight dams were sacrificed at Recovery and the remainder were permitted to undergo parturition, which was accomplished successfully by four of the five dams. Fetal offspring that underwent a portion of their gestation (about 20%) in space were morphologically intact and appeared on a gross level to have grown and developed properly. The Cosmos 1514 flight fetuses weighed approximately 10% less than control fetuses at recovery and flight dams weighed approximately 18% less than control dams (Serova *et al.*, 1993). Analyses of fetal tissue revealed mitotic figures in cortical locations that were interpreted as signs of retarded cellular development and migration (Keefe, 1985). The failure of pregnant Flight dams to gain weight normally compromised interpretation of any seemingly anomalous outcomes in the offspring. Nevertheless, the fact that space-flown fetuses were intact and appeared to be well developed was itself an extraordinary finding and it left open the door to additional explorations.

Pregnant dams on NIH.R1 and NIH.R2

The NIH.R1 and NIH.R2 were experiments flown in mid-deck lockers of the Space Shuttle. Rats were housed in Animal Enclosure Modules (AEMs), which are sealed so the rats are inaccessible to the crew, except for visual access through one Lexan surface. As shown in Fig. 2, these missions were 11- and 9-days-long, and were timed to impose microgravity conditions during the second and third weeks of the rats' 22-day pregnancy.

Weight gains in pregnant dams during the NIH.R1 and NIH.R2 flights were identical to or very close to that of the ground controls (Burden *et al.*, 1997; Ronca and Alberts, 2000a), which enhanced interpretation of the measures made in the offspring, compared to the earlier results from the Cosmos mission.

NIH.R1 was the first time video data were collected systematically and analyzed as part of an experiment protocol, and the effort paid great dividends for several investigators. It also provided valuable information that will contribute to future experiments as well as to hardware design. Even in these short, 7- to 15-min-long video segments, Flight dams were observed to eat, drink, groom and interact with one another. Species-typical behaviors and movements were observed in both Flight and Synchronous control dams. It was clear that under flight conditions, the dams ambulated about the full volume of the AEM, using their paws and legs on the walls and surfaces to propel themselves throughout the habitat.

Although Flight dams appeared to adjust to weightlessness, we found that their behavior during space flight differed from that of the ground controls. This was seen when we applied a self-referencing movement notation system, by which we classified each movement of the dams in relation to the posture of the body at an immediately preceding time-point (Alberts and Ronca, 1997). The results of this analysis indicated that movements involving pitch and yaw were about equivalent in Flight and Control dams. In contrast, Flight dams

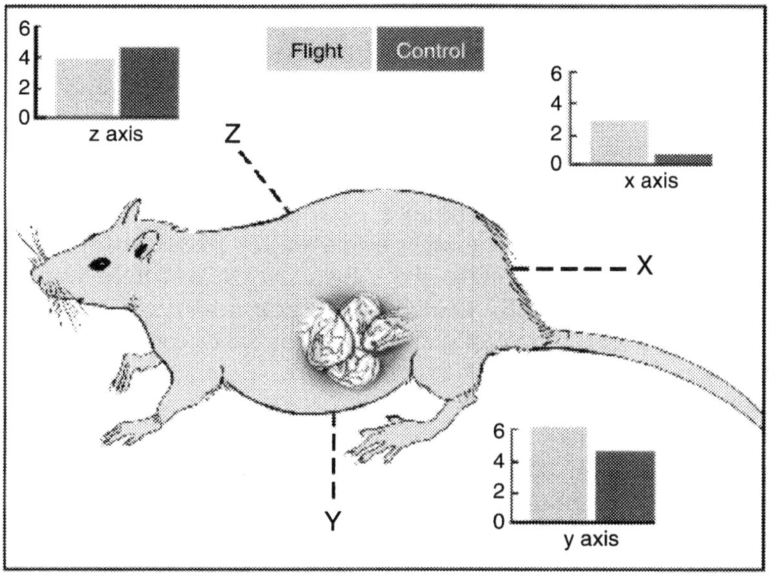

Fig. 3. Relative rates of body movements, measured in self-referenced transitions from an immediately previous timepoint, and expressed on three vectors (X, Y, Z) of movement. Data were from time-matched samples of behavior of pregnant dams housed in the AEM during orbital flight and from Synchronous Control dams housed in an environmental chamber that was yoked to temperature and humidity conditions of the Space Shuttle middeck.

displayed about seven times more rolling movements than did Controls. This difference, we think, was a consequence of the increased numbers of surfaces available in microgravity for walking and crawling. Many of the rats' movements from surface to surface involved rolling movements on the longitudinal body axis (see Fig. 3).

These flight-induced alterations in maternal movements changed the pattern of prenatal vestibular input to fetuses *in utero* and were associated with striking changes in fetal vestibular morphology and perception (Ronca *et al.*, 2000).

Fetuses from NIH.R1 and NIH.R2

Several investigators who analyzed fetal tissues from NIH.R1 and R2 discovered differences between space flight- and Earth-gestated fetuses. Among the noteworthy findings were:

(a) R1 fetuses showed a six-fold increase in atrial natriuretic peptide (ANP), a cardiovascular-related hormone that regulates sodium and water excretion, and vasodilatation (Davet *et al.*, 1999). This observation is similar to reports of increased ANP in the astronauts, presumably related to inflight fluid shifts.

(b) R1 fetuses also showed delayed development of the choroid plexus, responsible for cerebrospinal fluid (CSF) secretion and involved in brain homeostasis (Mani-Ponset *et al.*, 1997). Significant immunocytochemical changes were observed in the distribution of proteins involved in cell differentiation and CSF production within the choroidal epithelial cells.
(c) R2 flight fetuses evinced changes in the developing superchiasmatic nucleus (SCN), which is the reception area for photic information that entrains circadian rhythms to the light/dark cycle. Flight fetuses SCN appeared developmentally delayed (Murakami *et al.*, 1997), a finding that fits well with the host of disturbances in the circadian timing system observed in astronauts (Dijk *et al.*, 2001).
(d) Sonnenfeld and colleagues (Sonnenfeld *et al.*, 1998) reported no changes in immune responses of R1 flight fetuses, despite significant immune suppression in the dams that was characteristic of that seen in non-pregnant adult animals.
(e) R2 Flight and Control fetuses did not differ in key proteinases important in bone development and remodeling (Davis *et al.*, 1998).

NIH.R2 included a study designed to assess the functional status of the fetus' vestibular system prior to re-adaptation to 1-***g***. This was accomplished by applying the procedures described earlier in which live fetuses are externalized from the mother's body and tested for sensory responsivity (cf. Ronca and Alberts, 1994). Flight fetuses were tested about 2 h after Recovery from orbital space flight. They showed magnified responses to vestibular stimulation (rolling through a 70° arc) compared to Control fetuses tested identically. This finding was particularly surprising because previous data (from NIH.R1) had prepared us to expect an attenuated response to vestibular stimuli, as evidenced by the water immersion righting responses of the neonates. Nevertheless, we were better able to understand the dramatic hyper-responsivity of Flight fetuses to angular accelerations after we analyzed the video recordings of the Flight dams and Control dams (Alberts and Ronca, 1997). The video data revealed differences in the mother's behavior that created contrasting adaptive challenges for the fetuses.

Postnatal pups that experienced prenatal space flight

Cosmos-1514, NIH.R1 and NIH.R2 together provide a broad, preliminary picture of how space or microgravity adaptation may proceed during early, formative phases of the body, brain, and behavior. Four dams from Cosmos-1514 delivered litters that were used for postnatal tests. The satellite flew from G13 to G17.5, so that gestation in these dams continued for about as long as the mothers and fetuses had been exposed to microgravity. For NIH.R1, newborn rats were derived from ten dams that gave birth vaginally on G22 following unilateral hysterectomy (removal of one of the dam's paired uterine horns) on G20. For NIH.R2, newborn rats were available from six of the ten intact Flight dams. They were allowed to undergo vaginal deliveries.

In all three missions, litter sizes were similar between the Flight and Control conditions. The pups' external appearance was appropriate for their calendar ages. Average birth weights of Flight pups did not differ from Synchronous controls. In the Cosmos rats, however, pup weights were strikingly variable in the Flight group. Body weights of NIH.R2 pups were modestly but significantly less than Synchronous controls until the 14th postnatal day (Hoban-Higgins *et al.*, 1999; see also Ronca, 2003). This outcome may have been related to the minor (5%) reduction on the dam's body weights of Recovery. No differences in body weight were observed thereafter (through P90) for either NIH.R1 or NIH.R2 offspring.

Postnatally, the emergence and development of locomotion and gait appeared equivalent for Flight and Control pups through P 81, when testing was terminated (Wong and DeSantis, 1997).

We (Ronca and Alberts, 2000b) analyzed vestibular-mediated responses from Day of birth (P0) through P5. The contact righting reflex, composed of stereotyped movements that rotate the body from supine to prone on a solid surface, did not differ in Flight and Control pups. This proved to be important evidence, confirming that Flight pups could perform the coordinated movements involved in righting the body to prone. Such tests of "surface righting" present to the pup abundant tactile and proprioceptive cues, in addition to the vestibular stimulus, when the pup is held on its back. We therefore included tests involving "water immersion righting." In this preparation, the tactile and proprioceptive cues are eliminated and the challenge to the pup becomes much more purely vestibular, presumably testing its perception of linear acceleration—or gravity. Here the pup is rotated to supine, held, and then released just below the surface of a warm waterbath. The pup's buoyancy allows slow descent and its "diving reflex" prevents aspiration of water. This is designed to be a test of vestibular sensitivity of gravity.

Flight newborns were deficient in responding to the vestibular perturbation on P1 and P3 whereas Control pups righted themselves quickly and reliably in the immersion tests. The same Flight pups were demonstrably capable of surface righting, so we were certain that their failures in the immersion test were not due to motor deficits. We believe the Flight pups returned from space relatively insensitive to gravitational cues. Their deficit was profound, but transient, as evidenced by complete response recovery on P5.

Collectively, the emergent picture is enticing. Removal of gravitational cues during the early phases of vestibular development can eliminate sensitivity to gravity. At the same time, sensitivity of angular accelerations may be magnified. We are aware of two, non-mutually exclusive contributions to such a phenomenon. First, the pups' habitat (the uterus) is affected profoundly by the mothers' weightlessness. The fetuses are without forces of linear acceleration (gravity). But, because the mothers are active in space, the fetuses are nonetheless subjected to forces of angular acceleration. Moreover, and importantly, the Flight dams *moved differently* than did the Controls. Their movements included much more rolling, in addition to pitch and saw movements. Thus the fetuses' space must adapt to deprivation of gravity cues and an *enrichment* of inertial cues. We believe we observed evidence of precisely this unique form of adaptation.

The second kind of contribution to the pups' behavioral adaptation was presented on the level of neural growth.

Gravid on the gravity continuum

Gravity is a graded stimulus. Yet, for many years, space flight opportunities have strongly influenced the kinds of data that have been available and thus have led many to focus on comparisons of Earth-normal gravity (1-***g***) and its absence, as approximated by orbital space flight (0-***g*** or micro-***g***). Rigorous understanding of gravity as a biological factor will be attained, however, only after we can describe and explain the form of biological processes, as they are over a range of gravitational forces, ranging from 0-***g*** through fractional ***g*** levels below 1.0 and above it, say to 2-***g***, which happens to encompass the range most pertinent to human and animal space travel.

In recent years, ground-based studies of early mammalian development have been conducted using centrifugation at varying ***g***-loads in excess of 1-***g***. Two common functions can describe many (but not all) of the observed responses to deviations from 1-***g***: some physiological and behavioral systems respond to hypo- and hypergravity in opposite directions, thus providing a somewhat continuous function from 0-***g*** to higher (>1-***g***) levels (Phillips, 2002). A classic example comes from data on antigravity muscles, such as the soleus, that shows a linear increase in volume with increasing ***g***-load (Vasques *et al.*, 1998.). In contrast, there are other cases in which physiological or behavioral systems respond to deviations from 1-***g*** with similar responses, regardless of whether the deviation is above or below 1-***g***. For example, when infant rats are placed in the supine position near the top of a heated water-bath and then released, they tend to rotate their bodies to a proneposition. Neonatal rats that underwent gestation during space flight were impaired in their ability to perform this vestibular-based response postflight, but the response recovered several days later (Ronca and Alberts, 2000b). We recently repeated this experiment at 2-***g*** and observed a similar pattern of compromised responses (Ronca *et al.*, unpublished observations).

Analyses of differing biological responses across a range of gravity vectors will ultimately lead to the establishment of general principles and the development of regression equations that will help us further delineate relationships between effects of hypo- and hypergravity. In this way, we can begin to make initial, limited predictions for specific responses in microgravity from those studied in hypergravity.

Gravid at greater gravities

Oyama and colleagues (Oyama and Platt, 1967; Oyama *et al.*, 1985) conducted some of the initial studies of hypergravity-rearing in rodents. In those studies, young female rats and mice were adapted to either 2.16-***g*** or 3.14-***g*** centrifugation, and then mated. Rats that were impregnated and gave birth during centrifugation were reported to be "less maternal" and neonatal survival was greatly diminished relative to 1-***g*** controls. The period around the time of birth was reported to be highly vulnerable to hypergravity exposure, with extensive neonatal losses occurring during this time. This led Oyama's group to interrupt centrifugation for approximately sixteen hours each day beginning at birth and throughout the first few postnatal days. Offspring survival rates declined precipitously as ***g***-load increased (Oyama and Platt, 1967; Oyama *et al.*, 1985; Baer *et al.*, 2000). It is interesting to note that mice born and reared during centrifugation were somewhat less affected by hypergravity exposure than were rats.

Exposure during pregnancy and birth to even modest increases in ***g***-load (1.5-***g***) exerts immediate effects on dams, and is likely to affect the growth and development pups *in utero*. Initially, body mass declines, stabilizing at about 8–15% less than 1-***g*** controls (Ronca *et al.*, 2000, 2001). Food and water intake (adjusted per 100 g dam body mass) were reduced in hypergravity-exposed dams relative to controls. For the first four days of centrifugation, hypergravity-exposed pregnant dams were approximately 25% less active that controls. Within just a few days, dams begin to show signs of adaptation to centrifugation. After nine days of centrifugation, late pregnant (G20/21 of the rats' 22-day pregnancy) dams in the 1-***g*** condition began to reduce their previous levels of activity, but dams in the 1.5-***g*** condition did not (Ronca *et al.*, 2000). The augmented activity of hypergravity-exposed dams during late pregnancy led us to examine specific behaviors of the dams during this period. Time spent feeding, drinking, and self-grooming was comparable in the *1.5-**g*** and *1-**g*** dams, regardless of circadian cycle. In contrast, the late pregnant hypergravity-exposed dams spent three times more time engaged in nest-building behavior. This latter observation suggested to us that changes in patterns of maternal care might play an important role in neonatal losses during exposure to hypergravity.

We tested the hypothesis that maternal reproductive experience determines neonatal outcome following gestation and birth under hypergravity conditions (Ronca *et al.*, 2001). Primigravid (first pregnancy) and bigravid (second pregnancy) female rats were exposed to 1.5-***g*** centrifugation from G11 throughout birth and the first postnatal week. On the day of birth, litter sizes were identical across gravity and parity conditions although significantly fewer live neonates were observed among hypergravity-reared litters born to primigravid dams as compared to bigravid dams (82% and 94%, respectively; 1.0-*g* controls, 99%).

Within the hypergravity groups, neonatal mortality was comparable across parity conditions from the first to the seventh postnatal day at which time litter sizes stabilized. These results indicate that prior pregnancy and birth can reduce neonatal losses in hypergravity during the first 24 h after birth, but not on subsequent days.

In seeking to explain the hypergravity-related neonatal losses, we analyzed the dams' postpartum maternal behavior. Similar to the results of the space flight studies, there were no observable changes in the mothers' behavior during birth. The behavior of primigravid hypergravity mothers differed from the other conditions in that these dams tended to disrupt nursing bouts and pups within the huddle by frequently digging within the nest and rearranging the pups. This pattern was highly correlated with neonatal mortality ($R^2 = 0.99$, $p < 0.02$).

Our data suggest that during the period around the time of birth, the maternal–offspring system is particularly sensitive to relatively modest (0.5-*g*) increments in the Earth's gravitational field. Maternal reproductive experience appears to be a major determinant of postpartum survival in hypergravity. The newborn rat is profoundly vulnerable to changes in the Earth's gravitational field, in part through changes in the mother's behavior.

The mother as habitat

When the infant mammal emerges from the birth canal it suddenly and irreversibly enters a dramatically different world. Before birth the fetus is influenced indirectly by the mother's behavior. After birth, however, the mother's actions are more specifically directed toward the young and she becomes a central environmental factor, essentially defining the newborn rat's postnatal habitat. As each pup is born, the mother licks it, both cleaning the pup and providing it with tactile stimulation that facilitates its newborn physiology and behavior, including pulmonary respiration (Ronca and Alberts, 1995a,b), thermogenesis (Alberts *et al.*, 1992), and suckling (Ronca *et al.*, 1996; Abel *et al.*, 1998). Without such early postnatal tactile stimulation, survival is compromised.

The mother's body and the nest that she builds, become the buffer between the vulnerable newborn and the outside world. No longer embedded within her body, the altricial infant rat depends on the conductive heat and insulation provided by contact with the dam. During the first few postnatal days, the rat dam spends 60–80% of the time in the nest and in direct physical contact with the pups. Most of this time is spent nursing. Within minutes of birth, the rat fetus metamorphoses into a veritable suckling machine. The dam provides milk during nursing bouts that are punctuated by "letdowns" of milk that occur simultaneously to all 12 nipples, even if there are fewer than 12 suckling pups. These milk letdowns are stimulated by the contraction-inducing effects of oxytocin, which is released by the maternal pituitary prior to each milk ejection (Lincoln *et al.*, 1973). The letdowns occur every 3–5 min and there may be 8–10 letdowns per nursing bout. Mother's milk is the rat pup's sole source of food and fluid for the first 17 days and pups continue to attach to nipples and suckle until day 28 or later (Thiels *et al.*, 1990).

Pups are equipped with a host of specializations for acquiring and utilizing milk as a source of nutrition, water, electrolytes, and immune competence. The shape of the pups' face and the neuromuscular patterns of facial muscle are organized around suckling (Westneat and Hall, 1992). With the mother as habitat and her milk representing the pups' sole substrate, the pup's gastrointestinal system is adapted to utilization of mother's milk. This is particularly evident in the specialized cells of the small intestine which are shaped to hold fat globules, typical of the milk diet, and also of the preponderance of lactase, the enzyme used to digest the major carbohydrate in milk. As can be seen in Fig. 6, lactase levels essentially disappear as weaning progresses and milk is replaced by solid food.

Rat pups can arouse a rat dam with their ultrasonic vocalizations (Farrell and Alberts, 2002a,b). When a dam approaches her litter of pups, it is the tactile stimulation that they deliver to her ventrum that provokes from her the arched "kyphosis" posture above the litter (Stern, 1996), depicted in Fig. 1B. Beginning *in utero*, rat pups learn the identity of chemical cues (Pedersen and Blass, 1981) which, upon birth, arouse them into nipple-searching activities (Teicher and Blass, 1977). The aroused pup moves its head from side-to-side, scanning along the dam's ventrum until it encounters a nipple. Nipple apprehension involves coordinated head and mouth movements and seems to rely on stereotyped "rooting reflexes." Once attached, the pup's mouth forms a tight, salivary seal at the base of the nipple. Suckling begins. Tactile stimulation to the dam's nipples exerts direct excitatory effects on hypothalamic neurons which summate and trigger the pituitary release of oxytocin and the subsequent milk ejection. Pups can detect the intramammary pressure associated with milk letdown and they display a vigorous extensor reflex in response to the letdown cues.

Before and between nursing bouts, the dam licks each pup, focusing her oral activity at the anogenital (AG) region. The licking behavior is highly stereotyped. Typically, the dam grasps a pup with her forepaws and manipulates it into position, which is ventrum up with the infant's head below her chin and its caudal region in front of her mouth and snout, thus exposing the anogenital region. Pups placed in this supine position would otherwise exhibit a vigorous "righting reflex" and rotate themselves to prone (Pellis *et al.*, 1996), but the tactile stimulation to the pup's ventrum inhibits this reflex and the pup remains supine and still, but for a leg-extension response that appears to augment the dam's access to the AG region (Moore, 1992).

The mother's licking triggers the pups' micturation reflex, whereupon it urinates and/or defecates. The infant cannot void spontaneously. Maternal licking clearly provides the infant rat with needed stimulation, helps with hygiene, and serves the infants in other ways (Gubernick and Alberts, 1983). But the dam does not only stimulate the needed release of pup urine; she consumes the urine. In fact, by licking the pups and consuming their urine, the dam recycles about 2/3 of the water in the previous day's milk. This coordinated exchange between the mother and the pup is vital to the dam's water and electrolyte balance (Friedman *et al.*, 1981).

If this weren't enough, there is a sex difference in how much pups are licked. Males receive about three times more licking of their anogenital regions than do their female littermates. Equivalent amounts of licking are received on the rest of the body (Moore, 1992). Male rats provide an androgen-dependent chemical cue from the preputial gland that augments the dam's AG licking. Interestingly, the effect of the AG licking is to masculinize the recipient. Thus, the male rat hormones not only masculinize its brain and body, but the same hormones affect the mother's behavior in a way that delivers additional, masculinizing stimulation to them.

We relate these research stories concerning maternal licking because they emphasize nicely the important point that the pups' adaptations to the mother (and vice versa) create a true, integrated system of dam-and-offspring that is functionally intertwined. When these individuals or the entire system is challenged to function in a novel environment, it is vital to appreciate the multiple levels of integration if one is to interpret accurately the changes that may be observed.

Maternal behavior and offspring adaptation: effects and responses along the gravity continuum

When a broad, contextual cue such as gravity is manipulated, it is often necessary to parse the consequences into two broad kinds: we must distinguish *effects* from *responses*, especially when interpretations of adaptation may be affected. To borrow from an example used by G.C. Williams (1966), a biologist whose writing has been incisive and important in evolutionary theory, consider a flying fish leaping out of the water and splashing back in. If the fish does not return to the water, he noted, it won't survive. Williams argues emphatically, however, that it is wrong to invoke biological adaptation as an explanation to account for the fish's return to the sea. In this case, returning to the water can be explained as a mere "effect" of gravity. No specialization is needed to account for the phenomenon. In contrast, an adaptive "response" is an active, organized biological process, something that has probably been shaped by natural selection. A kidney that enables a fish to return to salt water after living in fresh water is a candidate for an adaptation.

Now, the kinds of phenomena faced by contemporary gravitational biologists who might witness the very first occasion *ever* that a particular species lives in microgravity clearly cannot be considered as direct results of natural selection. Nevertheless, our task is to parse whether a phenomenon that we observe is a mere effect of weightlessness versus an alteration that involves a system that has been shaped or evolved by natural selection in a way that has incorporated gravitational forces. How that system reacts and possibly adjusts to altered gravity during the animal's lifetime (i.e., without invoking a change in genotype) may be its somatic adaptation.

We believe that developmental biologists investigating gravity–related phenomena that involve maternal-offspring systems will face the basic challenge of recognizing active responses and distinguishing them from mere "effects" of altered gravity. This was heralded by some of the pioneering efforts on the Neurolab mission. For the first time, lactating rats and their suckling litters were placed in the microgravity conditions of orbital flight. Many pups did not survive, especially in the groups launched at the youngest ages (Maese and Ostrach, 2002). Does this automatically mean that gravity is essential for the developing system to survive? No, it does not. A more direct and parsimonious explanation (the preferred type in an empirically based understanding) is that some of the effects of weightlessness in the cages used for the flight made it difficult for the mothers and pups to remain together. Absent were some of the "passive" effects of gravity that normally afford specific adaptations such as the lactational system, to transfer milk to pups, who, in turn, are applying adaptive behaviors such as nipple search and sucking responses to ingest and utilize milk. Hardware design is thus paramount in such studies, and every flight (including the so-called failures) will be a source of valuable information with which our science can be honed.

There have been too few opportunities to study mothers and offspring during the important early postnatal period. Consequently, there is little to report. The prospects are bright, however, for the earliest efforts suggest that our mammalian model is so beautifully and robustly organized, that it is capable of adjusting adaptively to microgravity and solving the kinds of passive problems that may arise. We know, for example, that weightless, suckling rat pups, given sufficient support, can attach to nipples and suckle milk. Initially, this was determined by one of us (JRA) in studies conducted on NASA's KC-135 aircraft, used for creating brief periods (approximately 25 s) of weightlessness during parabolic maneuvers. During a sequence of such microgravity episodes, milk-deprived rat pups, aged 5-, 10-, or 15-days were held near the exposed ventrum of their anesthetized mother and allowed to attach to a nipple. Then, by appropriately timing an intra-peritoneal injection of oxytocin, it was possible to induce a letdown of milk. Weightless pups exhibited the typical stretch-reflex to milk letdown (Lincoln *et al.*, 1973) and maintained their oral grasp of the nipple! This observation was extremely informative and encouraging, because it demonstrated that a weightless pup *can* perform its role in the mother–pup suckling sequence.

Video analysis of brief inflight footage of lactating dams and litters taken during the NIH.R3 flight provide evidence that, even in weightlessness, maternal nursing and maternal care (pup retrieving and licking) are retained in space (Daly and Ronca, 2002). This was true despite the observation that, the overall coherence and dynamics of the nest were drastically changed.

Our task, as investigators, will be to create nests that will properly afford the mother the opportunity to care for the young under various gravitational challenges.

Neurolab experiences and observations are consistent with the view that the *R. norvegicus*' mother–litter system is highly robust and capable. Many of the views of the mothers and pups during that flight revealed unconstrained pups and chaotic arrangements of bodies. Nevertheless, the survival of most of the pups under what turned out to be suboptimal conditions is testimony to the species' ability to cope with novel challenges. With additional attention and experience in hardware design, it is likely that appropriately supportive environments can be designed that will enable us to study gravitational *responses* of the mother–offspring system.

Though speculative, we can raise a couple of examples of promising areas involving the postnatal pup living in the maternal habitat that might reveal adaptive effects related to gravity. An appropriate nest environment for microgravity studies involving nursing mothers and their developing offspring is one that will assist (in the absence of Earth-normal gravity) the coherence of the litter group, for the dam normally nurses them as a coherent unit. Within this nest, pups should be able to respond to the episodic appearances of the dam and her maternal behavior, including presentation of her ventrum for nursing. Figure 4 illustrates the head movements made by pups during their "nipple search" behaviors. In Earth-gravity, this behavior is the source of both linear and angular forces to the developing vestibular apparatus. It is probably important that such stimuli will be linked, proprioceptively, to efferent activity of the infant's neck, shoulder and forelimb musculature. These scanning movements bring the pup's sensitive perioral skin in contact with the nipple protuberance, and stimulate the grasp reflex. There follows vigorous suckling and treading, both of which contribute to the release of oxytocin from the dam's pituitary and milk letdown. It will be instructive to observe and measure such movements under various gravitational loads. We will learn about the control and regulation of such adaptive behaviors and, importantly, we will better understand the quantity and qualities of the stimulation sustained by the developing vestibular system. Nipple-shifting, mentioned earlier, will also be important to understand under novel gravitational conditions, for this may help in studies such as hippocampal development and spatial learning (Cramer, 1998).

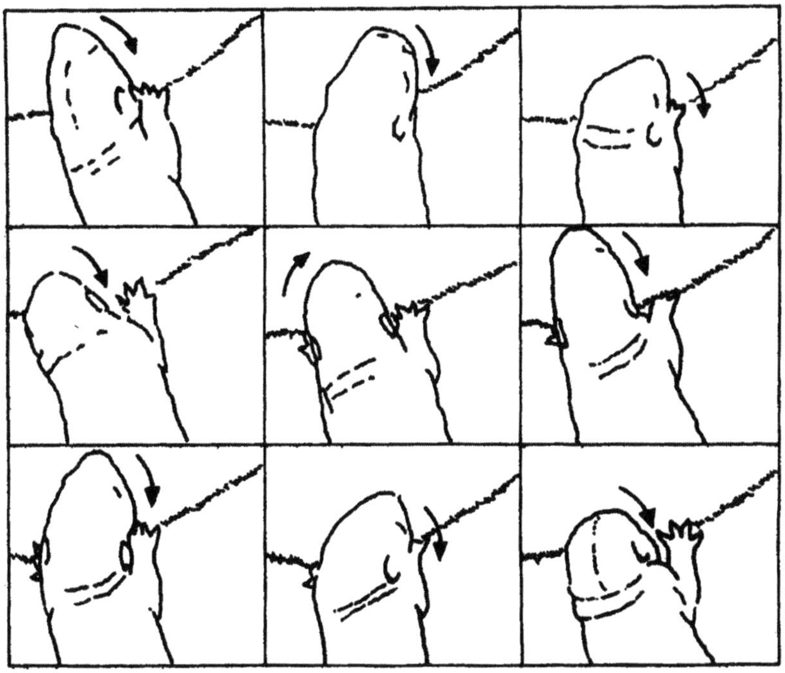

Fig. 4. Scanning movements of a 7-day-old rat pup in the process of locating a nipple of an anesthetized dam. Each panel represents a single frame from a videorecording (Adapted from Pedersen and Blass, 1981).

A related challenge to accurate analysis of the effects of gravity on development is separating effects of gravity on mothers vs. offspring. Because the maternal–fetal system is so highly intertwined in mammals, differentiation among these "direct" and "indirect" effects of gravity on the developing young is a critical, yet extremely difficult task. As we have previously discussed (Alberts and Ronca, 1999), "direct" effects of gravity are those that operate through a primary relation with the recipient organism, tissue or cell. In this case, applications of fractional *g*-loads would be expected to produce fractional differences from the *1-g* phenotype. In contrast, "indirect" effects of gravity are those expressed through avenues of the mammalian system. For example, if mothers cannot stabilize their bodies in the nursing posture or if pups cannot retain metabolic heat because their huddling behavior is disrupted (see below), the altered growth effect would clearly be an indirect consequence of weightlessness.

The huddle as habitat

As the pups develop, the mother begins to make more frequent and longer excursions from the nest. As this occurs there is, in effect, a habitat shift.

Compared to the birth transition, this one is gradual and subtle, but the effect is profound. The huddle of littermates becomes habitat (Fig. 1C) and the pups interact in important and adaptive ways.

There is great thermal significance to the huddle (Alberts, 1978). Though their thermogenic capabilities are limited and their insulation is meager, pups can generate metabolic heat, mostly through activation of brown adipose tissue, a uniquely mammalian specialization. By huddling with their littermates there is less heat-losing surface area exposed to the environment and there is the opportunity for conductive heat exchanges. Pup behavior augments this. A huddle is not a pile of passive bodies. Pups move and exchange positions frequently. In some ways, the huddle behaves like a single organism and displays "group behavioral regulation" whereby the huddle regulates its own surface: mass ratio and hence its thermal properties (Alberts, 1978). The key to this feat is the behavior of individual pups, some of which involves diving down and moving up through the mass of bodies in a nest. Other behavioral factors include movements that bring pups together and maintain contact. Responding to the activity state of other pups is also key to the development of a more integrated, responsive group behavior (Schank and Alberts, 1997). The huddle becomes habitat, exclusive of the dam, for significant portions of each day beginning around day 5. It gains in prominence through about day 15, when longer and more frequent egressions from the nest begin.

In addition to the important thermal benefits of huddling, pups also derive cutaneous and proprioceptive stimuli from the contact behavior. This is probably a significant feature of the huddle as habitat. Rats are considered a "contact species" because even their adult social behavior is dominated by physical contact among conspecifics.

Anticipating the gravitational aspects of the huddle as habitat

Hardware design will again be critical to the science. Huddling should be enabled, but not necessarily forced. Ideally, the environment will allow self-regulation by individuals and the group.

It is possible, based on observations such as those made during parabolic flight studies, that weightless pups will react to weightlessness by becoming even more responsive to tactile cues than they are on Earth. That is, during the "0-***g***" phases of parabolic flight, pups tended to grab and tenaciously hold onto grids, wires, each other, even their own tail (Alberts, unpublished observations). If such tendencies persisted during continuous exposures to microgravity, the pups would create a novel tactile environment for themselves and this would require incorporation into the subsequent interpretation of outcomes.

Figure 5 is from a study (conducted at 1-***g***) of pup movements in a huddle that was contained in a bowl-shaped nest. Gravity was at work, of course. But so were the pups. The arrows superimposed on the drawings depict

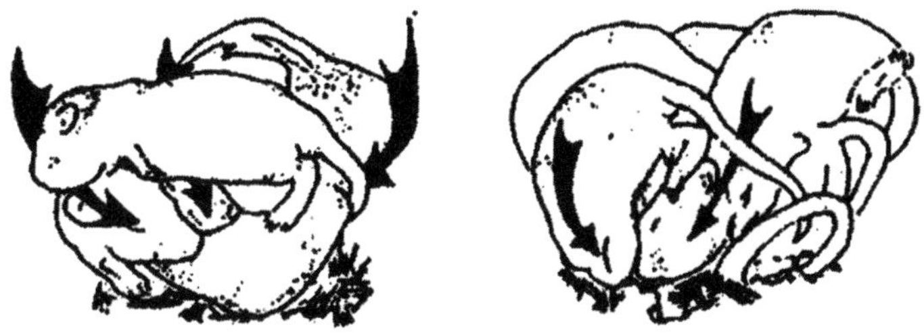

Fig. 5. Drawings of a huddle of 10-day-old rat pups in a concave nest. The arrows depict the direction of "pup flow" when individuals burrow and dive into the group.

the movements of the pups as they burrowed *down* into the huddle. Their individual movements in the huddle create a phenomenon termed "pup flow" and this changes systematically with changes in the micro-environment. Under the conditions used in this study, the pups' behavior was motivated by thermal conditions, for when the nest temperature was increased, the warmed pups reversed directions and pup flow was *up*. How will this be executed in different gravities? How will the resulting self-stimulation of the musculature, proprioception, and the vestibular system be expressed later in life? Such questions, we think will be key to developmental insights emerging from this habitat.

Some of the pups in Neurolab studies flew during the developmental phase that we are discussing here. The conditions experienced by the Neurolab pups launched on postnatal day 8 might have potential associations with a number of postnatal impairments or delays. For example, undernourishment in early life causes significant delays in physical growth of the young and long-lasting morphological changes, particularly in brain areas undergoing cell proliferation (e.g., Fish and Winick, 1969). Reduced neocortical dendrogenesis, number of dendritic spines, and reduced synapse-to-neuron ratio have been reported following neonatal malnourishment. These changes are frequently correlated with reflexive and long-term behavioral abnormalities, including aberrant locomotor patterns (see Ronca, 2003 for further discussion).

In an effort to identify critical periods in peripheral vestibular system development, plastic changes in the organization of the vestibular efferent network of the rat utricle were studied in the postnatal rats launched on day 9 of the Neurolab flight (Dememes *et al.*, 2001). Immunofluorescence experiments were performed with a specific biochemical marker of the efferent system, the calcitonin gene-related peptide (CGRP). The utricles were analyzed by confocal microscopy. Maturation of the vestibular efferent system was similar in space flight and control rats. The absence of changes in the development of vestibular efferents is encouraging in view of the dramatically reduced body weights of the pups studied.

The coterie as habitat

Though the nest diminishes as the site for behavioral interactions beginning around day 15–19, the pups remain highly social and interactive with each other and with the dam, as well as with other conspecifics in a colony or colony-like setting. The pups are on the verge of weaning at this stage, i.e., they will begin to eat and drink independently and gradually diminish suckling and milk intake. At this stage, interactions with littermates continue but they occur outside the nest. In a real sense, the social cohort or coterie becomes habitat. Huddling, mutual grooming, and play fighting are observed in a variety of situations. Feeding is a social behavior, and much is known about the onset of food intake and the formation of dietary preferences and other socially mediated choices (Galef, 1981).

Just as we noted that the suckling pup was exquisitely equipped to utilize milk as diet, the weanling demonstrates corresponding adaptations to a diet of solid food, usually higher in carbohydrates and protein than was mother's milk. Figure 6 depicts some of the dramatic changes that enable the weanling to leave milk, eat food, and maintain a trajectory of growth and development. We present this graph as a representation of the multileveled adaptive changes that the developing pup displays. These changes in digestive enzymes are beautifully coordinated with changes in morphology, sensory function, and behavior (e.g., Alberts, 1994).

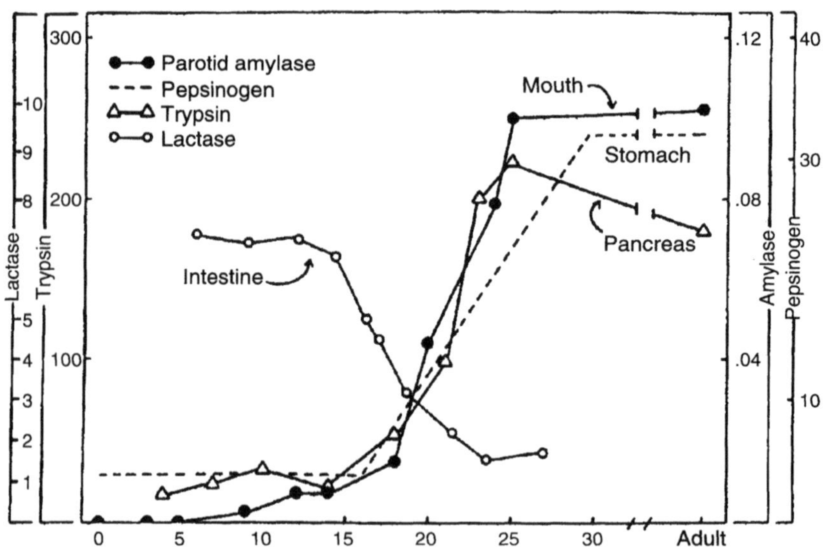

Fig. 6. An overview of selected physiological developments that enable adaptation to suckling (intestinal lactase) and to the utilization of solid food, as measured in rat intestine, mouth, stomach, and pancreas (from Alberts, 1994).

During this phase of life, pups are establishing a more independent existence, though they remain highly social and contactful, at least on Earth. Sexual maturation is proceeding and, in the course of such changes, sexual differences begin to emerge. These are seen in physical developments as well as in behaviors, including grooming and play.

The coterie and gravity

Again, the kinds of habitats in which animals develop are likely to contribute to the outcomes seen. A deep understanding of the many dimensions of adaptation to the social and physical habitat during this phase will be important in our ability to understand gravitational adaptations.

Bone and muscle develop importantly during this phase of growth and maturation. What the animals do in different gravitational fields, as well as how much they do it, will be key to data interpretation.

It is especially enticing to take note of a number of findings from the Neurolab mission derived form pups flown in the AEMs at ages that involved the coterie as habitat. A sample of these are noted herein. It is valuable to keep in mind the benefits that will be derived from having at hand a full view of what these animals experienced in the coterie, as a basis of understanding the outcomes.

The myosin heavy chain (MHC) molecule is the principal structure and regulatory protein that serves as the molecular motor to control the intrinsic contractile properties of muscle (Adams *et al.*, 2000). The separate and combined effects of space flight and thyroid deficiency on myosin heavy chain (MHC) gene expression (protein and mRNA) were examined in muscles of the Neurolab pups. It was found that space flight markedly reduced the expression of the slow, type I MHC gene by approximately 55%, whereas expression of the fast IIx and IIb MHCs in antigravity skeletal muscles was enhanced. In fast muscles, space flight caused subtle increases in the fast IIb MHC relative to the other adult MHCs.

Central pathways for oxytocin and vasopressin have been implicated in the neurobiology of anxiety and social behaviors. Garcia-Ovejero *et al.* (2001) studied oxytocinergic and vasopressinergic magnocellular hypothalamic neurons of prepubertal rats of the 15-day-old rat pups flown on the Neurolab mission either at landing or 18 weeks postlanding. Enhanced transcriptional and biosynthetic activity was observed in magnocellular supraoptic neurons of flight animals on the day of landing as compared to control rats, including increased c-Fos expression, enlarged nucleoli and cytoplasm, and increased volume in the neuronal perikaryon of mitochondriae, endoplasmic reticulum, Golgi apparatus, lysosomes, and cytoplasmic inclusions (nematosomes). Vasopressin levels, cytoplasmic volume and c-Fos expression returned to control levels by 18 weeks after landing, whereas other changes did not normalize. Together, the results of this study suggest that space flight during the preweanling period may induce irreversible modifications in the regulation of oxytocinergic neurons. Central pathways for oxytocin and vasopressin have been implicated in the neurobiology of anxiety and social behaviors.

A study of the postnatal day 15 pups flown on the Neurolab mission suggests that development in microgravity leads to a constellation of interesting changes in the number and morphology of cortical synapses and does so in a laminar-specific manner (DeFelipe *et al.*, 2002). In the layers II/III and Va, the synaptic cross-sectional lengths were significantly larger in flight animals than in ground control animals. Flight animals also showed significantly lower synaptic densities in layers II/III, IV, and Va. The greatest difference was found in layer II/III, where there was a difference of 344 million synapses per mm^3 (15.6% decrease). After a four-month period of re-adaptation, some changes disappeared (i.e., the alterations were transient), while conversely, some new differences also appeared. For example, significant differences in synaptic density in layers II/III and Va after re-adaptation were no longer observed, whereas in layer IV the density of synapses increased notably in flight animals (a difference of 185 million synapses per mm^3 or 13.4%). In addition, all the changes were in asymmetrical synapses, known to be excitatory. These results suggest that gravity may be an important environmental parameter for normal cortical synaptogenesis.

Temple *et al.* (2002) studied spatial learning in the postnatal day 9 or 15 Neurolab pups using several different tasks. Performance and search strategies evaluated for each task revealed remarkably few differences between the flight groups and their Earth-bound controls at both ages. Together, these data suggest that development in an environment without gravity has minimal long-term impact on spatial learning and memory abilities. Nevertheless, we must more fully understand the ways and rates at which adaptation to micro-***g*** and re-adaptation to 1-***g*** occur before we can fully interpret such findings.

Developmental biology as a core discipline for space biology and biomedicine

Throughout the present chapter we have noted, even celebrated, patterns of development that are unique to and universal among mammals. From the standpoint of evolutionary biology, such commonality of process represents the common ancestry shared by contemporary mammalian species. It reinforces the unity and continuity among mammalian forms and is the foundation for cross-species generalizations and detailed comparisons of differences. From the standpoint of fundamental biology and biomedicine, because mammals share so many biological processes, species such as the Norway rat are valid models of human function and disease. This been explored and validated in ground-based science; the principles apply equally well to contemporary concerns in the space life sciences.

How is rodent development pertinent to space biology and space biomedicine? We believe that rigorous study of the gravitational biology of vertebrate development is a valuable pursuit, directly and perhaps uniquely relevant to some of the most pressing issues of astronaut well being in an era of extended missions and possible space exploration.

First, *adaptation* is the most general and pervasive question regarding human health during long-term missions. Fundamentally, as we have shown, *development is adaptation.* Each insight gained into the roles of gravity in development is an increment of understanding better the grand issue of adaptation to the space environment. Among potential advantages of studying space adaptation via development rather than adaptation in adults, are rapidity and magnification. To be accurate, studies of adults over time are also developmental studies, for there really is no fixed or "final" adult stage. Nevertheless, changes during adulthood typically proceed more slowly and subtly than during phases of rapid growth and reorganization. Generally, during phases of rapid growth and reorganization, such as the postnatal phase from birth to weaning in mammals, the changes are rapid and dramatic. The underlying processes in this phase of life are the same as those at other phases, but they occur rapidly and are often expressed more dramatically (magnification). We have endeavored to show how the somatic plasticity evident during development can be understood as adaptive change. Many of these processes are gravity sensitive. Thus, we can conclude that developmental preparations are relevant to and efficient for space studies.

Another way in which some investigators use developmental analysis is as a window into understanding how a system is "built" or assembled during maturation. This approach can be used to gain insight into how complex systems are organized and how it may be that some components are differentially sensitive to perturbation. Again, this is an adaptive issue and developmental analyses can be the key that unlocks the passageway to deeper understanding.

With such considerations in mind, we see a fruitful and profitable future for space research that includes developmental analysis in its agenda. To the extent that such studies have been supported in the past, the results have been rewarding and varied. Knowledge and expertise is accumulating and the pace at which new knowledge can be generated will increase further with continued support.

Acknowledgment

Preparation of this chapter and the original research was supported by grant MH-28355 from the National Institute of Mental Health to Jeffrey R. Alberts, grant NCC 2-870 from the National Aeronautics and Space Administration to Jeffrey R. Alberts and April E. Ronca, grant MH-46485 from the National Institute of Mental Health to April E. Ronca and Jeffrey R. Alberts, and Grant 121-1040 from the National Aeronautics and Space Administration to April E. Ronca.

References

Abel, R.A., Ronca, A.E. and Alberts, J.R. (1998) Perinatal stimulation facilitates suckling onset in newborn rats. Developmental Psychobiology 32, 91–99.

Adams, G.R., Haddad, F., McCue, S.A., Bodell, P.W., Zeng, M., Qin, L., Qin, A.X. and Baldwin, K.M. (2000) Effects of space flight and thyroid deficiency on rat hindlimb development. II. Expression of MHC isoforms. Journal of Applied Physiology 88, 904–916.

Alberts, J.R. (1978) Huddling by rat pups: Group behavioral mechanisms of temperature regulation and energy conservation. Journal of Comparative and Physiological Psychology 92, 231–245.

Alberts, J.R. (1984) Sensory-perceptual development in the Norway rat: A view toward comparative studies. *In* Kail, R. and Spear, N. (eds.). *Comparative Perspectives On Memory Development*. pp. 65–101. Plenum Press, New York.

Alberts, J.R. (1994) Learning as adaptation of the infant. Acta. Pediatrica, Suppl. 397, 77–85.

Alberts, J.R. and Cramer, C.P. (1988) Ecology and experience: Sources of means and meaning of developmental change. *In* Blass, E.M. (ed.). *Handbook of Behavioral Neurobiology: Behavioral Ecology and Developmental Psychobiology*. Vol. 9, pp. 1–39. Plenum Press, New York.

Alberts, J.R. and Gubernick, D.J. (1985) Early learning as ontogenetic adaptation for ingestion by rats. Learning and Motivation 15, 334–359.

Alberts, J.R. and Ronca, A.E. (1996) Fetal behavior in developmental psychobiology. Developmental Psychobiology 29, 185–190.

Alberts, J.R. and Ronca, A.E. (1997) Rat pregnancy and parturition survive microgravity challenge: New considerations of developmental consequences. Journal of Gravitational Physiology 4, P55–P58.

Alberts, J.R. and Ronca, A.E. (1999) Effects of gravity on development: The importance of behavior. Journal of Gravitational Physiology 6, P5–P8.

Alberts, J.R., Ronca, A.E. and Blumberg, M.S. (1992) Thermal imaging of postpartum rat pups upon delivery into the nest. Proceedings of the International Society for Developmental Psychobiology, Newport Beach, CA.

Alberts, J.R., Serova, L.V., Keefe, J.R. and Apanasenko, Z. (1985) Early postnatal development of rats gestated during flight of Cosmos 1514. The Physiologist 28, 81–82.

Bradley, R.M. and Mistretta, C.M. (1975) Fetal sensory receptors. Physiological Reviews 55, 352–382.

Baer, L.A., Ronca, A.E. and Wade, C.M. (2000) Survival and growth of developing rats during centrifugation at 2-g. Journal of Gravitational Physiology 7, 17–22.

Burden, H.W., Zary, J.T., Lawrence, I.E., Jonnalagadda, P., Davis, M. and Hodson, C.A. (1997) Effects of space flight on ovarian-hypophyseal function in postpartum rats. Journal of Reproducitive Fertility 109, 193–197.

Cramer, C.P. (1998) Experience during suckling increases weight and volume of rat hippocampus. Brain Research 470, 151–155.

Daly, M.E. and Ronca, A.E. (2002) Lactating rats retain nursing behavior and maternal care in space. Developmental Psychobiology 39, 75.

Davet, J., Fagette, S., Mani-Ponset, L., Bayard, B., Dumars, P., Reiss-Bubenheim, D., Guell, A., Gharib, C. and Gabrion, J. (1999) Cardiac atrial natriuretic peptide (ANP) in rat dams and fetuses developed in space (NIH-R1 and NIH-R2 experiments). Life Sciences 64, 1533–1541.

Davis, B.A., Sipe, B., Gershan, L.A., Fiacco, G.J., Lorenz, T.C., Jeffrey, J.J. and Partridge, N.C. (1998) Collagenase and tissue plasminogen activator production in developing rat calvariae: normal progression despite fetal exposure to microgravity. Calcified Tissue International 63, 416–422.

Dawes, G.S. (1968) Fetal and Neonatal Physiology. Year Book Medical Publishers, Chicago.

Dawes, G.S. (1974) Breathing before birth in animals and man. New England Journal of Medicine 290, 557–559.

Decasper, A.J. and Fifer, W.P. (1980) Of human bonding: Newborns prefer their mothers' voices. Science 2083, 1174–1176.

DeFelipe, J., Arellano, J.I., Merchan-Perez, A., Gonzalez-Albo, M.C., Walton, K. and Llinas, R. (2002) Space flight induces changes in the synaptic circuitry of the postnatal developing neocortex. Cerebral Cortex 12, 83–91.

Dememes, D., Dechesne, C.J., Venteo, S., Gaven, F. and Raymond, J. (2001) Development of the rat efferent vestibular system on the ground and in microgravity. Developmental Brain Research 128, 35–44.

Dijk, D.J., Neri, D.F., Wyatt, J.K., Ronda, J.M., Riel, E., Ritz-De Cecco, A., Hughes, R.J., Elliott, A.R., Prisk, G.K., West, J.B. and Czeisler, C.A. (2001) Sleep, performance, circadian rhythms, and light-dark cycles during two space shuttle flights. American Journal of Physiology: Regulatory, Integrative and Comparative Physiology 281, 1647–1664.

Drachman, D.B. and Sokolov, L. (1966) The role of movement in embryonic joint development. Developmental Biology 14, 401–420.

Farrell, W.J. and Alberts, J.R. (2002a) Maternal responsiveness to infant rat ultrasonic vocalizations during the maternal behavior cycle and following steroid and experiential induction regiments. Journal of Comparative Psychology 116, 286–296.

Farrell, W.J. and Alberts, J.R. (2002b) Stimulus control of maternal responsiveness to rat pup ultrasonic vocalizations. Journal of Comparative Psychology 116, 297–307.

Fifer, W.P. and Moon, C. (1988) Auditory experience in the fetus. *In* Smotherman, W.P. and Robinson, S.R. (eds.). *Behavior of the Fetus*. Telford Press, New Jersey.

Fish, I. and Winick, M. (1969) Effect of malnutrition on regional growth of the developing rat brain. Experimental Neurology 25, 534–540.

Friedman, M.I., Bruno, J.P. and Alberts, J.R. (1981) Physiological and behavioral consequences of water recycling during lactation. Journal of Comparative and Physiological Psychology 95, 26–35.

Galef, B.G., Jr. (1981) The ecology of weaning: parasitism and the achievement of independence by altricial mammals. *In* Gubernick, D.J. and Klopfer, P.H. (eds.). *Parental Care in Mammals*. Plenum Press, New York.
Garcia-Ovejero, D., Trejo, J.L., Ciriza, I., Walton, K.D. and Garcia-Segura, L.M. (2001) Space flight affects magnocellular supraoptic neurons of young prepuberal rats: transient and permanent effects. Developmental Brain Research 130, 191–205.
Gottlieb, G. (1971) Ontogenesis of sensory function in birds and mammals. *In* Tobach, E., Aronson, L.R. and Shaw, E. (eds.). *The Biopsychology of Development*. pp. 67–128. Academic Press, New York.
Gould, S.J. (1977) Ontogeny and Phylogeny. Belknap Press, Cambridge.
Gubernick, D.J. and Alberts, J.R. (1983) Maternal licking of young: Resource exchange and proximate controls. Physiology and Behavior 31, 593–601.
Hoban-Higgins, T.M., Murakami, D.M., Tang, I.H., Fuller, P.M. and Fuller, C.A. (1999) Development of circadian rhythms in rat pups exposed to microgravity during gestation. Journal of Gravitational Physiology 6, 71–79.
Hofer, M.A. (1981) *The roots of human behavior: An introduction to the psychobiology of early human development*. W.H. Freeman, San Francisco.
Hubel, D.H. and Wiesel, T.N. (1998) Early exploration of the visual cortex. Neuron 20, 401–412.
Keefe, J.R. (1985) Final Report of the NASA Mammalian Developmental Biology Working group. *In* Souza, K.A. and Halstead, T.W. (eds.). *NASA Developmental Biology Workshop*. pp. 46–63. NASA Technical Memorandum-86756, Washington D.C.
Kenny, P.A. and Turkewitz, G. (1986) Effects of unusually early visual stimulation on the development of homing behavior in the rat pup. Developmental Psychobiology 19, 57–66.
Lickliter, R. (1993) Timing and the development of perinatal perceptual organization. *In* Turkewitz, G. and Devenny, D.A. (eds.). *Developmental time and timing*. pp. 105–123. Lawrence Erlbaum, Hillsdale, NJ.
Lincoln, D.W., Hill, A. and Wakerly, J.B. (1973) The milk-injection reflex of the rat; An intermittent function not abolished by surgical levels of anesthesia. Journal of Endocrinology 57, 459–476.
Liggins, G.C. (1982) The fetus and birth. *In* Austin, C.R. and Short, R.V. (eds.). *Reproduction in Mammals*. Cambridge University Press, New York.
Lott and Dale, F (1973) Parental behavior. *In Perspectives on animal behavior*. Scott, Foresman, and Co..
Maese, A.C. and Ostrach, L.H. (2002) Neurolab Final Report. NASA Technical Memorandum-0073543. Washington D.C.
Mani-Ponset, L., Masseguin, C., Davet, J., Herbute, S., Maurel, D., Ghandour, M.S., Reiss-Bubenheim, D., Guell, A. and Gabrion, J. (1997) Effects of an 11-day space flight on the choroid plexus of developing rats. Developmental Brain Research 99, 187–200.

Moessinger, A.C. (1983) Fetal akinesia deformation sequence: An animal model. Pediatrics 72, 857–863.

Moore, C.L. (1992) The role of maternal stimulation in the development of sexual behavior and its neural basis. Annals of the New York Academy of Sciences 662, 160–177.

Murakami, D.M., Hoban-Higgins, T.M., Tang, I.-H. and Fuller, C.A. (1997) The effect of spaceflight on retino-hypothalamic tract development. Journal of Gravitational Physiology 4, 67–70.

Narayanan, C.H., Fox, M.W. and Hamburger, V. (1971) Prenatal development of spontaneous and evoked activity in the rat (Rattus norvegicus albinus). Behaviour 40, 100–134.

Oliver, R.E. (1981) Of labor and the lungs. Archives of Diseases in Childhood 56, 59–62.

Oppenheim, R.W. (1981) Ontogenetic adaptations and retrogressive processes in the development of the nervous system and behavior: A neuroembryological perspective. *In* Connelly, K. and Prechtl, H. (eds.). *Maturation and development: Biological and psychological perspectives*. Spastica Society Publications, London.

Oyama, J. and Platt, W.T. (1967) Reproduction and growth of mice and rats under conditions of simulated increased gravity. American Journal of Physiology 212, 164–166.

Oyama, J., Solgaard, L., Corrales, J. and Monson, C.B. (1985) Growth and development of mice and rats conceived and reared at different g-intensities during chronic centrifugation. The Physiologist 28, 83–84.

Pedersen, P.E. and Blass, E.M. (1981) Olfactory control over suckling in albino rats. *In* Aslin, R.N., Alberts, J.R. and Petersen, M.R. (eds.). *Development of perception: Psychobiological perspectives*. Vol. 1. Academic Press, New York.

Pellis, V.C., Pellis, S.M. and Teitelbaum, P. (1996) A descriptive analysis of the postnatal development of contact righting in rats (Rattus norvegicus). Developmental Psychobiology 24, 237–267.

Phillips, R. (2002) Gravity: It's the law. Journal of Gravitational Physiology 9, P15–16.

Pittendrigh, C.S. (1958) Adaptation, natural selection, and behavior. *In* Roe, A. and Simpson, C.G. (eds.). *Behavior and evolution*. pp. 390–401. Yale University Press, New Haven.

Raff, R.A. (1996) *The Shape of Life*. The University of Chicago Press, Chicago.

Reynolds, G.D. and Lickliter, R. (2004a) Effects of redundant and nonredundant bimodal sensory stimulation on heart rate in bobwhite quail embryos. Developmental Psychobiology 43, 304–310.

Reynolds, G.D. and Lickliter, R. (2004b) Modified prenatal sensory stimulation influences postnatal behavioral and perceptual responsiveness in bobwhite quail chicks (Colinus virginianus). Journal of Comparative Psychology 118, 172–178.

Ronca, A.E. (2003) Mammalian development in space. *In* Marthy, H.J. (ed.). *Advances in Space Biology and Medicine*. Vol. 9. Development in Space. Elsevier, Netherlands.
Ronca, A.E., Abel, R.A. and Alberts, J.R. (1996) Perinatal stimulation and adaptation of the neonate. Acta Pediatrica 415, 8–15.
Ronca, A.E. and Alberts, J.R. (1990) Heart rate development and sensory-evoked cardiac responses in perinatal rats. Physiology and Behavior 47, 1075–1082.
Ronca, A.E. and Alberts, J.R. (1994) Sensory factors associated with gestation and parturition evoke cardiac and behavioral responses in fetal rats. Psychobiology 55, 270–282.
Ronca, A.E. and Alberts, J.R. (1995a) Simulated uterine contractions facilitate fetal and newborn respiratory behavior in rats. Physiology and Behavior 58, 1035–1041.
Ronca, A.E. and Alberts, J.R. (1995b) Cutaneous induction of breathing in perinatal rats. Psychobiology 23, 261–269.
Ronca, A.E. and Alberts, J.R. (2000a) Physiology of a Microgravity Environment Selected Contribution: Effects of spaceflight during pregnancy on labor and birth at 1 G. Journal of Applied Physiology 89, 849–854.
Ronca, A.E. and Alberts, J.R. (2000b) Effects of prenatal spaceflight on vestibular responses in neonatal rats. Journal of Applied Physiology 89, 2318–2324.
Ronca, A.E., Baer, L.A., Daunton, N.G. and Wade, C.M. (2001) Maternal reproductive experience enhances birth outcome and early postnatal survival following gestation and birth of rats in hypergravity. Biology of Reproduction 65, 805–813.
Ronca, A.E., Baer, L.A., Mills, N.A., Sajdel-Sulkowska, E.M. and Wade, C.E. (2000) Body mass, food and water intake, and activity of pregnant and lactating rat dams during 1.5-g centrifugation. Journal of Gravitational Physiology 7, P131–132.
Ronca, A.E., Fritsch, B., Bruce, L. and Alberts, J.R. (2000) Invited review: The effects of microgravity on vestibular development of function in rats: Genetics and environment. Korean Journal of Biological Science 4, 215–221.
Ronca, A.E., Lamkin, C.A. and Alberts, J.R. (1993) Maternal contributions to sensory experience in the fetal and newborn rat (Rattus norvegicus). Journal of Comparative Psycholology 107, 61–74.
Schank, J.C. and Alberts, J.R. (1997) Self-organized huddles of rat pups modeled by simple rules of individual behavior. Journal of Theoretical Biology 189, 11–25.
Serova, L.V. (1993) Changing gravity level and the development of animals. The Physiologist 36, S31–3.
Serova, L.V., Denisova, L.A., Lavrova, E.A., Makeyeva, V.F., Natochin, Yu, V., Pustynnikova, A.M. and Shakhmatova, E.I. (1993) Parameters of the reproductive function of the animals: Fetal and placental characteristics. *In* Gazenko, O.G. (ed.). *Ontogenesis of Mammals in Microgravity*. pp. 35–36. NASA Technical Memorandum-103978, Washington D.C.

Sonnenfeld, G., Foster, M., Morton, D., Bailliard, F., Fowler, N.A., Hakenewerth, A.M., Bates, R. and Miller, E.S., Jr. (1998) Spaceflight and development of immune responses. Journal of Applied Physiology 85, 1429–1433.
Stern, J.M., (1996) Somatosensation and maternal care in Norway rats. In: Rosenblatt, J.S., Snowden, C.T. (eds.). Parental Care: Evolution, Mechanisms and Adaptive Significance. Advances in the Study of behavior, Vol 25..
Teicher, M.H. and Blass, E.M. (1977) First suckling response of the newborn albino rat: The roles of olfaction and amniotic fluid. Science 19, 635–636.
Temple, M.D., Kosik, K.S. and Steward, O. (2002) Spatial learning and memory is preserved in rats after early development in a microgravity environment. Neurobiology of Learning and Memory 78, 199–216.
Thiels, E., Alberts, J.R. and Cramer, C.P. (1990) Weaning in rats: II. Pup behavior patterns. Developmental Psychobiology 23, 495–510.
Uexkull, J. Von (1909) *Umwelt and Innenwelt der Tiere*. Springer-Verlag, Berlin.
Vasques, M., Lang, C., Grindeland, R.E., Roy, R.R., Daunton, N., Bigbee, A.J. and Wade, C.E. (1998) Comparison of hyper- and microgravity on rat muscle, organ weights and selected plasma constituents. Aviation, Space and Environmental Medicine 69, A2–8.
West, M.J., King, A.P. and Arberg, A.A. (1988) The Inheritance of Niches. *In* Blass, E.M. (ed.). *Handbook of Behavioral Neurobiology*, Vol. 9., Behavioral Ecology and Developmental Psychobiology. Plenum Press, New York.
Westneat, M.W. and Hall, W.G. (1992) Ontogeny of feeding motor patterns in infant rats: an electromyographic analysis of suckling and chewing. Behavioral Neuroscience 106, 539–554.
Williams, G.C. (1966) Adaptation and natural selection. Princeton University press, Princeton.
Wong, A.M. and DeSantis, M. (1997) Rat gestation during space flight: outcomes for dams and their offspring born after return to Earth. Integrative Physiological and Behavioral Sciences 32, 322–342.
Yeh, K.-Y. and Moog, F. (1974) Intestinal lactase activity in the suckling rat: Influence of hypophysectomy and thyroidectomy. Science 182, 77–99.

Experimentation with Animal Models in Space
G. Sonnenfeld (editor)
2005. Published by Elsevier B.V.
DOI: 10.1016/S1569-2574(05)10008-2

Use of Animal Models to Study Skeletal Effects of Space Flight

Stephen B. Doty[1], Laurence Vico[2], Thomas Wronski[3] and Emily Morey-Holton[4]

[1]*Hospital for Special Surgery, USA*
[2]*INSERM E0366, Paris, France*
[3]*University Florida, Gainsville, FL, USA*
[4]*NASA Ames Research Center, Moffett Field, CA, USA*

Introduction

Space flight results in many physiological changes in humans and animals which includes cardiovascular, immunological, muscular, hormonal, and skeletal effects. In one of the earliest publications on the subject, Morey and Baylink (1978) showed that following 19.5 days in space, the skeleton responded to space flight by a reduction in bone formation. This study involved a Russian biosatellite (Cosmos 782) carrying adult male Wistar rats which had tetracycline as a fluorescent label in their skeletal tissues. The factors which made this experiment so meaningful were: the small size of the experimental animal, the *in vivo* aspects of this study of bone formation rate, the ability of tetracycline to label newly formed bone, and the opportunity to sacrifice animals at the end of the flight and at specific times post-flight. All of these factors are critical to accurately determine how and where the skeletal effects are experienced during space flight or during non-weight-bearing conditions. Eventually, the objective behind the use of animals for space flight studies is to determine the mechanism(s) behind this reduction of bone mass and eventually to prevent this loss from occurring. Animals rather than humans have been the subject of choice because the number of animals which can be flown are so much greater than crew size. The animals most commonly used are rats so that genetic variation is greatly reduced and skeletal response better understood, and of course examination of an entire animal skeleton is much more informative than a single human bone biopsy, even if a biopsy were available. As we will see from several studies presented below, the bone response to space flight is not uniform throughout the skeleton.

However, even with these advantages of using animals as models for space flight studies, most of the actual flights are not very long in time compared to the animal age, thus significant skeletal changes have sometimes been difficult to document. According to Tipton (2003) the Bion/Cosmos biosatellite series flew for variable time periods, ranging from 5.0–21.5 days. The US shuttle flights were limited to a maximum of 16–17 days and often were considerably shorter. The NASA/MIR flights were of longer duration but they studied calcium and bone changes in the flight crew, and no animals models were in use. Van Loon *et al.* (1996) indicated that the longest flights were in the 1973 Skylab series, which went out to 84 days. However, these skeletal studies were limited to the crew and no animal models were in use at the time. From this information it is apparent that many different approaches for the study of bone loss in space flight have been taken and different outcomes have been measured. Only in the case where animal studies have been carried out is there any attempt to discern mechanisms behind the bone loss.

In this chapter, we have arranged the results from *in vivo* animal model studies according to different areas of interest relating to skeletal physiology. There are no human studies included, nor any cell culture studies since the use of animal models is the subject of this review. There are no published studies of embryonic bone development in space which would indicate that space flight has any significant effect on embryonic development, so this area will be left for future reviews. In general, rats have been the animal of choice, although some data exists from monkeys, which will be presented. Any data collected and noted as coming from a flight group will usually be compared to Vivarium Controls (animals housed under standard conditions but not subjected to any factors associated with launch, re-entry or other space flight conditions) and Synchronous Controls (housed similarly to flight animals and exposed to launch and re-entry forces, vibrations, and any anomalies during flight which can be reproduced on the ground). In many cases, a Basal group is sacrificed at the time of launch so that a baseline of information will be collected prior to any space flight related activities.

Reduction in bone quantity due to a change in bone formation

Morey and Baylink (1978) (Cosmos 782) showed by microscopy that there was an arrest line present along the endosteum and periosteum in male rats following a 19.5 day space flight, suggesting that bone formation had ceased. The increased extent of the arrest lines in these animals was an indication of complete cessation of periosteal bone formation during space flight but results also indicated that the formation process could return to normal during the postflight period. Spengler *et al.* (1979, 1983) and Turner *et al.* (1979, 1985) (Cosmos 936) found a 45% decrease in periosteal new bone formation which was not corrected by centrifugation during flight. The in-flight centrifugation also did not improve the mechanical properties of the flight bone. Their results did show that the centrifuged rats began correcting their bone loss during recovery at 1-*g* (Earth's gravity) at a faster rate than the non-centrifuged flight animals. Wronski and Morey (1983a, b) and Wronski *et al.* (1980) (Cosmos 1129) in a study of rats flown for 18.5 days found that the tibia contained an increase in marrow fat, a decreased periosteal bone formation and evidence of arrest lines. These various studies which showed as much as 40–45% decrease in periosteal bone formation did not demonstrate any affect on cortical bone resorption (shown by marrow cavity size). The rib bones, which are non-weight-bearing did not reflect any significant bone formation change compared to the weight bearing tibia.

In Cosmos 1667, Vico *et al.* (1988) showed in Wistar rats after 7 days space flight that there was reduced trabecular volume and density in the tibia metaphysis but not in the femur metaphysis. There was also a reduced osteoid surface in the tibia, measured by morphometry, but not a significant change in the greater trochanter of the femur, in an area containing Sharpey's fibres (i.e., an attachment site). This was one of the earliest studies to demonstrate that the reduced bone formation due to space flight was somewhat site dependent, at least for relatively short duration space flights. This same study concluded there was no change in osteoclast numbers in either tibia or femur in flight animals. There are several studies, exemplified by Wronski *et al.* (1987) which have shown that a space flight of 7 days or less is often not long enough to produce recognisable skeletal changes in rats using morphometric techniques.

Jee *et al.* (1983) (Cosmos 1129) showed after 18.5 days of space flight that there was a reduced bone mass in the metaphysis and an associated increase in marrow fat. This increase in marrow fat has been seen in osteoporotic conditions and as a result of increased corticoids. The cause of this marrow response is unknown. This was a morphometric study using male Wistar rats.

In general, static indices of bone formation such as osteoblast numbers or osteoid surface measurements were found to be decreased in flight rats (Vico *et al.*, 1988; Turner *et al.*, 1995; Zerath *et al.*, 1996, 2000). This inhibition of bone formation was also seen in dynamic measurements based (unconventionally) on measurement of a single preflight fluorochrome label (Turner *et al.*, 1995; Zerath *et al.*, 2000). In contrast, space flight studies have not shown much evidence of effects on bone resorption utilising static measurements of osteoclast numbers or resorptive surfaces. This will be discussed further in a following section.

The vertebral bodies seem to respond differently than the long bones, but a response to space flight has occasionally been documented. In the Cosmos 1667 report by Vico *et al.* (1988) there was no significant bone loss in the lumbar or thoracic vertebrae. This was a short 7-day flight and if the bone turnover in vertebrae is less than the long bones, then it will take a longer time in space to find significant changes. Eurell and Kazarian (1983) studied the lumbar vertebrae following an 18.5-day space flight and concluded that the vertebrae show reduced turnover during flight. The most significant mineral loss occurred from this tissue at 6 days post-flight. By 29 days post-flight the vertebrae had recovered and were the same as the controls. The tissue content was analysed by histology and staining for proteoglycans, collagen and mineral, which are not the most sensitive indicators of tissue change. Another study by Vico *et al.* (1987a) from a 5-day space flight (Cosmos 1514) made note of the fact that vertebrae are not weight-bearing bones in the rat (they are weight-bearing in humans) and their 5-day flight was again not a long enough duration to produce change in bone formation. In this study, there was increased bone resorption in flight animals, but these rats were also into late pregnancy so that this bone loss result cannot be precisely related to space flight conditions.

Bion 11 which was a 13.7-day space flight takes on added importance because two adult male rhesus monkeys were the subjects of study. A post-flight analysis of iliac crest biopsies (Zerath *et al.*, 2002) found a reduction in bone volume associated with reduced osteoid, reduced mineralisation and reduced bone formation rate. Oganov *et al.* (2000) collected data from 6 monkeys, flown on different space flights, and compared acoustical data from *in vivo* analysis to determine bone mineral density. They documented a significant difference in mid diaphysis compared to the metaphyseal ends of the long bones and this difference decreased during space flight suggesting that microgravity either delayed bone growth or caused a change in mineralisation of the long bones. These results were similar to many of the findings from studies of rat long bones.

Reduction in bone quantity due to a change in bone resorption

Spengler *et al.* (1979, 1983) and Turner *et al.* (1979, 1985) (Cosmos 936) did not find a change in endosteal bone resorption resulting from an 18.5-day flight, based on morphometric analysis of rat tibias. Wronski and Morey (1983a, b) and Wronski *et al.* (1980) (Cosmos 1129) also did not note any effect of 18.5 days of space flight on the endosteal bone resorption, using morphometric techniques. Cann *et al.* (1980) and Cann and Adachi (1983) (Cosmos 1129) substituted calcium-40 for dietary calcium during an 18.5-day space flight which meant that the calcium excreted during the flight which was stable calcium-48 had to be the calcium released from pre-existing bone. From this study it was determined that bone "resorption" (i.e., loss of calcium from bone) was reduced during flight to about 75–80% of normal. In the male rat, decreased bone formation rather than increased bone resorption seems to be the primary affect of space flight effects on the skeleton. Female rats and their estrogen status may result in differing results. This will be discussed below in the section on cytokines.

Reduction in bone quality due to space flight

Vailas *et al.* (1990) (Cosmos 1887) found that although there was no difference in the length of humerus following 12.5 days of space flight, there was a significant loss of bending stiffness in the flight animals compared to controls. And in the vertebral body, there was less compressional stiffness compared to the synchronous and vivarium controls. This is interesting since in general, there were few quantitative bone changes found in the rats flown on Cosmos 1887. Doty *et al.* (1990) (Cosmos 1887) using electron microscopy, noted that the flight animals had some osteocytic death in the long bones, probably due to vascular occlusions which were seen in blood vessels within the diaphyseal bone of the tibia. Such vascular changes could be responsible for a change in bone quality without showing any corresponding quantitative changes. Doty *et al.* (1992) also showed some similar vascular inclusions (lipid deposits) in the blood vessels beneath the periosteal bone surface in flight animals from Cosmos 2044. The question was raised whether these vascular changes occured in space or perhaps might have occurred during re-entry or the resumption of weight bearing upon re-entry.

Loss of mineral from bone

Cann *et al.* (1980) and Cann and Adachi (1983) (Cosmos 1129) demonstrated that a reduced loss of calcium from bone occurred during space flight compared to ground controls. Although calcium excretion occurs through the intestine and kidney, this experiment was able to distinguish calcium derived directly from bone and showed reduced turnover rates. Mechanic *et al.* (1990) (Cosmos 1887) biochemically measured a reduction in calcium in diaphyseal rat bone following a 12.5-day space flight. These results were complicated by the finding that the adrenal glands were enlarged in these animals, suggesting a possible stress effect.

Simmons *et al.* (1990) (Cosmos 1887) found a lower bone ash in flight animals and also a lower specific gravity of the mineral fractions from calvaria and vertebrae. He concluded that the space flight conditions caused a reduced rate of maturation of the mineral and matrix components of bone.

Site specificity of bone effects

Tran Van and Mailland (1979) (Cosmos 1129) worked with alveolar bone, a non-weight-bearing bone, and showed that space flight reduced bone formation specifically in those areas not attached to muscle. As was also noted by other investigators in the long bones, the periosteum appeared to be the most affected bone surface.

Roberts *et al.* (1981, 1986) (Cosmos 1129) showed in the rat periodontal ligament that the cell number was decreased following 18.5 days of space flight. This technique involved measuring cell nuclei size as well as cell numbers within the periodontal ligament. This decrease in pre-osteoblasts was attributed to either a defect in proliferation or a delay in cell differentiation. It is suggested that this effect could also explain the reduced bone formation due to space flight. In Cosmos 1887, animal recovery was delayed for 55 h. This probably explains the results of Garetto *et al.* (1990) who measured a decrease in osteoprogenitor cells with a concomitant increase in pre-osteoblasts. This would indicate a rapid recovery during the 55 h recovery delay following the 12.5-day space flight.

Simmons *et al.* (1983) (Cosmos 1129) found that space flight did not affect periosteal bone formation in the non-weight-bearing ribs or some areas of the mandible. Bone formation was reduced in the mandible at sites where there was no muscle attachment, again suggesting a bone saving effect due to muscle forces. Nevertheless, they did find some change in the maturation of bone and mineral, in the mandibles, suggesting that microgravity may affect all skeletal tissues in some fashion.

Space flight effect on growth plate and bone length

Duke *et al.* (1990) and Montufar-Solis and Duke (1991) (Cosmos 1187) noted that the proliferative zone increased in cell number and the hypertrophic/degenerate cells decreased in number in the growth plate following this 12.5-day flight. They suggested that the 55 h delay in recovery at 1-*g* would be enough to cause this increased cell response. However, in Cosmos 2044 (Montufar-Solis *et al.*, 1992) they found the same effect and 2044 was designed to repeat the 1187 flight but without the delay in recovery. Nevertheless, there is always some delay in the animal recovery and flight 2044 had 6–10 h at 1-*g* which in the cartilage growth plate could have influenced the final results.

Cann studied the skeletal growth of two monkeys (*Macaca mulatta*) on the Cosmos 1887 flight and using radiographs of long bones, concluded that there was some slight inhibition of growth. One of these monkeys had a food supply problem and lost weight during the flight, so that the final results were biased towards a conclusion of reduced growth.

In the rat model, because of the prolonged growth phase of the long bones, impaired bone growth could contribute to the decrease in mass of the cancellous bone near the growth plates. The growth plate and its role in forming the primary trabeculae, if it were inhibited, could result in fewer or smaller trabeculae within the cancellous region of the long bone. The results of Duke *et al.* (1990) and Montufar-Solis *et al.* (1992) suggest that there is a flight effect on the growth plate. However, direct measurements of the rate of longitudinal bone growth in rats using fluorochrome labels indicate that space flight and weightlessness does not impair this process (Westerlind and Turner, 1995; Zerath *et al.*, 2000; Sibonga *et al.*, 2000). Therefore the cancellous osteopenia observed in flight rats does not appear to be a consequence of a decline in longitudinal bone growth.

Bone recovery following space flight

Spengler *et al.* (1979, 1983) and Turner *et al.* (1979, 1985) (Cosmos 936) found that inflight centrifugation did not help preserve bone during flight but did speed up recovery after animals were brought back to Earth. In the 14-day flight of SLS-2, young male rats showed reduced bone volume in the tibial metaphyses associated with thinning and loss of trabeculae. After two weeks of recovery at 1-*g*, bone loss had partially recovered and only the loss of trabeculae was in evidence. (Lafage-Proust *et al.*, 1998). There was also some indication that bone resorption had returned to normal whereas bone formation was greater than the preflight value. Morey and Baylink (1978) (Cosmos 782) and Holton (1982) showed that by 25 days, following a 19.5-day space flight, there had occurred a significant rebound in bone formation. Similar findings were reported in rhesus monkeys following the Russian Bion 10 mission (Zerath *et al.*, 1996).

Cytokine effects on bone and space flight

Bikle *et al.* (1994) (STS-54) extracted mRNA from long bones of Sprague Dawley rats housed individually on a 6 day space flight. Their prior studies led them to believe that expression of Insulin-like Growth Factor (IGF-1), IGF-IR (receptor for IGF), alkaline phosphatase, and osteocalcin would be useful indicators of bone formation and osteoblast differentiation. Following the flight, they found an increase in mRNA for alkaline phosphatase, IGF-1 and IGF-1R with a concomitant decrease in osteocalcin expression. They interpreted this as indicating a decrease in osteoblast maturation and/or an inability of this cell to respond to normal cytokine levels during space flight. It is interesting that Patterson-Buckendahl *et al.* (1987) also found a reduced osteocalcin content in humerus and vertebrae following a 7-day flight on Spacelab 3. This reduction was associated with reduced mineral content of bone and a decrease in mechanical strength of these bones. In addition, the long bones from this same Spacelab flight also demonstrated reduced mechanical strength (Shaw *et al.*, 1988).

Bateman *et al.* (1998) investigated the ability of infused IGF-1 to increase bone formation in an attempt to prevent bone loss in male Sprague-Dawley rats during a 10-day flight on STS-77. Alzet pumps were implanted prior to flight to infuse either IGF-1 or saline during the flight. Animals were housed in the Animal Enclosure Module (AEM), began the flight at 150 g body weight, and were recovered within 6 h after flight. The IGF increased bone mineral density, but had a very limited effect on new bone formation. These results and those from Bikle *et al.* (1994) would suggest that the normal effect of IGF on osteoblast population may be influenced by the presence or absence of mechanical load on the bone cells. This relationship needs to be further studied, especially as this relates to space flight.

A 14-day space flight on STS-62 was investigated using ovariectomised Fisher 344 rats (Cavolina *et al.*, 1997). mRNA was extracted for a subunit of collagen type I, osteocalcin, TGF-beta, and IGF-1. Although there was a decrease in periosteal bone formation and decreased collagen subunit activity, there was almost no change in calculated bone formation rate. However, bone resorption was greatly enhanced. The conclusions from this study is that estrogen and/or ovariectomy greatly alters the bone response to mechanical loading. And since we know that estrogen has a significant role in the regulation of osteoclastic activity, we need to be aware that female and male response to space flight may be quite different. Most of the skeletal bone loss studies have utilised male animal models.

Zhang and Turner (1998) analysed the mRNA from bone following a 14-day space flight on STS-62 (Physiological Systems Experiment-4). Female Fisher 344 rats, 12 weeks old, were housed in a group housing unit (AEM). Although the hypothesis suggested that IL-1 and IL-6 might control bone resorption through some kind of estrogen regulation, they found that space flight had no effect on IL mRNA. Since bone resorption in females is greater than in male rats, an estrogen and IL relationship was strongly suggested prior to flight. However, this flight study and subsequent work indicated that the presence or absence of estrogen had no effect on IL mRNA. This study did show that IL-1beta (interleukin) and INF (interferon gamma) mRNA was increased when ovariectomy was combined with space flight. These findings could potentially explain the increased bone resorption which occurred but further space flight data will be needed to support this hypothesis.

Stress and animal housing

In several reports, the enlarged adrenals of flight animals (and sometimes the controls) are indicative of stress to the animal during the experimental period and therefore the skeletal changes have been questioned (Jee *et al.*, 1983; Mechanic *et al.*, 1990). This makes the report by Wronski *et al.* (1998) especially valuable. Their study used adrenalectomised male Sprague-Dawley rats flown on STS-78 for 17 days. Half the animals were adrenalectomised and half were flown with intact adrenals. The adrenalectomised flight animals were given implants containing corticosterone and aldosterone to maintain normal hormone levels during flight. The recovery showed that the intact animals had enlarged adrenals. However none of the flight animals, with or without adrenals, demonstrated any significant bone loss. Although these were growing animals (6 weeks old, 165 g) the fact that they were group-housed during flight was thought to help protect against the bone loss. It also throws into question, the importance of adrenal hypertrophy, at least in young growing animals.

Some investigators have failed to observe the expected loss of cancellous bone and significant inhibition of bone formation in rats placed in earth orbit for as long as 10–17 days (Bateman *et al.*, 1998; Wronski *et al.*, 1998). Although there are several possible explanations, two studies by Vico *et al.* (1988, 1993) are very instructive. In both studies, the experimental animals were male Wistar rats that were 15 weeks of age at launch. Thus, animal gender, age and strain

Table 1

Bone changes during spaceflight in singly housed animals

Flight	Age at launch (days)	Flight length (days)	Strain	Bone formation rate (% Δ control)		
				Site	% Change	*p*
SLS2-Holton 1993	38	14	Harlan S/D	TFJ*	−11%	<.05
				Humerus	−10%	<.05
SL3 (young)	56	7	Teconic S/D		Not done	
SL3 (adult) 1985	84	7	Teconic S/D	TFJ	−34%	<.05
SLS1 1991	58	9	Teconic S/D	TFJ	−27%	<.05
Cosmos 782 1975	63	19.5	Munich Wistar	TFJ	−40%	<.05
Cosmos 936 1977	63	18.5	Munich Wistar	TFJ	−37%	<.05
Cosmos 1129	83	18.5	Munich Wistar	TFJ	−44%	<.05
				Humerus	−23%	<.05
				Rib	−18%	NS

*TFJ, Tibiofibular Junction, S/D, Sprague-Dawley.
Note: Data in Tables 1 and 2 provided by Dr Emily Morey Holton.

have been controlled in these two flights. In one study (Vico *et al.*, 1988) the flight rats exhibited obvious decreases in tibial cancellous bone mass and osteoid surface after only 7 days of space flight, whereas these histomorphometric indices were not significantly altered by 14 days in space in the second study (Vico *et al.*, 1993). The most obvious difference between the two studies was the difference in housing during space flight. In the study which demonstrated obvious bone changes, the rats were housed in individual cages. In the latter study with minimal bone changes, the flight rats were group-housed in a single cage. This comparison suggested that group housing of rats might be an important factor in space flight results.

A study of housing of animals during space flight was carried out by Morey-Holton, *et al.* (2000) (Tables 1 and 2). Male Sprague-Dawley rats (285 g), were group housed (AEM) or housed singly (RAHF; Research Animal Holding Facility) on STS-40 (SLS-1 experiment) for a 9-day flight. It was found that the group housed rats experienced a 6% decrease in bone formation (not significant) whereas the singly housed rats showed a 30% fall in bone formation. These two modes of housing in Earth's gravity and without the experience of space flight, did not result in any bone loss. This suggests that there is some factor present during space flight, when rats are placed into single housing environments, that results in significant bone loss. Although metabolic stress would be a natural suggestion, the earlier work cited concerning adrenals and stress, would indicate that stress is not a factor in bone loss during space flight. The analysis of this housing effect needs additional animal studies. The use of the AEM housing also resulted in a decrease in mechanical properties of bone (Vajda *et al.*, 2001) following a 17-day flight on STS-78.

Table 2

Bone changes during spaceflight in group (AEM) housed animals

Flight	Age at launch (days)	Flight length (days)	Strain	Bone formation rate (% change AEM control)		
				Bone	% Change	*p* value
PARE 3 1993	34	9	Harlan Holtz. S/D	TFJ*	−6%	NS
				Femur	−13%	NS
PSE-1 1990	40	4	Charles River	Femur	+3%	NS
				Humerus	+5%	NS
NIH-R 1996	40	18	Teconic S/D	TFJ	−3%	NS
PSE-2 1992	42	10	Teconic S/D	Humerus	−10%	NS
SLSI 1991	58	9	Teconic S/D	TFJ	−8%	NS
PSE-3 1993	60	11	Teconic F344	Femur	−35%	<.05
PSE-4 1994	84–91	14	Teconic F344	TFJ	−25%	<.05

*TFJ, Tibiofibular Junction

Conclusions

Animal models for the study of bone loss due to space flight has been limited largely to rats and the occasional monkey. The rat is not ideal for bone loss studies because the bone is structured in a lamellar pattern and is not osteonal bone as found in humans and larger animals. In addition, the rat does not walk in an upright stance and is not a bipedal animal, which causes differences in weight bearing, especially on the vertebrae. Also, the rat tends to grow throughout much of its life span and the growth plate in the long bones tends to remain open for extended periods. Martin *et al.* (2003) have shown that the growth plate of the tibia closes in male rats at 8 months of age and at 10 months of age in the female (Harlan Sprague-Dawley rats). This too is a significant difference between the rat and humans or other mammals. If we want to mimic the human physiology more closely, then the rats flown in space should probably be a year old or older. And if we assume that age, as well as housing, strain and flight conditions, all impact on the skeletal response, then aged animals might well produce a different response than the younger rats used in all previous space flight experiments. However when one considers all the differences which occur in the overall physiology of an animal or human in space, there probably is no animal which can accurately model all the events which change during space flight. The rat does have the advantage that its skeletal physiology has been studied for many years and under many different physiological and pathological conditions. Therefore most structural or metabolic changes which may occur during space flight can be related to known events from the long history of skeletal studies of these animals. For example, the fracture healing process may be affected by the space flight environment and this is a cause of concern for human flight and potential space related medical problems. There is a tremendous amount of information regarding fracture healing in rats on earth, which can be a basis for the space studies. This background information will aid in the space flight studies and move the research along at a much faster rate. In addition, there are advantages in use of these smaller animals in that the numbers of subjects can be utilised to reach significance in statistical analysis. For the future, knock out rats or mice can be flown which will have specific characteristics or disabilities, and will provide insight into mechanisms behind skeletal change. In the final analysis, animal models will have to be studied to understand microgravity effects on bone loss and to obtain information on the cellular and molecular mechanisms involved. This will continue to be true as we extend our time in space and increase our human exposure to its dangers. However these questions are not only relevant for space flight and the safety of our astronauts but also increase our understanding of fundamental processes in bone physiology. Using these animal models can provide links to the pathophysiology of bone loss and benefit those of us, the majority of us, who will never leave the Earth's gravitational field.

References

*Adams, G.R., McCue, S.A., Bodell, P.W., Zeng, M. and Baldwin, K.M. (2000) Effects of spaceflight and thyroid deficiency on hindlimb development I. Muscle mass and IGF-I expression. Am. J. Physiol. 88, 894–903.

*Arnaud, S.B., Buckendahl, P., Durnova, G., Bromage, T. and Yamauchi, M. (2000) Bone biochemistry in rat diaphysis after spaceflight. J. Gravit. Physiol. 7, 7–15.

*Bateman, T.A. and Countryman, S. (2002) Osteoprotegerin and bone loss associated with spaceflight. Drug Discov. Today 7, 456–457.

Bateman, T.A., Zimmerman, R.J., Ayers, R.A., Ferguson, V.L., Chapes, S.K. and Simske, S.J. (1998) Histomorphometric, physical, and mechanical effects of spaceflight and insulin-like growth factor-I on rat long bones. Bone 23, 527–535.

Bikle, D.D., Harris, J., Halloran, B.P. and Morey-Holton, E. (1994) Altered skeletal pattern of gene expression in response to spaceflight and hindlimb elevation. Am. J. Physiol. 267, E822–E827.

*Bikle, D.D., Sakata, T. and Halloran, B.P. (2003) The impact of skeletal unloading on bone formation. Grav. Space Biol. Bull. 16, 45–54.

Cann, C.E., Adachi, R.R. and Holton, E.M. (1980) Bone resorption and calcium absorption in rats during spaceflight. Physiologist Supl. 23, S83–S86.

Cann, C.E. and Adachi, R.R. (1983) Bone resorption and mineral excretion in rats during spaceflight. Am. J. Physiol. 244, R327–R331.

Cavolina, J.M., Evans, G.L., Harris, S.A., Zhang, M., Westerlind, K.C. and Turner, R.T. (1997) The effects of orbital spaceflight on bone histomorphometry and messenger ribonucleic acid levels for bone matrix proteins and skeletal signaling peptides in ovariectomized growing rats. Endocrinology 138, 1567–1576.

Doty, S.B., Morey-Holton, E.R., Durnova, G.N. and Kaplansky, A.S. (1990) Cosmos 1887: Morphology, histochemistry, and vasculature of the growing rat tibia. FASEB Journal 1, 16–23.

Doty, S.B., Morey-Holton, E.R., Durnova, G.N. and Kaplansky, A.S. (1992) Morphological studies of bone and tendon. Am. J. Physiol. 73, 10S–13S.

Duke, P.J., Durnova, G. and Montufar-Solis, D. (1990) Histomorphometric and electron microscopic analysis of tibial epiphyseal plates from Cosmos 1887 rats. FASEB Journal 4, 41–46.

Eurell, J.A. and Kazarian, L.E. (1983) Quantitative histochemistry of rat lumbarvertebrae following spaceflight. Am. J. Physiol. 244, R315–R318.

*Evans, G.L., Morey-Holton, E. and Turner, R.T. (1998) Spaceflight has compartment- and gene-specific effects on mRNA levels for bone matrix proteins in rat femur. J. Appl. Physiol. 84, 2132–2137.

Garetto, L.P., Gonsalves, M.R., Morey, E.R., Durnova, G. and Roberts, W.E. (1990) Preosteoblast production 55 h after a 12.5-day spaceflight on Cosmos 1887. FASEB Journal 4, 24–28.

*Grindeland, R.E., Popova, I.A., Arnaud, S.B. and Vasques, M. (1990) Cosmos 1887 mission overview: effects of microgravity on rat body and adrenal weight and plasma constituents. FASEB J. 4, 105–109.

*Howe, E.E., Wronski, T.J., Halloran, B.P. and Miller, S.C. (2000) Histologic changes in the adrenal cortex from young rats following spaceflight. Aviat. Space Environ. Med. 71, 1039–1044.

Jee, W.S., Wronski, T.J., Morey, E.R. and Kimmel, D.B. (1983) Effects of spaceflight on trabecular bone in rats. Am. J. Physiol. 244, R310–314.

Johnson, R.B. (1998) The bearable lightness of being: bones, muscles, and spaceflight. Anat. Rec. 253, 24–27.

*Kaplansky, A.S., Durnova, G.N., Sakharova, Z.F. and Ilyina-Kakuyeva, Ye.I. (1987) Histomorphometric analysis of rat bones after spaceflight aboard Cosmos-1667 biosatellite. Moscow Kosmicheskaya Biologiya I Aviakosmicheskaya Meditsina 21, 25–31In Russian..

*Kawashima, K., Yamaguchi, A., Shinki, T., Noji, S., Yokose, S., Yamaai, T., Endo, H., Yoshiki, S., Abe, E. and Suda, T. (1995) Microgravity generated by spaceflight has little effect on the growth and development of chick embryonic bone. Biol. Sci. Space 9, 82–94.

Lafage-Proust, M.H., Collet, P., Dubost, J.M., Laroche, N., Alexandre, C. and Vico, L. (1998) Space-related bone mineral redistribution and lack of bone mass recovery after reambulation in young rats. Am. J. Physiol. Regulatory Integrative Comp. Physiol. 274, R324–334.

*Li, M., Shen, Y., Halloran, B.P., Baumann, B.D., Miller, K. and Wronski, T.J. (1996) Skeletal response to corticosteroid deficiency and excess in growing male rats. Bone 19, 81–88.

Martin, E.A., Ritman, E.L. and Turner, R.T. (2003) Time course of epiphyseal growth plate fusion in rat tibia. Bone 32, 261–267.
Mechanic, G.L., Arnaud, S.B., Boyde, A., Bromage, T.G., Buckendahl, P., Elliott, J.C., Katz, E.P. and Durnova, G.N. (1990) Regional distribution of mineral and matrix in the femurs of rats flown on Cosmos 1887 Biosatellite. FASEB Journal 4, 34–40.
Montufar-Solis, D. and Duke, P.J. (1991) Effect of strain, diet and housing on rat growth plates: a Cosmos '87-Spacelab 3 comparison. Physiologist 34, s183–s184.
Montufar-Solis, D., Duke, P.J. and Durnova, G. (1992) Spaceflight and age affect tibial epiphyseal growth plate histomorphometry. J. Appl. Physiol. 73, s19–s25.
Morey, E.R. and Baylink, D.J. (1978) Inhibition of bone formation during spaceflight. Science 201, 1138–1141.
Morey-Holton, E.R., Halloran, B.P., Garetto, L.P. and Doty, S.B. (2000) Animal housing influences the response of bone to spaceflight in juvenile rats. J. Appl. Physiol. 88, 1303–1309.
Oganov, V.S., Bakulin, A.V., Novikov, V.E. and Murashko, L.M. (2000) Change and recovery of bone mass and acoustic properties of rhesus monkeys after Bion 11 spaceflight. J. Gravit. Physiol. 7, S163–S168.
Patterson-Buckendahl, P., Arnaud, S.B., Mechanic, G.L., Martin, R.B., Grindeland, R.E. and Cann, C.E. (1987) Fragility and composition of growing rat bone after one week in spaceflight. Am. J. Physiol. Feb. 252, R240–246.
*Portugalov, V.V., Savina, E.A., Kaplansky, A.S., Yakovelva, V.I., Plakhuta-Plakhuta, G.I., Pankova, A.S., Katunyan, P.I., Shubich, M.G. and Buvailo, S.A. (1976) Effect of spaceflight factors on the mammal: experimental morphology study. Aviat. Space Environ. Med. 47, 813–816.
Roberts, W.E., Mozsary, P.G. and Morey, E.R. (1981) Suppression of osteoblast differentiation during weightlessness. Physiologist Suppl. 24, S75–S76.
Roberts, W.E., Weinreb, M.M., Gonsalves, M.R. and Morey, E.R. (1986) Nuclear volume assay of osteogenic potential in ulna primary spongiosa of rats flown in space for 18.5-days. Anatomical Record 214, 109A.
*Schneider, V., Oganov, V., LeBlanc, A., Rakmonov, A., Taggart, L., Bakulin, A., Huntoon, C., Grigoirev, A. and Varonin, L. (1995) Bone and body mass changes during spaceflight. Acta. Astronaut. 36, 463–466.
*Shaw, S.R., Vailas, A.C., Grindeland, R.E. and Zernicke, R.F. (1988) Effects of a 1-week spaceflight on morphological and mechanical properties of growing bone. Am. J. Physiol. 254, R78–R83.
Sibonga, J.D., Zhang, M., Evans, G.L., Westerlink, K.C., Cavolina, J.M., Morey-Holton, E. and Turner, R.T. (2000) Effects of spaceflight and simulated weightlessness on longitudinal bone growth. Bone 27, 535–540.
Simmons, D.J., Grynpas, M.D. and Rosenberg, G.D. (1990) Maturation of bone and dentin matrices in rats flown on the Soviet Biosatellite Cosmos 1887. FASEB Journal 4, 29–33.

Simmons, D.J., Russell, J.E., Winter, F., Tran Van, P., Vignery, A., Baron, R., Rosenberg, G.D. and Walker, W.V. (1983) Effect of spaceflight on the non-weight-bearing bones of rat skeleton. Am. J. Physiol. Mar. 244, R319–R326.

Spengler, D.M., Morey, E.R., Carter, D.R., Turner, R.T. and Baylink, D.J. (1983) Effects of spaceflight on structural and material strength of growing bone. Proceedings of the Society for Experimental Biology and Medicine 174, 224–228.

Spengler, D.M., Morey, E.R., Carter, D.R., Turner, R.T. and Baylink, D.J. (1979) Effects of spaceflight on bone strength. Physiologist Suppl. 22, S75–S76.

*Suda, T. (1998) Lessons from the space experiment Sl-J/FMPT/L7: the effect of microgravity on chicken embryogenesis and bone formation. Bone May 22, 73S–78S.

Tipton, C.M. (2003) Animals and biosatellites in space: A historical perspective. J. Gravitational Physiol. 10, P1–P4.

Tran Van, P. and Mailland, M. (1979) Early response of periodontal ligament and alveolar bone to hypofunction in the rat. Journal of Dental Research 38 (Special Issue D), 2261.

Turner, R.T., Bell, N.H., Duvall, P., Bobyn, J.D., Spector, M., Holton, E.M. and Baylink, D.J. (1985) Spaceflight results in formation of defective bone. Proceedings of the Society for Experimental Biology and Medicine 180, 544–549.

Turner, R.T., Morey, E.R., Liu, C. and Baylink, D.J. (1979) Altered bone turnover during spaceflight. Physiologist Suppl. 22, S73–S74.

*Turner, R.T. (2000) Invited review: What do we know about the effects of spaceflight on bone? J. Appl. Physiol. 89, 840–847.

Turner, R.T., Evans, G.L. and Wakley, G.K. (1995) Spaceflight results in depressed cancellous bone formation in rat humeri. Aviat. Space. Environ. Med. 66, 770–774.

Vailas, A.C., Zernicke, R.F., Grindeland, R.E., Kaplansky, A., Durnova, G.N., Li, K.C. and Martinez, D.A. (1990) Effects of spaceflight on rat humerus geometry, and biochemistry. FASEB Journal 4, 47–54.

Vajda, E.G., Wronski, T.J., Halloran, B.P., Bachus, K.N. and Miller, S.C. (2001) Spaceflight alters bone mechanics and modeling drifts in growing rats. Aviat. Space Environ. Med. 72, 720–726.

Van Loon, J.W.A., Veldhuijzen, J.P. and Burger, E.H. (1996) Bone and space flight: An overview. *In* Moore, D., Bie, P. and Oser, H. (eds.). *Biological and Medical Research in Space*, pp. 259–299, Springer-Verlag, Berlin.

Vico, L., Bourrin, S., Genty, C., Palle, S. and Alexandre, C. (1993) Histomorphometric analyses of cancellous bone from COSMOS 2044 rats. J. Appl. Physiol. 75, 2203–2208.

*Vico, L., Chappard, D., Alexandre, C., Palle, S., Minaire, P., Riffat, G., Novikov, V.E. and Bakulin, A.V. (1987a) Effects of weightlessness on bone mass and osteoclasts number in pregnant rats after a five-day spaceflight (Cosmos 1514). Bone 8, 95–103.

*Vico, L., Chappard, D., Bakulin, A.V., Novikov, V.E. and Alexandre, C. (1987b) Effect of 7-day spaceflight on weight-bearing and non-weight-bearing bones in rats (Cosmos 1667). Physiologist 30, S45–S46.

Vico, L., Chappard, D., Palle, S., Bakulin, A.V., Novikov, V.E. and Alexandre, C. (1988) Trabecular bone remodeling after seven days of weightlessness exposure (BIOCOSMOS 1667). Am. J. Physiol. 255, R243–R247.

*Vico, L., Hinsenkamp, M., Jones, D., Marie, P.J., Zallone, A. and Cancedda, R. (2001) Osteobiology, strain and microgravity. Part II. Studies at the tissue level. Calcif. Tissue Int. 68, 1–10.

*Vico, L., Novikov, V.E., Very, J.M. and Alexandre, C. (1991) Bone histomorphometric comparison of rat tibial metaphysis after 7-day tail suspension vs. 7-day spaceflight. Aviat. Space Environ. Med. 62, 26–31.

Westerlind, K.C. and Turner, R.T. (1995) The skeletal effects of spaceflight in growing rats: tissue-specific alterations in mRNA levels of TGF-β. J. Bone Miner. Res. 10, 843–848.

Wronski, T.J. and Morey, E.R. (1983a) Alterations in calcium homeostasis and bone during actual and stimulated spaceflight. Medicine and Science in Sports and Exercise 15, 410–414.

Wronski, T.J. and Morey-Holton, E.R. (1983b) Effect of spaceflight on periosteal bone formation in rats. American Journal of Physiology 244, R305–R309.

Wronski, T.J., Li, M., Shen, Y., Miller, S.C., Bowman, B.M., Kostenuik, P. and Halloran, B.P. (1998) Lack of effects of spaceflight on bone mass and bone formation in group-housed rats. J. Appl. Physiol. 85, 279–285.

Wronski, T.J., Morey-Holton, E. and Jee, W.S. (1980) Cosmos 1129: Spaceflight and bone changes. Physiologist Supl. 23, S79–S82.

Wronski, T.J., Morey-Holton, E.R., Doty, S.B., Maese, A.C. and Walsh, C.C. (1987) Histomorphometric analysis of rat skeleton following spaceflight. Am. J. Physiol. 252, R252–R255(Regulatory Integrative Comp Physiol 21).

*Yagodovsky, V.S., Triftanidi, L.A. and Gorokhova, G.P. (1976) Spaceflight effects on skeletal bones of rat (light and electron microscope examinations). Aviat. Space Environ. Med. 47, 734–738.

*Yamada, G., Sugimura, K., Nakamura, S., Yamada, M.O., Tohno, Y., Maruyama, I., Kitajimi, I. and Minami, T. (1997) Trace element composition and histological analysis of rat bones from the space shuttle. Life Sci. 60, 635–642.

Zerath, E., Godet, D., Holy, X., Andre, C., Renault, S., Hott, M. and Marie, P.J. (1996) Effects of spaceflight and recovery on rat humeri and vertebrae: histological and cell culture studies. J. Appl. Physiol. 81, 164–171.

Zerath, E., Grynpas, M., Holy, X., Viso, M., Patterson-Buckendahl, P. and Marie, P.J. (2002) Spaceflight affects bone formation in rhesus monkeys: a histological and cellular study. J. Appl. Physiol. 93, 1047–1056.

Zerath, E., Holy, X., Roberts, S.G., Andre, S., Renault, S., Hott, M. and Marie, P.J. (2000) Spaceflight inhibits bone formation independent of corticosteroid status in growing rats. J. Bone Miner. Res. 15, 1310–1320.
Zhang, M. and Turner, R.T. (1998) The effects of Spaceflight on mRNA levels for cytokines in proximal tibia of ovariectomized rats. Aviat. Space Environ. Med. Jul. 69, 626–629.

**These references are relevant to the subject but are not cited in the text.*

Experimentation with Animal Models in Space
G. Sonnenfeld (editor)

2005. Published by Elsevier B.V.
DOI: 10.1016/S1569-2574(05)10009-4

Responses across the Gravity Continuum: Hypergravity to Microgravity

Charles E. Wade

Life Sciences Division, NASA – Ames Research Center, Moffett Field, CA 94035, USA; and United States Army Institute of Surgical Research, Fort Sam, Houston, TX 78234, USA

Introduction

Life on Earth has evolved with a single environmental constant, gravity (1 ***g***). As changes in gravity are not readily observed in Earth's environment, unlike effortlessly observed differences such as temperature or barometric pressure, the study of altered gravitational fields is a relatively new science. One of the first observations of the impact of gravity was the "gravitic" response of plants (Knight, 1806). Plants, grown on a rotating centrifuge to simulate hypergravity (> 1 ***g***), oriented to the resultant gravitational field rather than that of Earth's 1 ***g***. The impact of increased acceleration and the resulting increased gravitational loads spurred new interest in the area and the study of gravity's influence of biological processes. Most of these efforts were oriented towards defining the limits to acute hypergravity (> 1 ***g***) exposure. It was not until the late 1950s with the advent of the possibility of space flight and exposure to microgravity (< 1 ***g***) that interest developed in chronic exposure to alteration in the level of gravity. Early investigators viewed the alteration of the gravitational field as a continuum (Oyama and Platt, 1965; Smith and Burton, 1967; Smith, 1976; Pitts, 1977; Burton and Smith, 1996, 1999; Phillips, 2002). This approach was based in an extensive history in environmental sciences where the spectrum on either side of homeostatic set points i.e., hypergravity > 1 ***g*** and hypo-gravity < 1 ***g***, elicited opposite responses.

While responses to the reduced gravity environment of space flight have stimulated intense interest, little work has addressed the gravitational continuum. Smith and Oyama with their colleagues (Oyama and Platt, 1965; Smith and Burton, 1967; Smith, 1976; Pitts, 1977; Burton and Smith, 1996, 1999) were early proponents who investigated responses to hypergravity using centrifugation and extrapolated the findings to the microgravity environment of space flight. Neither of these early investigators had the opportunity to personally test their theories in the microgravity environment of space. However their work using centrifugation, and that of others, clearly demonstrates a dose response across gravity levels greater than the 1 ***g*** environment of Earth (Oyama and Platt, 1965; Smith and Burton, 1967; Smith, 1976; Pitts, 1977; Burton and Smith, 1996, 1999; Phillips, 2002).

There are two physical principles that provide the basis for this approach. The first is the "theory of equivalence" postulated by Ernst Mach and Albert Einstein. The principle states that "acceleration acting on a mass can not be distinguished from the attracting force of gravity". Thus, if a mass is accelerated, on a centrifuge for example, the resultant force will be the same as an increase in the level of gravity. The second principle is that of "continuity". Simply stated, gravitational fields are continuous above and below the gravitational field of Earth, and biological responses to changes across the spectrum of gravity exhibit a similar continuity. The "principle of continuity" has not yet been rigorously tested and validated across the gravitational continuum. However, there is data suggesting that this principle is valid. These data are the focus of the present review. This review postulates that gravity is a continuum that can be demonstrated by responses to space flight and exposure to hypergravity during centrifugation. Furthermore, what is observed in acclimation to hypergravity may be extrapolated to the micro-gravity of space flight.

Centrifugation

The normal gravity of Earth can be summed with acceleration forces to induce hypergravity. Changing the direction of motion or the linear rate of motion induces acceleration. In the case of centrifugation, acceleration is the rate of rotation. Additionally, the housing of the specimens within the centrifuge should consider the posture/position. In the case of rodents, gravitational forces are normally applied from the back through the feet. In order to maintain this orientation, the housing units on a centrifuge must be gimbaled such that the resultant force is through the floor and thus the normal direction across the animal, from the posterior to the feet.

The use of centrifuges serves two differing objectives. The first is alteration of the gravitational field as an experimental manipulation to study the response of the specimen. The second is the use of centrifugation as a countermeasure to ameliorate or attenuate the adverse effects of space flight/microgravity. As humans consider space flight missions of longer duration, the ability to maintain function and fitness has become increasingly important. Rotation of the spacecraft to induce a gravitational field has been proposed (Young, 1999; Clement and Pavy-Le Traon, 2004). Additionally, periodic exposure to centrifugation is being investigated. The confounding factors occurring with the use of centrifugation as a countermeasure must be investigated and understood to adequately employ this manipulation.

Factors in the study of hypergravity

In the study of hypergravity induced by centrifugation, there are a number of factors to be considered: scaling effect, impact of the rotational effects on a moving living system (Coriolis forces), duration of the exposure and resultant level of gravity.

Scaling effects

Simply stated, the bigger the mass of an animal, the greater the impact of alterations in gravity. Scaling was first identified in observations of survival of animals in hypergravity (Wunder, 1960, 1962; Burton and Smith, 1965; Oyama and Platt, 1967; Oyama, 1975; Amtmann *et al.*, 1976; Doden *et al.*, 1978; Economos, 1979). Survival of a species is the ultimate test of acclimation to a new environment. The question as to the size of animal and its response has been considered in a number of environments with the primary end point being survival. Earlier works suggest that there is a decrease in the magnitude of the gravity load that an animal can survive is relative to its body mass (Fig. 1) (Wunder, 1960, 1962; Burton and Smith, 1965; Oyama and Platt, 1967; Oyama, 1975; Amtmann *et al.*, 1976; Doden *et al.*, 1978; Economos, 1979). Wunder *et al.* (Wunder, 1960, 1962) found that mice were able to acclimate to hypergravity levels up to 7 ***g*** and beyond this level, death occurred. The larger sized rats had reduced surival above 4.5 ***g*** (Oyama and Platt, 1967). Thus, there appears to be a difference in ability of rodents to acclimate dependent upon the size of the animal. Scaling effects should be considered in

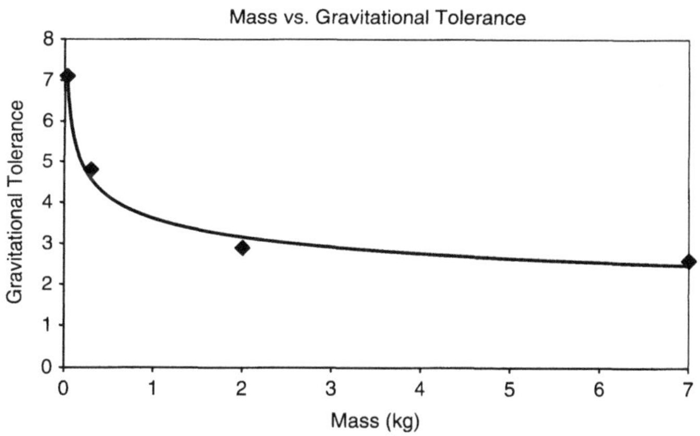

Fig. 1. The influence of body mass on the gravity level at which mortality is reported (Relevant references: Wunder, 1960, 1962; Burton and Smith, 1965; Oyama and Platt, 1967; Oyama, 1975; Amtmann *et al.*, 1976; Doden *et al.*, 1978; Economos, 1979).

cross comparison of responses to changes in the level of gravity between species and genders.

Rotation

The rate of rotation should be considered, as the same gravitational field may be produced employing differing rates of rotation and lengths of the centrifuge arm (radius). The impact of rotation can be addressed by the use of rotational controls. Placing the specimens on the center point of the centrifuge such that they sense the rate of rotation with a minimal change in gravitational force does this. Another approach to evaluate the impact of rotation is to study the gravity level at different arm lengths. A third method is to make the centrifuge as large as possible to necessitate the slowest possible rate and to minimize the effect of rotation. Rotational effects are confounded when the specimen is allowed to change orientation within the gravitational field. An example of this is the sensation of motion experienced with the acceleration of a car. Once the body achieves the same speed as the vehicle the motion is no longer detected. However, if one moves about in the environment, the direction of motion is sensed as orientation of the body varies in relationship the direction of acceleration.

Duration of exposure

The classic definitions in the study of environmental change are applied to the study of altered gravity. That is, there are acute and chronic changes. As noted by Burton and Smith (1965, 1996) many of the early (i.e., acute) responses to altered gravity are the same as those seen on exposure to any change in environment. In time (i.e., chronic) and if the gravity level is low enough, the organism establishes a new homeostasis and acclimates to the change. The final stage is adaptation to the altered environment where genetic traits of preference are passed from one generation to the next. Thus, the duration of the exposure to a change in the gravity field should also be considered.

In many studies it is necessary to periodically stop the centrifuge to care for the specimens or to collect samples. This periodic stopping of the centrifuge led Smith (Smith and Burton, 1967; Burton and Smith, 1996) to propose the concept of accumulated exposure. Acclimation is a function of the total duration at a gravity level rather than total time of the experiment. In addition, there is also the function of dual adaptation. That is, the impact of transitioning from one environment is influenced by prior exposure to the environment. An example of this is the drop in body temperature of rats upon exposure to hypergravity. Daily movement from 1–2 ***g***, following the initial exposure and acclimation, no longer elicits the response of reduced body temperature (Oyama *et al.*, 1971; Holley *et al.*, 2003). This acclimation may last for weeks after reexposure to 1 ***g***.

Responses to Altered Gravity

Species

A large number of species including *C. elegans*, fruit flies, turtles, rodents, chickens and non-human primates, have been acutely exposed to hypergravity. While there is an extensive literature on the response of human to acute increases in gravity associated with the advent of high performance aircraft, information relating to chronic exposures is lacking (Burton and Smith, 1996). Of interest among species are not the differences in responses, but the similarities. While much of the literature deals with data obtained using rodents, other species will be mentioned in this review as they provide insights into possible mechanisms involved in acclimation to altered gravity.

Reproduction

Adaptation of a species to a novel environment may be determined based on the ability of the animals to reproduce. In response to attempts to mate animals, mice exposed to 3.5 ***g*** had no pregnancies (Janer and Duke, 1984; Moore and Duke, 1988) while a similar finding was noted in rats at 1.87 ***g*** (Ishay and Barr-Nea, 1977). Oyama and Platt (1967) successfully bred rats at 3.6 ***g*** but no neonates survived. At 2.5 ***g*** similar rates of pregnancy occurred in rats and mice, but only 14% of the rat neonates survived whereas, 28% of the mice neonates survived. Exposure of rats to hypergravity results in a dramatic reduction in the survival of offspring above gravity levels of 1.3 ***g*** (Baer *et al.*, 2000). At 2 ***g*** less than 30% of the rat offspring survive. In mice, a similar reduction in survival was noted in the pups but at higher gravity levels (Wunder, 1962).

In response to reductions in gravity the data is inconclusive as to survival of neonates (Ronca, 2003). Most of the space flights have been of relatively short durations (< 30 days), and limitations appear to be related to engineering problems rather than the ability of the animals to acclimate to the space flight environment (Reichhardt, 1998). Thus, the ability of rodents to survive alterations in ambient gravity level (1 ***g***) and to procreate appears to be related to the size of the animal. Due to larger size, it appears that rats are more sensitive to gravitational changes than mice.

Body mass

The question has been raised as to the size of an animal and its response to exposure to altered gravity. The larger the animal, the greater the response and thus the scaling effect. Furthermore, in response to hypergravity, the magnitude of the reduction in body mass is closely correlated to the gravity level to which the animals are exposed and the initial mass of the animal (Smith and

Burton, 1967; Burton and Smith, 1996; Wade *et al.*, 1997; Wade *et al.*, 2000). Smith and Burton (1967) stated, "Animals raised under acceleration fields stronger than Earth-gravity have a reduced mature body size—and the reduction is rectilinearly related to the field strength". With exposure to hypergravity there is a reduction in body mass that persists for over 400 days (Oyama and Platt, 1967; Pitts *et al.*, 1972; Pitts, 1977, 1982; Oyama *et al.*, 1985). At 2 ***g*** the magnitude of the reduction in body mass when compared to control animals is on the order of 10% in rats and 3% in mice (Oyama and Platt, 1967; Pitts *et al.*, 1972; Pitts, 1977, 1982; Oyama *et al.*, 1985; Yuwaki and Okuno, 2003). In a recent experiment we assessed the effect of exposure to 2 ***g*** on mice and rats of different ages and of both gender. There was an effect of species. In rats the decrease in body mass was greater (6–14%) than that of mice (2–9%) with no significant effect of age. There was a gender effect with males having a greater reduction of body mass than females reflective of a greater body mass. In rat neonates the magnitude of the reduction in body mass is similar to that of adults (Ronca *et al.*, 2001). In weanling rats with a similar body mass as that of mice, a decrease in body mass was on the order of 13%. Thus, the change in body mass in response to hypergravity may be dependent on the species rather than solely on the initial mass of the animal. Smith and Burton (1967) noted that the final mature body size may be the determining factor.

The reduced body mass response to the microgravity of space flight appears to be influenced by the fact that upon reexposure to Earth's gravity mammals have pronounced diuresis that contributes to a reduction in body mass (Wade and Morey-Holton, 1998). However, in the majority of flights the body mass of rats upon return to Earth is similar to that of ground controls (Wade *et al.*, 1997, 2000). Only one study of space flight of mice resulted in a significant difference between controls and flight animals of about −12%. Furthermore, the flight animals lost weight (−5%) from preflight levels (32). This reduction may have been due to engineering issues (i.e., increased ambient temperatures). Thus, the comparison of mice to rats cannot be made. The only study where rat body mass was determined during space flight indicated a significant increase of 9% (Wade *et al.*, 2000).

In response to hypergravity, it appears that the lowering of body mass compared to controls is greater and more consistent in rats. This observation led Pitts (ref) to propose a series of mathematical models to predict the response of body mass to alterations in gravity. The models consider the species, age and gender. For rats, there is a close relationship with body mass across the spectrum of gravity (Wade *et al.*, 2000; Moran *et al.*, 2001) (Fig. 2). During the microgravity of space flight, body mass was increased (9%) and response to hypergravity reduced (15–20%) compared to controls. While a lower body mass is detectable in mice exposed to hypergravity, an increase in the number of animals is required because of increased variability in body mass. Furthermore, the response of mice to space flight is ill-defined as they have not been intensively studied in microgravity.

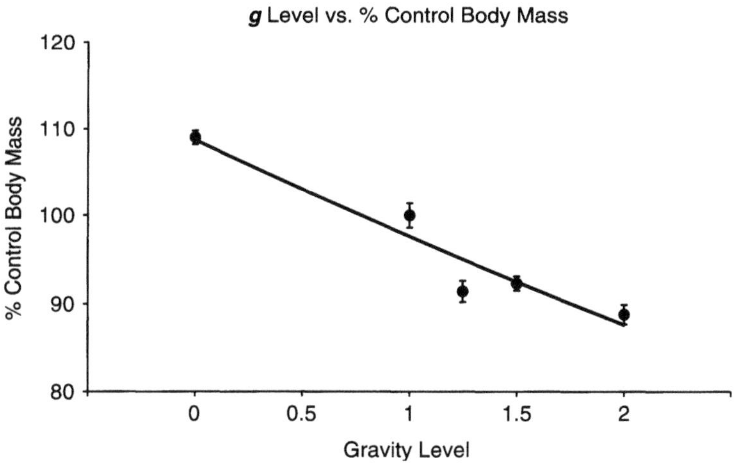

Fig. 2. The response of body mass of rats compared to control values at various gravity levels (Relevant References: Wade *et al.*, 2000; Moran *et al.*, 2001).

Growth/Body composition

The acute reduction in body mass in both rats and mice during centrifugation is subsequently associated with normal rates of weight gain after an initial period of acclimation to hypergravity. Thus, growth does not appear to be affected (Keil, 1969; Pitts *et al.*, 1972; Pitts, 1982; Oyama *et al.*, 1985; Wade *et al.*, 1997; Fuller *et al.*, 2000; Wade *et al.*, 2000; Warren *et al.*, 2001; Harrison *et al.*, 2003; Yuwaki and Okuno, 2003). However, there are significant changes in body composition. In mice there is a significant increase in lean mass and reduction in fat mass (Keil, 1969; Fuller *et al.*, 2000). Fat mass as a percentage of body mass at 2 ***g*** is reduced by 7% and lean mass is increased proportionally, with no significant difference in total body mass (Fuller *et al.*, 2000). Keil *et al.* (1969) reported a decrease in fat mass from 10 to 3% in both male and female mice exposed to 4.7 ***g***. Lean mass was also reduced resulting in a net reduction in total body mass. In rats, a similar response is noted. Pitts *et al.* (1972) reported fat mass of rats at 4.15 ***g*** to be reduced from 19 to 10%. At 2.76 ***g***, percent fat mass was 15%. The same group found that space flight increased the percent of body fat (Pitts *et al.*, 1983). The response of lean body mass to altered gravity in rats is a slight increase. Warren *et al.* (2001) reported that lean mass decreased by 5% while fat mass decreased by 50% in rats exposed to 2 ***g*** for 8 weeks. Thus, at low alterations of gravity (<2.1 ***g***), mice appear to adjust their body composition by increasing lean mass to compensate for the decrease in fat mass. Rats at similar levels of gravity have a decrease in both lean and fat mass resulting in a net reduction in total body mass.

Data as to the contribution of changes in total body water (TBW) to the alteration in body mass are variable. Pitts *et al.* (1972) found TBW to be reduced with high levels of gravity (4.1 ***g***). However, others noted no change or an increase in the proportion of TBW to the reduction in body mass (Ortiz and Wade, 2000). If the shift in body composition induced a decrease in fat mass and increase in lean mass then an increase in TBW would be expected.

Evidence from space flight regarding the response of growth during space flight for mice or rats has been limited. Rats gain weight at a normal rate during space flight (Wade *et al.*, 2000). The data on mice is confounded as noted above for body mass. Reports on space flight effects on body composition are also limited. Fast *et al.* (1985) found no change in body composition of young or old rats flown in space for 7 days. Pitts *et al.* (1972) and Ushakov *et al.* (1980) reported an increase in percent body fat of rats exposed to space flight for 18.5 days. Thus, the duration of exposure to the microgravity environment of space flight may affect the response of body composition.

Energy stores

At the onset of hypergravity exposure, there is a period of hypophasia that is compensated for by mobilisation of fat stores. This decrease in body fat is not replaced. Thus, there is a reduction in percent body fat and the mass of fat pads (Oyama and Platt, 1965; Pitts *et al.*, 1972; Pitts, 1982; Moran *et al.*, 2001; Warren *et al.*, 2001). This decrease is dependent upon the level of gravity to which the animals are exposed. Moran *et al.* (2001) reported the level of gravity to be closely related to the decrease in fat pad mass and this was correlated with the reduction in plasma leptin concentration. Recent work by Warren *et al.* (1997) suggested an increase in plasma leptin in rats flown in space. However, an accompanying increase in indices of body fat was not reported. The flight data on leptin, coupled with the decrease during centrifugation, demonstrate a continuum across the spectrum of gravity reflective of the change in body fat.

Energy intake

A period of reduced food consumption occurs at the onset of exposure to hypergravity. The duration and magnitude of the reduction in food intake is proportional to the ***g*** load (Pitts *et al.*, 1972; Moran *et al.*, 2001; Warren *et al.*, 2001). The greater the ***g*** load, the greater the cumulative reduction in food intake. There is conflicting data that relate the level of energy intake after acclimation to an increase in gravity. However, when adjusted for differences in the absolute body mass no change in food intake after acclimation from controls is noted.

With space flight, there is extensive literature regarding nutrition problems in astronauts. Negative energy balance is due to a reduction in food intake in astronauts, however this does not appear to be the case in rats flown in space. Wade *et al.* (2002) found food consumption of rats to be similar in flight to that of ground controls housed under the same conditions and provided the same diet. Furthermore, the caloric intake of space flight rats was similar to rats housed and fed under laboratory conditions. Thus, space flight does not appear to adversely affect the energy intake of rats. Across the continuum of gravity, food intake in relation to body mass is sustained.

Muscle

Alterations of muscles in rodents in response to space flight have been evaluated in over 100 studies (Stevens *et al.*, 1993; Fejtek and Wassersug, 1999; Adams *et al.*, 2003). Despite studies employing animals of different strains, ages and gender with variations in the duration of exposure, the response in the weight bearing muscles studied have been relatively consistent. Recent space flight data suggest significant muscle atrophy occurs in both mice and rats (Stevens *et al.*, 1993; Harrison *et al.*, 2003; Nikawa *et al.*, 2004). With space flight, there is a reduction in muscle mass of the limbs associated with weight bearing. The muscle atrophy is the result of a reduction in cross-sectional area and changes in fiber type. These changes lead to a decrease in muscle function such as contraction strength and duration.

In response to increases in gravity, acclimated animals have increased muscle mass and alterations in fiber type associated with an increase in oxidative metabolism (Burton *et al.*, 1967; Martin and Romond, 1975; Martin, 1978, 1980; Roy *et al.*, 1996; Frey *et al.*, 1997; Picquet *et al.*, 1997; Vasques *et al.*, 1998; Picquet *et al.*, 2002; Stevens *et al.*, 2003; Bozzo *et al.*, 2004). This change is specific to muscle associated with weight bearing. In 1967, Burton *et al.* proposed a reduction in muscle mass of extensor muscles during space flight based on increases in muscle mass observed in chickens during exposure to hypergravity levels of 1.5, 2.0 and 3.0 *g* for up to one year. They postulated a reduction in muscle mass during space flight and an increased dependence on carbohydrates as an energy source but did not have the opportunity to test their hypothesis. In a subsequent comparison of hypergravity and space flight, Vasques *et al.* (1998) found the size of individual muscles was a function of the gravity load (Fig. 3). For muscles associated with posture, the relationship is stronger; it appears that in hypergravity, the level of activity is reduced confounding the interpretation of data (Holley *et al.*, 2003, Keil, 1969). The responses relating to fiber type shifts and muscle function can thus be confounded by the level of acclimation of the animals (Martin, 1978). In rats born and reared in hypergravity, the weight bearing soleus muscle shifted to a slower fiber type. These results were based on changes in both myosin heavy chain isoform expression and kinetic parameters that contributed to an increase in relative maximum tension (Burton *et al.*, 1967; Martin and Romond, 1975; Martin, 1978; 1980; Stevens *et al.*, 1993; Roy *et al.*, 1996; Frey *et al.*, 1997;

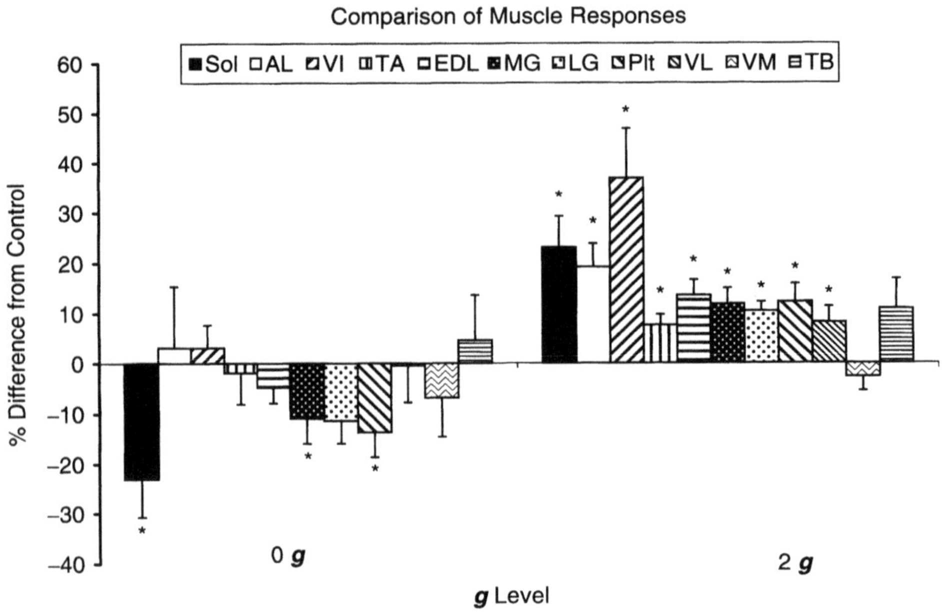

Fig. 3. Alteration of muscle mass as a function of body mass in response to hyper-gravity or the microgravity of space flight (Relevant reference: Vasques *et al.*, 1998) (Sol—soleus; MG—medial gastrocnemius; LG—lateral gastrocnemius; Plt—plantaris; EDL—extensor digtorium longus; TA—tibialis anterior; AL—adductor longus; VL—vastus lateralis; VI—vastus intermedius; VM—vastus medialis; TB—triceps brachii).

Picquet *et al.*, 1997; Vasques *et al.*, 1998; Fejtek and Wassersug, 1999; Picquet *et al.*, 2002; Stevens *et al.*, 2003; Adams *et al.*, 2003; Bozzo *et al.*, 2004; Nikawa *et al.*, 2004). Thus, alterations in muscle mass, fiber type and function of weight bearing muscles appear to respond across the gravity continuum, especially if the behavior of the animal is considered.

Bone

Loss of bone mass is one of the classic observations following space flight and potentially a critical limiting factor for long-term human exploration (Turner, 2000). The alteration in bone mass is a function of changes in the mechanical load experienced. Thus, in rats exposed to reduced mechanical load associated with the microgravity of space flight, a decrease in bone mass and density has been postulated. This hypothesis appears to be supported by findings but may be confounded by the use of rats of varying strains, ages and genders from different vendors (Morey-Holton *et al.*, 2000, Turner, 2000). In addition, the type of housing used in studies also appears to impact bone results (Morey-Holton *et al.*, 2000).

Rats reared in hypergravity have alterations in bone structure, growth, mineral composition and density (Smith, 1975; Jaekel *et al.*, 1977; Wunder *et al.*, 1979; Pace *et al.*, 1985; Wunder *et al.*, 1987). In mice exposed to hypergravity there is an increase in mineral content in young animals compared to age-matched controls with no difference detected in mature animals (Wunder, 1962; Fosse *et al.*, 1974; Keil and Evans, 1979). The changes in bone appears to be related to the load to which they are exposed, which is a function of body weight (body mass × gravity level). Thus, the larger the animal, the greater the response to increases in gravity. Despite the extensive evaluation of bone responses to space flight, the absence of the use of modern techniques to study bone tissues in animals exposed to hypergravity make comparison across the spectrum of gravity difficult.

Immune system

Numerous immune aberrations have been reported in response to alterations in the level of gravity. Early work by Burton and Smith (1965) noted many of these response were similar to general responses to stress. Furthermore, they found that if the change was not rectified the animals subsequently died. Changes in the organs important in immune function such as cellular responsiveness were reported to respond to alterations in gravity. In mice, hypergravity led to a reduction in the mass of the spleen and thymus (Goldstein and Ishay, 1998; Gridley *et al.*, 2002). In rats, no reduction was observed (Oyama and Platt, 1965). There are acute responses in cell type distribution that do not persist in mice or rats exposed to hypergravity (Gridley *et al.*, 2002). With space flight the data have been inconsistent, and appear to be related to mission length, ambient environment (i.e., temperature, carbon dioxide, etc.) and the type of assays used (Gmunder and Cogoli, 1996; Sonnenfeld, 1998; Sonnenfeld *et al.*, 1998; Gridley *et al.*, 2003; Pecaut *et al.*, 2003). Specifically, recent publications on mice report a transient increase in ambient temperature, pronounced loss of body mass, and reduced water consumption (Stevens *et al.*, 1993; Gridley *et al.*, 2002; Pecaut *et al.*, 2003). These types of factors encountered during space flight confound results, making comparisons between species difficult. Furthermore, persistent alteration across the continuum of gravity has not been demonstrated in either mice or rats. Sonnenfeld *et al.* (1998a, 1998b) have conducted extensive studies on the alterations of the immune system during space flight. They reported alterations in leukocyte subset distribution, inhibition of bone marrow cells to stimulating factors and decreases in modulating factors such as interferon and cytokines. When this group studied rats under similar conditions to space flight but in a hypergravity environment of 2 *g*, no significant alterations were found (Sonnenfeld *et al.*, 1995). The authors also noted the absence of immune changes even during the acute phase. The absence of immune responses across the gravity continuum raises a number of possible explanations (Scibetta *et al.*, 1984; Sonnenfeld *et al.*, 1995). First, the changes during space flight are due to other environmental factors such as exposure to radiation or the increase in ambient carbon dioxide. Second, the immune alterations may be threshold phenomena where the set point is at gravity levels below 1 ***g***. These questions await further investigation.

Red blood cells

With space flight, there is a well-defined reduction in red cell mass in rats and humans (Alfrey *et al.*, 1996; Allebban *et al.*, 1996). This decrease was attributed to a reduction in bone marrow production of red cells. In contrast, Burkovskaya and Krasnov (1991) reported that rats reared in hypergravity at 2 ***g*** showed an increase in red blood cell production. This conclusion was based on the finding that the bone marrow of centrifuged rats had a decrease in neutrophyles and increases in eosinophyles and erythrocaryocytes. Furthermore, there was an increase in circulating reticulocytes. Thus, the limited findings on red blood cells in hypergravity animals suggest a response opposite to that reported for space flight.

Cardiovascular system

The impact of the space flight on the cardiovascular system of humans has been extensively studied. One of the major contributing factors in the response of humans is the height of their hydrostatic column, which necessitates a continual response to Earth's gravitational field each time the body is reoriented. In rodents, the orthostatic column is minimal except when the animal rears up on their hind legs. Thus, the responsiveness of the cardiovascular system of rodents to altered gravity may be limited. Investigations relating to the response of the cardiovascular system of rats to space flight have been limited and conflicting (Zhang, 2001). There appears to be a decrease in the heart rate of rats during space flight and an alteration in vascular responsiveness upon return to Earth (Fuller, 1985; Zhang, 2001). Chronic exposure to centrifugation at 2 ***g*** has been associated with a persistent increase in the heart rate of rats (Sudon and Ikawa, 1988; Fuller *et al.*, 2002). This may be the result of an increase in sympathetic tone, as urinary levels of catecholamines are increased in centrifuged rats (Moran *et al.*, 2001). The alteration in heart rate could contribute to morphological changes associated with alterations in the level of gravity (Goldstein *et al.*, 1992, 1998).

Body temperature

Acute exposure to hypergravity induces a decrease in body temperature. This reduction is a function of the level of gravity to which the animal is exposed. Horowitz and colleagues (Horowitz and Horwitz, 1978; Fuller, 1994) have expressed the decrease in body temperature and the rectification in a variety of mathematical considerations. Holly *et al.* (2003) recently found a dose response with variations in the magnitude of the initial decrease, as well as the rate of recovery. During space flight, Fuller *et al.* (2002) demonstrated a mild increase in body temperature upon the initial period of exposure to microgravity.

Following the initial changes in body temperature, normal mean daily values are attained within 5–7 days. However, differences in the diurnal pattern during space flight and hypergravity have been suggested (Schertel *et al.*, 1980).

Vestibular system

Alterations in the function of the vestibular system result in a number of aberrant behaviors. In response to a hypergravity of 2 ***g***, mice have shown balance disturbances, less rearing activities, an inability to right and difficulties in swimming behavioral tests (Dalton *et al.*, 1991; Fox *et al.*, 1998; Sondag *et al.*, 1995; Sondag *et al.*, 1996; Wubbels *et al.*, 2002a; Wubbels *et al.*, 2002b). Of note was that hamsters reared in hypergravity, when dropped underwater, had extreme difficulty in locating the surface. Similar behavioral alterations have been reported in rats (Dalton *et al.*, 1991; Fox *et al.*, 1998; Wubbels *et al.*, 2002a).

These changes in behavior have been attributed to morphological changes in the vestibular system, specifically the otolith organs which are responsible for the transduction of forces resulting from linear accelerations. Gravity is such a force. In hypergravity animals, there is a change in the size distribution and density of otoconia (Sondag *et al.*, 1995, 1996). Exposure to hypergravity (2.5 ***g***) for a month resulted in no effect on the calcium content, size or shape of otoconia, but the total area of the utricular macula with small otoconia was increased. The change in otoconia distribution would contribute to a decrease in the sensory transduction of the otolith organ upon transition from one gravity field to another. That is, the organ functions normally for the gravity level in which the animal is reared but contribute to abnormal behavior in an altered gravity.

Changes in the sensing organs could contribute to abnormal behavior. With exposure to space flight, there is a change in the distribution of hair cell types and in the innervation of the cell bodies (Ross, 1998, 2000). With space flight, there is an increase of 100% in the number of synapses on type II hair cells. In contrast, hypergravity at 2 ***g*** resulted in a significant reduction, of 30% (Dalton *et al.*, 1991). Thus, alteration in the vestibular system across the spectrum of gravity may in part contribute to abnormal behaviors upon transitioning from one gravitational field to another. However, acclimation of vestibular apparatus to a specific gravitational field allows the animal to function and survive.

Mammary metabolism

The series of studies on the effects of altered gravity on mammary metabolisms by Plaut and coworkers (1999, 2003) clearly define the influence of gravity on a physiological system. Pregnant animals were exposed to space flight (microgravity; 0 ***g***), control conditions at 1 ***g***, and centrifugation at gravity levels of 1.5, 1.75 and 2.0 ***g*** for eleven days (days 11 to 20 of gestation). On the twentieth day of gestation, glucose oxidation into carbon dioxide and incorporation into lipids was measured in mammary glands. There was a strong negative correlation between metabolic rate and the level of gravity (Fig. 4).

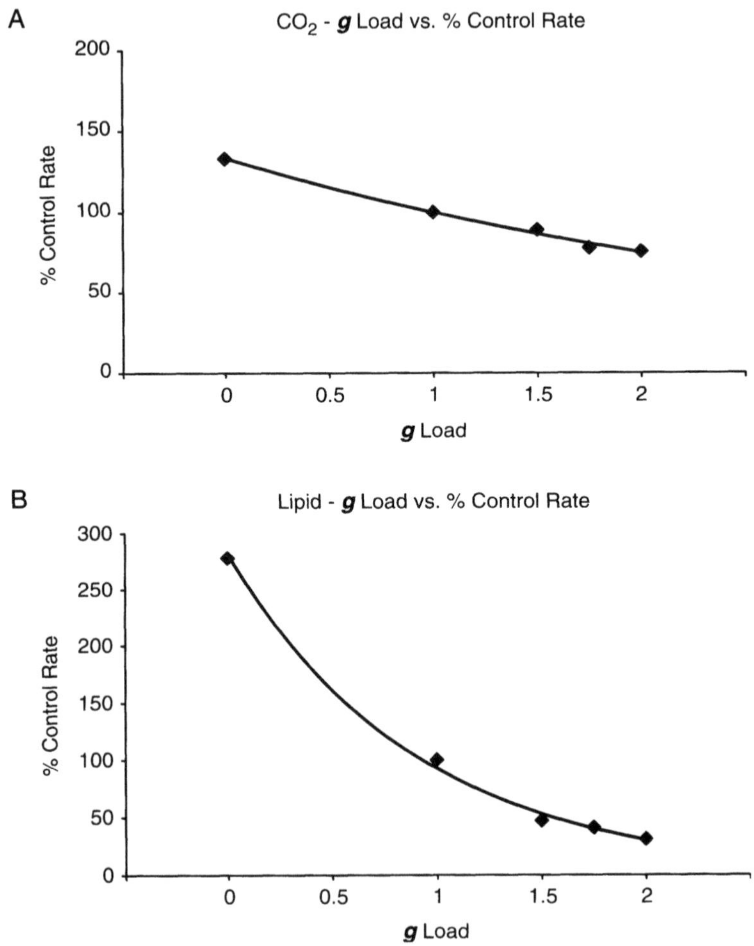

Fig. 4. Changes in glucose oxidation into carbon dioxide (A) or incorporation into lipids (B) across gravity levels (Relevant reference: Plaut *et al.*, 2003).

Approximately 99% of the variance in glucose incorporation into lipids or oxidation could be accounted for by differences in gravity load. These findings demonstrated a clear alteration in metabolism across the gravity continuum, and the utility of using hypergravity as a means of predicting responses to space flight.

Summary

In response to hypergravity, it appears that the larger the animal, the greater the response, if present. Therefore, the response of a rat exceeds that of a mouse in the same hypergravity environment. When investigated in the microgravity environment of space flight, this appears to hold true. The lack of definitive data obtained in space for either species makes the extrapolation of the continuum to levels below Earth-gravity problematic. However, in systems where responses are detected for both space flight and acceleration by centrifugation, a gravitational continuum is present supporting the "principle of continuity". For those and similar systems, it appears that the use of hypergravity could be used to predict responses to space flight.

References

Adams, G.R., Caiozzo, V.J. and Baldwin, K.M. (2003) Skeletal muscle unweighting: spaceflight and ground-based models. J. Appl. Physiol. 95, 2185–2201.

Alfrey, C.P., Udden, M.M., Leach-Huntoon, C., Driscoll, T. and Pickett, M.H. (1996) Control of red blood cell mass in spaceflight. J. Appl. Physiol. 81, 98–104.

Allebban, Z., Gibson, L.A., Lange, R.D., Jago, T.L., Strickland, K.M., Johnson, D.L. and Ichiki, A.T. (1996) Effects of spaceflight on rat erythroid parameters. J. Appl. Physiol. 81, 117–122.

Amtmann, E., Oyama, J. and Fisher, G.L. (1976) Effect of chronic centrifugation of the musculoskeletal system of the dog. Anat. Embryol. (Berl). 149, 71–78.

Baer, L.A., Ronca, A.E. and Wade, C.E. (2000) Survival and growth of developing rats during centrifugation at 2G. J. Gravit. Physiol. 7, 17–22.

Bozzo, C., Stevens, L., Bouet, V., Montel, V., Picquet, F., Falempin, M., Lacour, M. and Mounier, Y. (2004) Hypergravity from conception to adult stage: effects on contractile properties and skeletal muscle phenotype. J. Exp. Biol. 207, 2793–2802.

Burkovskaya, T.E. and Krasnov, I.B. (1991) Blood and bone marrow of rats born and grown under hypergravity. Physiol. 34, S139–140.

Burton, R.R. and Smith, A.H. (1965) Chronic Acceleration Sickness. Aerosp. Med. 36, 39–44.

Burton, R.R., Besch, E.L., Sluka, S.J. and Smith, A.H. (1967) Differential effects of chronic acceleration on skeletal muscles. J. Appl. Physiol. 23, 80–84.

Burton, R.R. and Smith, A.H. (1996) Adaptation to acceleration environments. *Handbook of Physiology Environmental Physiology*, pp. 943–976, American Physiological Society, Bethesda, MD.

Burton, R.R. and Smith, A.H. (1999) Can centrifugation at increased G be used to predict results of 0G? J. Gravit. Physiol. 6, P1–4.

Clement, G. and Pavy-Le Traon, A. (2004) Centrifugation as a countermeasure during actual and simulated microgravity: a review. Eur. J. Appl. Physiol. 92, 235–248.

Dalton, N.G., Ross, M.D., Fox, R.A., Corcoran, M.L., Cutler, L.D. and Wu, L.C. (1991) Effects of chronic hyper-gravity on the righting reflex and vestibular endorgan in the rat. Neurosc. Abs. 17, 316.

Doden, E., Oyama, J. and Amtmann, E. (1978) Effect of chronic centrifugation on bone density of the dog. Anat. Embryol. (Berl). 153, 321–329.

Economos, A.C. (1979) Gravity, metabolic rate and body size of mammals. Physiol. 22, S71–72.

Fast, T., Grindeland, R., Kraft, L., Ruder, M., Vasques, M., Lundgren, P., Scibetta, S., Tremor, J., Buckendahl, P., Keil, L. *et al.* (1985) Rat maintenance in the Research Animal Holding Facility during the flight of Space Lab 3. Physiologist 28, S187–188.

Fejtek, M.B. and Wassersug, R.J. (1999) Survey of studies on how spaceflight affects rodent skeletal muscle. Adv. Space Biol. Med. 7, 1–30.

Fosse, G., Gat, H., Holmbakken, N. and Kvinnsland, S. (1974) Bone atrophy and hypergravity in mice. Growth 38, 329–342.

Fox, R.A., Daunton, N.G. and Corcoran, M.L. (1998) Study of adaptation to altered gravity through systems analysis of motor control. Adv. Space Res. 22, 245–253.

Frey, M., von Kanel-Christen, R., Stalder-Navarro, V., Duke, P.J., Weibel, E.R. and Hoppeler, H. (1997) Effects of long-term hypergravity on muscle, heart and lung structure of mice. J. Comp. Physiol. [B] 167, 494–501.

Fuller, C.A. (1985) Homeostasis and biological rhythms in the rat during spaceflight. Physiol. 28, S199–200.

Fuller, C.A. (1994) The effects of gravity on the circadian timing system. J. Gravit. Physiol. 1, P1–4.

Fuller, P.M., Warden, C.H., Barry, S.J. and Fuller, C.A. (2000) Effects of 2-G exposure on temperature regulation, circadian rhythms, and adiposity in UCP2/3 transgenic mice. J. Appl. Physiol. 89, 1491–1498.

Fuller, P.M., Jones, T.A., Jones, S.M. and Fuller, C.A. (2002) Neurovestibular modulation of circadian and homeostatic regulation: vestibulohypothalamic connection? Proc. Natl. Acad. Sci. USA 99, 15723–15728.

Gmunder, F.K. and Cogoli, M.A. (1996) Effect of spaceflight on lymphocyte function and immunity. *Handbook of Physiology Environmental Physiology*, pp. 799–813, American Physiological Society, Bethesda, MD.

Goldstein, M.A., Edwards, R.J. and Schroeter, J.P. (1992) Cardiac morphology after conditions of microgravity during COSMOS 2044. J. Appl. Physiol. 73, S94–S100.

Goldstein, M.A., Cheng, J. and Schroeter, J.P. (1998) The effects of increased gravity and microgravity on cardiac morphology. Aviat. Space Environ. Med. 69, A12–16.
Goldstein, O. and Ishay, J.S. (1998) The white blood cell line: changes induced in mice by hypergravity. Adv. Space Res. 21, 1333–1343.
Gridley, D.S., Pecaut, M.J., Green, L.M., Miller, G.M. and Nelson, G.A. (2002) Hypergravity-induced immunomodulation in a rodent model: lymphocytes and lymphoid organs. J. Gravit. Physiol. 9, 15–27.
Gridley, D.S., Nelson, G.A., Peters, L.L., Kostenuik, P.J., Bateman, T.A., Morony, S., Stodieck, L.S., Lacey, D.L., Simske, S.J. and Pecaut, M.J. (2003) Genetic models in applied physiology: selected contribution: effects of spaceflight on immunity in the C57BL/6 mouse. II. Activation, cytokines, erythrocytes, and platelets. J. Appl. Physiol. 94, 2095–2103.
Harrison, B.C., Allen, D.L., Girten, B., Stodieck, L.S., Kostenuik, P.J., Bateman, T.A., Morony, S., Lacey, D. and Leinwand, L.A. (2003) Skeletal muscle adaptations to microgravity exposure in the mouse. J. Appl. Physiol. 95, 2462–2470.
Holley, D.C., DeRoshia, C.W., Moran, M.M. and Wade, C.E. (2003) Chronic centrifugation (hypergravity) disrupts the circadian system of the rat. J. Appl. Physiol. 95, 1266–1278.
Horowitz, J.M. and Horwitz, B.A. (1978) Thermoregulatory responses of unanesthetized rats exposed to gravitational fields of 1 to 4g. Life Sci. Space Res. 16, 77–82.
Ishay, J. and Barr-Nea, L. (1977) Effects of hypergravity on rat fertillity, pregnancy, parturition and survival. Experientia. 33, 244–246.
Jaekel, E., Amtmann, E. and Oyama, J. (1977) Effect of chronic centrifugation on bone density of the rat. Anat. Embryol. (Berl.). 151, 223–232.
Janer, L. and Duke, J. (1984) Effects of chronic centrifugation on mice. Physiol. 27, S75–76.
Keil, L.C. (1969) Changes in growth and body composition of mice exposed to chronic centrifugation. Growth 33, 83–88.
Keil, L.C. and Evans, J.W. (1979) Bone growth and composition in weanling and mature rats exposed to chronic centrifugation. Physiol. 22, S53–54.
Knight, T.A. (1806) On the direction of the radical and german during the vegitation of seeds. Philos. Trans. R. Soc. London 96, 99–100.
Martin, W.D. and Romond, E.H. (1975) Effects of chronic rotation and hypergravity on muscle fibers of soleus and plantaris muscles of the rat. Exp. Neurol. 49, 758–771.
Martin, W.D. (1978) Time course of change in soleus muscle fibers of rats subjected to chronic centrifugation. Aviat. Space Environ. Med. 49, 792–797.
Martin, W.D. (1980) Effects of chronic centrifugation on skeletal muscle fibers in young developing rats. Aviat. Space Environ. Med. 51, 473–484.
Moore, J. and Duke, J. (1988) Effect of chronic centrifugation on mouse breeding pairs and their offspring. Physiol. 31, S120–S121.

Moran, M.M., Stein, T.P. and Wade, C.E. (2001) Hormonal modulation of food intake in response to low leptin levels induced by hypergravity. Exp. Biol. Med. (Maywood) 226, 740–745.

Morey-Holton, E.R., Halloran, B.P., Garetto, L.P. and Doty, S.B. (2000) Animal housing influences the response of bone to spaceflight in juvenile rats. J. Appl. Physiol. 88, 1303–1309.

Nikawa, T., Ishidoh, K., Hirasaka, K., Ishihara, I., Ikemoto, M., Kano, M., Kominami, E., Nonaka, I., Ogawa, T., Adams, G.R., Baldwin, K.M., Yasui, N., Kishi, K. and Takeda, S. (2004) Skeletal muscle gene expression in space-flown rats. Faseb. J. 18, 522–524.

Ortiz, R.M. and Wade, C.E. (2000) Water balance in rats exposed to chronic centrifugation. J. Appl. Physiol. 89, 56–60.

Oyama, J. and Platt, W.T. (1965) Effects of prolonged centrifugation on growth and organ development of rats. Am. J. Physiol. 209, 611–615.

Oyama, J. and Platt, W.T. (1967) Reproduction and growth of mice and rats under conditions of simulated increased gravity. Am. J. Physiol. 212, 164–166.

Oyama, J., Platt, W.T. and Holland, V.B. (1971) Deep-body temperature changes in rats exposed to chronic centrifugation. Am. J. Physiol. 221, 1271–1277.

Oyama, J. (1975) Response and adaptation of beagle dogs to hypergravity. Life Sci. Space Res. 13, 11–17.

Oyama, J., Solgaard, L., Corrales, J. and Monson, C.B. (1985) Growth and development of mice and rats conceived and reared at different G-intensities during chronic centrifugation. Physiol. 28, S83–84.

Pace, N., Smith, A.H. and Rahlmann, D.F. (1985) Skeletal mass change as a function of gravitational loading. Physiol. 28, S17–20.

Pecaut, M.J., Nelson, G.A., Peters, L.L., Kostenuik, P.J., Bateman, T.A., Morony, S., Stodieck, L.S., Lacey, D.L., Simske, S.J. and Gridley, D.S. (2003) Genetic models in applied physiology: selected contribution: effects of spaceflight on immunity in the C57BL/6 mouse. I. Immune population distributions. J. Appl. Physiol. 94, 2085–2094.

Phillips, R.W. (2002) Gravity: It's the law. J. Gravit. Physiol. 9, P15–16.

Picquet, F., Stevens, L., Butler-Browne, G.S. and Mounier, Y. (1997) Contractile properties and myosin heavy chain composition of newborn rat soleus muscles at different stages of postnatal development. J. Muscle Res. Cell Motil. 18, 71–79.

Picquet, F., Bouet, V., Canu, M.H., Stevens, L., Mounier, Y., Lacour, M. and Falempin, M. (2002) Contractile properties and myosin expression in rats born and reared in hypergravity. Am. J. Physiol. Regul. Integr. Comp. Physiol. 282, R1687–1695.

Pitts, G.C., Bull, L.S. and Oyama, J. (1972) Effect of chronic centrifugation on body composition in the rat. Am. J. Physiol. 223, 1044–1048.

Pitts, G.C. (1977) Body size and chronic acceleration. Life Sci. Space Res. 15, 251–255.

Pitts, G.C. (1982) Effects of chronic acceleration on body composition. Physiol. 25, S13–S16.

Pitts, G.C., Ushakov, A.S., Pace, N., Smith, A.H., Rahlmann, D.F. and Smirnova, T.A. (1983) Effects of weightlessness on body composition in the rat. Am. J. Physiol. 244, R332–337.
Plaut, K., Maple, R., Vyas, C., Munaim, S., Darling, A., Casey, T. and Alberts, J.R. (1999) The effects of spaceflight on mammary metabolism in pregnant rats. Proc. Soc. Exp. Biol. Med. 222, 85–89.
Plaut, K., Maple, R.L., Wade, C.E., Baer, L.A. and Ronca, A.E. (2003) Effects of hypergravity on mammary metabolic function: gravity acts as a continuum. J. Appl. Physiol. 95, 2350–2354.
Reichhardt, T. (1998) Animal deaths turn shuttle into 'necrolab'. Nature 393, 4.
Ronca, A.E., Baer, L.A., Daunton, N.G. and Wade, C.E. (2001) Maternal reproductive experience enhances early postnatal outcome following gestation and birth of rats in hypergravity. Biol. Reprod. 65, 805–813.
Ronca, A.E. (2003) Mammalian development in space. Adv. Space Biol. Med. 9, 217–251.
Ross, M.D. (1998) Synaptic changes in rat maculae in space and medical imaging: the link. Otolaryngol. Head Neck Surg. 118, S25–28.
Ross, M.D. (2000) Changes in ribbon synapses and rough endoplasmic reticulum of rat utricular macular hair cells in weightlessness. Acta. Otolaryngol. 120, 490–499.
Roy, R.R., Roy, M.E., Talmadge, R.J., Mendoza, R., Grindeland, R.E. and Vasques, M. (1996) Size and myosin heavy chain profiles of rat hindlimb extensor muscle fibers after 2 weeks at 2G. Aviat. Space Environ. Med. 67, 854–858.
Schertel, E.R., Horowitz, J.M. and Horwitz, B.A. (1980) Effects of gravitational profiles on the rat's thermoregulatory response to cold. J. Appl. Physiol. 49, 663–668.
Scibetta, S.M., Caren, L.D. and Oyama, J. (1984) The unresponsiveness of the immune system of the rat to hypergravity. Aviat. Space Environ. Med. 55, 1004–1009.
Smith, A.H. and Burton, R.R. (1967) Influence of the ambient accelerative force upon mature body size. Growth 31, 317–329.
Smith, A.H. (1976) Physiological changes associated with long-term increases in acceleration. Life Sci. Space Res. 14, 91–100.
Smith, S.D. (1975) Effects of long-term rotation and hypergravity on developing rat femurs. Aviat. Space Environ. Med. 46, 248–253.
Sondag, H.N., de Jong, H.A., van Marle, J. and Oosterveld, W.J. (1995) Effects of sustained acceleration on the morphological properties of otoconia in hamsters. Acta. Otolaryngol. 115, 227–230.
Sondag, H.N., De Jong, H.A., Van Marle, J., Willekens, B. and Oosterveld, W.J. (1996) Otoconial alterations after embryonic development in hypergravity. Brain Res. Bull. 40, 353–356; discussion 357.
Sonnenfeld, G., Koebel, D.A. and Davis, S. (1995) Effects of hypergravity on immunologic function. Microgravity Sci. Technol. 7, 323–326.

Sonnenfeld, G. (1998) Immune responses in space flight. Int. J. Sports Med. 19 (Suppl. 3), S195–202; discussion S202–194.
Sonnenfeld, G., Foster, M., Morton, D., Bailliard, F., Fowler, N.A., Hakenewerth, A.M., Bates, R. and Miller, E.S. Jr. (1998) Spaceflight and development of immune responses. J. Appl. Physiol. 85, 1429–1433.
Stevens, L., Mounier, Y. and Holy, X. (1993) Functional adaptation of different rat skeletal muscles to weightlessness. Am. J. Physiol. 264, R770–776.
Stevens, L., Bozzo, C., Nemirovskaya, T., Montel, V., Falempin, M. and Mounier, Y. (2003) Contractile properties of rat single muscle fibers and myosin and troponin isoform expression after hypergravity. J. Appl. Physiol. 94, 2398–2405.
Sudon, M. and Ikawa, S. (1988) Response of rats to 3-week centrifugal accelerations. Jpn. J. Aerospace Environ. Med. 25, 36–74.
Turner, R.T. (2000) Invited review: what do we know about the effects of spaceflight on bone? J. Appl. Physiol. 89, 840–847.
Ushakov, A.S., Smirnova, T.A., Pitts, G.C., Pace, N., Smith, A.H. and Rahlmann, D.F. (1980) Body composition of rats flown aboard Cosmos-1129. Physiol. 23, S41–44.
Vasques, M., Lang, C., Grindeland, R.E., Roy, R.R., Daunton, N., Bigbee, A.J. and Wade, C.E. (1998) Comparison of hyper- and microgravity on rat muscle, organ weights and selected plasma constituents. Aviat. Space Environ. Med. 69, A2–8.
Wade, C.E., Harper, J.S., Daunton, N.G., Corcoran, M.L. and Morey-Holton, E. (1997) Body mass change during altered gravity: spaceflight, centrifugation, and return to 1 G. J. Gravit. Physiol. 4, 43–48.
Wade, C.E. and Morey-Holton, E. (1998) Alteration of renal function of rats following spaceflight. Am. J. Physiol. 275, R1058–1065.
Wade, C.E., Ortiz, R.M. and Baer, L.A. (2000) Increases in body mass of rats during spaceflight: models and measurements. Aviat. Space Environ. Med. 71, 1126–1130.
Wade, C.E., Miller, M.M., Baer, L.A., Moran, M.M., Steele, M.K. and Stein, T.P. (2002) Body mass, energy intake, and water consumption of rats and humans during space flight. Nutrition 18, 829–836.
Warren, L.E., Horwitz, B.A. and Fuller, C.A. (1997) Gravity and body mass regulation. J. Gravit. Physiol. 4, P89–92.
Warren, L.E., Horwitz, B.A., Hamilton, J.S. and Fuller, C.A. (2001) Effects of 2 G on adiposity, leptin, lipoprotein lipase, and uncoupling protein-1 in lean and obese Zucker rats. J. Appl. Physiol. 90, 606–614.
Wubbels, R.J., Sondag, H.N., van Marle, J. and de Jong, H.A. (2002a) Effects of hypergravity on the morphological properties of the vestibular sensory epithelium. I. Long-term exposure of rats after full maturation of the labyrinths. Brain Res. Bull. 57, 677–682.
Wubbels, R.J., van Marle, J., Sondag, H.N. and de Jong, H.A. (2002b) Effects of hypergravity on the morphological properties of the vestibular sensory

epithelium. II. Life-long exposure of rats including embryogenesis. Brain Res. Bull. 58, 575–580.
Wunder, C.C. (1960) Altered growth of animals after continual centrifugation. *In Proc. Iowa Acad. Sci.*
Wunder, C.C. (1962) Survival of mice during chronic centrifugation. I. Studies of male mice at different ages at onset of exposure to one field and those at different intensities of gravity for animals of the same age. Aeromed. Acta. 33, 866–870.
Wunder, C.C., Welch, R.C. and Cook, K.M. (1979) Femur strength as influenced by growth, bone-length and gravity with the male rat. J. Biomech. 12, 501–507.
Wunder, C.C., Cook, K.M., Watkins, S.R. and Moressi, W.J. (1987) Femur-bending properties as influenced by gravity: V. Strength vs. calcium and gravity in rats exposed for 2 weeks. Aviat. Space Environ. Med. 58, 977–982.
Young, L.R. (1999) Artificial gravity considerations for a mars exploration mission. Ann. NY. Acad. Sci. 871, 367–378.
Yuwaki, K. and Okuno, M. (2003) Changes in food intake and growth rate in mice under hypergravity. Biol. Sci. Space. 17, 219–220.
Zhang, L.F. (2001) Vascular adaptation to microgravity: what have we learned? J. Appl. Physiol. 91, 2415–2430.

Experimentation with Animal Models in Space
G. Sonnenfeld (editor)

DOI: 10.1016/S1569-2574(05)10010-0

Gravity Effects on Life Processes in Aquatic Animals

Eberhard R. Horn

University of Ulm, Germany

Introduction

Biological research using the space environment focuses on two aspects, space exploration and basic research. Further knowledge about the solar system will allow humans to extend their living area to other planets. This can only be achieved if effects of space radiation and weightlessness on living systems are known in order to reduce the risk of space flights and a life on other planets. Basic research in space deals with the question of how gravity has influenced the evolution of life. To achieve this goal, organisms have to be deprived from the influence of gravity during space flights and the induced changes have to be analyzed by means of any method including anatomy, physiology, behaviour, immunology, molecular biology or genetics.

The close relationship with humans has favoured the use of mammals such as monkeys, rats, dogs or mice to study the aspects of physiological risk estimation of a space flight. Parallel to applied research, basic biological research was included into space exploration. In particular, lower vertebrates such as fishes and amphibians and invertebrates such as insects, scorpions and worms were used or will be used as model species depending on the specific scientific questions. It is the feature of basic research that it also answers any human relevant question; thus all results have a chance to contribute to the risk estimation for manned space flights in future.

Early studies focused on the contribution of gravity to development. Following the classical line, aquatic vertebrates such as fish and amphibians, or aquatic invertebrates including sea urchins were used. The attractiveness of these species for developmental studies (which are still used in research on the molecular basis of development; see Gilbert, 2003) was their high reproduction rate, the easily available eggs, the high number and size of the eggs. These factors allow artificial cleavage, topological lesions, transplantation of parts of the eggs in recipient eggs or injections to knock down specific genes. In addition, a precise staging by means of external markers allows the description of physiological development at a high resolution. Extensive knowledge exists in aquatic vertebrate species about sensory, neuronal, motor and hormonal mechanisms (see Llinás and Precht, 1976) which serve as an excellent basis for physiological as well as morphological studies in space.

Due to the morphological and physiological similarities between the sensory vestibular systems of vertebrates, data from space research in aquatic lower vertebrates can be used for risk estimation of human space flight and for the establishment of countermeasures to overcome impaired postural control and impaired wellness caused by space sickness and kinetosis. These organ specific reasons were supplemented by the adaptive properties of the vestibular system called vestibular compensation. Mechanisms of vestibular compensation are common in all vertebrates; they are activated by lesions of the vestibular system and cause the normalisation of physiology and behaviour (for comparative aspects of vestibular compensation, see Schaefer and Meyer, 1974; Precht and Dieringer, 1985; for amphibian vestibular compensation, see Dieringer, 1995; Goto *et al.*, 2001).

While fish and amphibians were useful animal models to study developmental and neuronal processes, they were less useful to investigate microgravity effects on the skeletal, muscular and cardiovascular systems. These species are adapted to decreased weight effects during normal life conditions because buoyancy in their aquatic environment counteracts gravity-induced weight-loading on muscles and bones. Only a few studies showed modifications such as muscle loss (Mori *et al.*, 1994; Snetkova *et al.*, 1995) and observations on osteoporosis are unlikely. Some experiments on mineralisation processes in skeletons in the absence of gravity were performed in sea urchin during the IML-2 mission (Marthy *et al.*, 1996) and parabolic flights (Izumi-Kurotani and Kiyomoto, 2003) and revealed transient depression of mineralisation during short-term microgravity, but no effects during long-term microgravity.

Fluid shifts within the cardiovascular systems are strongly affected by gravity in land-living animals, in particular in upright walking species including man. Compensatory mechanisms such as venous pressure had been developed during evolution to avoid an accumulation of body fluid in lower portions of the body. In microgravity, these counter-mechanisms cause a fluid shift towards the upper body parts. Compensatory fluid-shifts towards the upper body parts are not needed in aquatic species. Thus, water-living species are also less useful for the study of physiological, homeostatic mechanisms during and after exposure to microgravity.

General features of aquatic animal models

Handling, life support and science are determining factors for the selection of the suitable animal model. Aquatic animals such as fish, amphibians, molluscs, sea urchins and medusae are in many instances appropriate for biological research in space. Some of them are classical models, while others were used only to answer specific questions. In most of these species, an atlas of standard development supplies scientists with a tool that delivers the basis for result comparisons between different laboratories (for *Xenopus laevis*: Nieuwkoop and Faber, 1967; for *Pleurodeles waltl*: Shi and Boucout, 1995; Gallien and Durocher, 1957; for *Oreochromis mossambicus*: Anken *et al.*, 1993; for *Danio rerio* : Kimmel *et al.*, 1995; for medaka fish *Oryzias latipes*: Anken and Bourrat, 1998).

Amphibians are divided in two subgroups, anurans and urodeles. Among anurans, various *Rana* species were used during the early period of space flight studies on Russian satellites and the manned American spacecraft Gemini. The clawed toad *Xenopus laevis* was the most frequently used anuran in gravitational biological research in later missions. *Xenopus* is a standard animal for all fields of morphology, physiology, and behaviour. Even epilepsy research takes advantage of the large oocysts to study properties of ion channels during this pathological hyperexcitation. Its general success as an aquatic experimental animal in space is probably due to its availability and easy handling.

Among the urodeles, *Pleurodeles waltl* and *Cynops pyrrhogaster* were used several times in space experiments, particularly in experiments on development and differentiation. They offer the advantage that the female retains the living sperm for several months in their cloacal pelvic gland; oocytes are fertilised during spawning. Thus, on-ground insemination can be performed and activation for egg deposition can be induced in-flight by hormonal injections (salamander, newt) or by an increase of the temperature in females that are sent into orbit in their hibernating status (newt). Compared to *Xenopus*, the rate of embryogenesis is slow; this fact facilitates studies with high stage resolution during embryonic development. They also regenerate lost extremities enabling scientists to study cell proliferation under weightlessness.

The list of fish species exposed to weightlessness during space flights is larger. *Fundulus heteroclitus*, zebrafish *Danio rerio* (cf. also *Brachyodanio rerio*), the Japanese medaka fish *Oryzias latipes* and the mouth-breeding cichlid fish *Oreochromis mossambicus* were all used in experiments on development. Adult goldfish *Carassius auratus* were used in studies on swimming. Specific questions on otolith formation were studied in the swordtail fish *Xiphophorus helleri* and neuronal activity in the toadfish *Ospsanus tau*.

Only a few invertebrate species were exposed to micro- or hypergravity. Sea urchins *Paracentrotus lividus* and *Sphaerechinus granularis* became useful for studies on mineralisation; the marine gastropod *Aplysia californica* and the freshwater pond snail *Biomphalaria glabrata* were used in otolith studies, the pond snail also in mineralisation studies, while locomotion characteristics and gravity receptor organs were studied in the scyphomedusa *Aurelia aurita*.

Development

Fertilisation and early developmental events in relation to gravity

One fundamental question in developmental biology is whether gravity is required for normal embryonic development, for axis and pattern formation, and for the subsequent morphogenesis and organogenesis. Most of the experiments at the beginning of the 20th century were done in frog eggs. Centrifugation that increases gravitational forces (hypergravity), clinostat rotation which produces a vector-free gravitational environment and, with progresses in the space flight techniques, orbital flights (weightlessness, microgravity) were applied to early developmental stages. All these experiments revealed that gravity influenced embryonic processes.

A typical feature of early development is the rotation of the egg inside the fertilisation membrane by which the animal-vegetal axis is aligned with gravity. The rotation is not a requirement for normal development since eggs prevented from rotation can develop normally. Generally, the direction of rotation determines the polarity of the embryonic axis. Eggs inclined with respect to gravity form the dorsal structures on the side of the egg's uppermost in the gravitational field. The answer to the question about the necessity of the gravity vector and morphogenetics in early development, however, needs the use of gravity deprivation during real space flights. Aquatic vertebrate (fish, frogs, salamander, newts) and invertebrate species (sea urchins) were the first-choice species in this experimental complex. Experiments with conditions modelling some aspects of the microgravity environment using the fast-rotating clinostat supplemented these studies and gave valuable hints to space flight experiments (Yokota *et al.*, 1994).

In the radial-symmetrical mature egg of *Xenopus laevis*, the polar animal-vegetal axis indicates roughly the embryo's main body axis. Pigment concentrating around the sperm-entry point marks the meridian that foreshadows the prospective ventral side. Since in most eggs the blastopore forms at the meridian about 180 degree away from the sperm-entry point, the embryo's general body pattern is established from that time on. However, the dorso-anterior and ventro-posterior polarities can still be altered by the influence of gravity and centrifugal forces which cause the rearrangement of yolk components. This suggests that gravity in conjunction with the sperm entry point establishes the dorso-ventral polarity.

In *Xenopus*, development in a clinostat where some components of the microgravity environment are reproduced, revealed no change of the cleavage rhythm. At the eight-cell stage, however, the location of the first horizontal cleavage furrow is shifted towards the vegetal pole and is completed earlier. Further modifications include that (1) the position of the blastocoel is more centered and the number of cell layers in the blastocoel roof is increased at the blastula stage, (2) a significant smaller blastocoel is formed, (3) the dorsal lip appeared closer to the vegetal pole at the gastrula stage and (4) the head and eye dimensions were enlarged at the hatching tadpole stage. Despite of these morphological changes, tadpoles at the feeding stage were largely indistinguishable from controls (Yokota *et al.*, 1994).

The first successful fertilisation in space was done during a ballistic rocket flight in 1988 using fully automated hardware (Ubbels, 1997). The experiment was successfully repeated on a sounding rocket flight in 1989 and two shuttle flights (IML-1 in 1992 and IML-2 in 1994). Due to the short duration of ballistic rocket flights, further development of these embryos occurred in 1*g*-conditions. In embryos raised in 1*g* after the MASER 3 rocket flight, development was slightly retarded compared to the ground embryos; microcephalisation and reduced tail formation was observed. In contrast, normally developed embryos were retrieved from the MASER 6 rocket flight and subsequent axis formation was normal (Ubbels, 1997; Ubbels *et al.*, 1995). A similar experiment was performed on Spacelab-J in 1995 (Souza *et al.*, 1995). All shuttle experiments differed from sounding rocket studies as embryonal development occurred for several days under microgravity conditions.

Similar to results obtained in the clinostat experiments, all sounding rocket and space flight studies with *Xenopus* embryos revealed a normal cleavage rhythm in microgravity but an increase in the number of cell layers of the blastocoel roof from 2 to 3 and a significant smaller blastocoel (Fig. 1, left

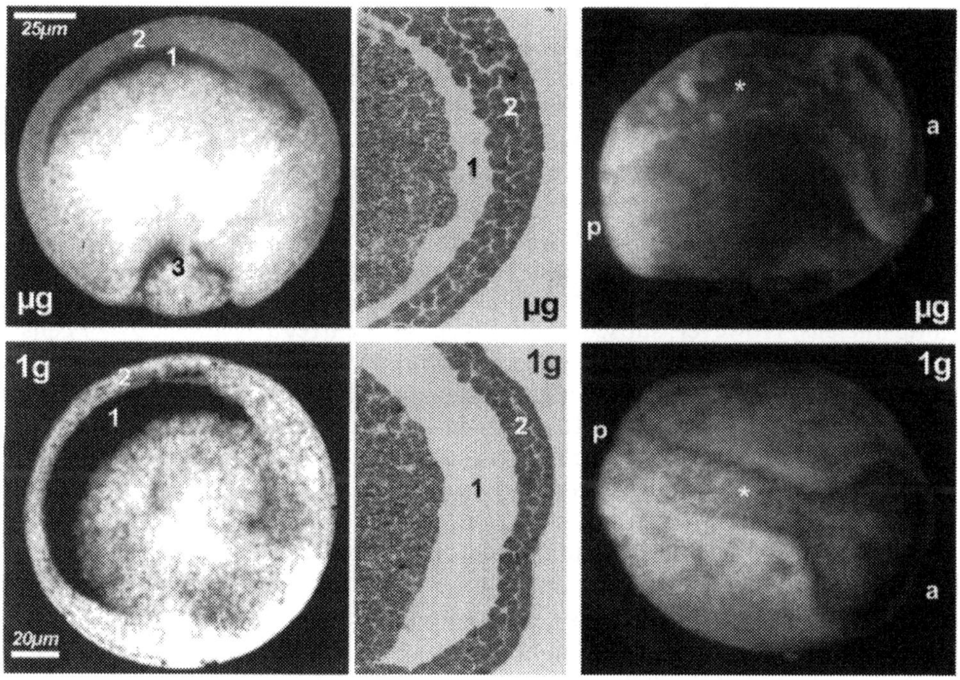

Fig. 1. Morphological malformations during embryonic periods of life in microgravity (μg)—Left and middle: Gastrulae from *Xenopus laevis* fixed in microgravity and 1G showing the thickening of the blastocoel roof by microgravity. 1, blastocoel, 2, blastocoel roof, 3, blastopore (from Ubbels *et al.*, 1995; see also Table 1)—Right: Disturbed neurulation (*) in the salamander *Pleurodeles waltl* by microgravity. Note the imcomplete closure of the neural tube in an embryo fixed in microgravity compared to the 1G control. a and p, anterior and posterior pole of the embryo, respectively (courtesy C. Dournon).

Table 1

Number of cell layers in the blastocoel roof of *Xenopus laevis* gastrulae—Note that 3 and 4 cell layers are more frequent in gastrulae exposed to microgravity than in the ground controls. Observations from the IML-1 mission (from Ubbels *et al.*, 1995).

	Cell layers (n)		
	2	3	4
microgravity	4	3	4
1G in-flight	1	–	2
1G ground	8	2	–

and middle; Table 1). Moreover, the blastocoel forms more centrally and vegetally due to an enlargement of the roof cells and not to a general increase of cell numbers. For all these microgravity conditions, these morphological differences disappear during further development to the late blastulae and early gastrulae (Ubbels *et al.*, 1995; Ubbels, 1997) and, after the space flight on Spacelab-J, normal tadpoles were retrieved (Souza *et al.*, 1995).

In-flight fertilisation was also performed using two urodele species, the salamander *Pleurodeles waltl* (FERTILE experiments on the Russian MIR space station in 1996 and 1998) and the newt *Cynops pyrrhogaster* (experiment Astronewt on IML-2 in 1994 with a repetition in 1995). In-flight video-recordings of early *Cynops* stages revealed normal morphological shapes of the late morula, early blastula, gastrula, neurula and tail bud stage up to the stage shortly before the first gill ramification appeared (Yamashita *et al.*, 2001). In *Pleurodeles*, 24 out of 25 microgravity-exposed eggs exhibited normal location of the first furrow, i.e., microgravity did not provoke an off-axis location of the zygotic nucleus. However, subsequent cleavages were irregular and 3, 5 or 7 cells were observed in the animal hemisphere. About 35% of microgravity-exposed eggs exhibited large unpigmented areas in the animal pole, and up to the morula stage movements of the pigment towards the animal pole were amplified. As in *Xenopus*, the blastocoel roof in microgravity-generated gastrulae was thicker than in 1***g***-gastrulae; but in contrast to *Xenopus*, it was always composed of two cell layers. Neurulation was also strongly affected by microgravity (Gualandris-Parisot *et al.*, 2002). During subsequent development in microgravity, all morphological changes were regulated to normal. In-flight video recordings revealed that the time between egg laying and hatching was identical in both microgravity-exposed and 1***g*** animals. Histological and immunohistochemical studies with larvae fixed 5 h after landing showed no microgravity specific effects in their central nervous system, eyes, somites, pronephros and gut (Dournon, 2003).

Studies in *Cynops pyrrhogaster* embryos and cultures from cells of the presumptive ectoderm make it likely that depression of apoptosis causes the increased thickness of the blastocoel roof. Under conditions that reproduced some aspects of the microgravity environment (clinostat rotation at 6 rpm for one day), *Cynops* gastrulae usually develop the thicker presumptive ectoderm compared to the controls as observed in *Xenopus* and *Pleurodeles*. In both gastrulae and cultures of presumptive ectoderm cells, TUNEL staining and electronmicroscopy revealed apoptotic cells, but their number was always smaller in clinostat-treated samples than in the controls (Komazaki, 2004).

In medaka fish (*Oryzias latipes*), ground-based studies revealed that gravity influences the position of the dorsoventral axis. Tilting or centrifugation (5***g***) affected the plane of bilateral symmetry and the orientation of the microtubules in the vegetal pole region of zygotes (Fluck *et al.*, 1998). In contrast, exposure to microgravity during the IML-2 mission (STS-65; 1995) caused no effects on further embryonic development. After the first successful mating under microgravity conditions during this mission (Ijiri, 1997, 1998), subsequent steps run similar to ground controls: newly laid eggs formed a cluster on the belly of the female fish; they left the body, and after detachment from the female's body, young fish hatched in microgravity. Post-flight examination of some space-borne fish revealed that neither the external appearance of embryos nor the formation of primordial germ cells was affected by microgravity. Off-springs from two remaining space originated fry one male and one female developed with no adverse effects (Ijiri, 2003). Similar observations were obtained earlier from guppies after the Russian Cosmos-1514 flight. Histological analysis of guppy embryos that performed organogenesis in space exhibited no abnormalities and also the second fish generation developed normal morphology (Cherdantseva, 1987).

Nervous system and neuron development

Neurulation is the first step in the formation of the nervous system. The FERTILE I, FERTILE II, and NEUROGENESIS experiments on Mir in 1996, 1998, and 1999, respectively, showed significant disturbances of neurulation in the salamander *Pleurodeles waltl.* During normal embryonic development, the closure of the neural folds bordering the neural plate occurs between stages 14 and 20. It starts in the median part of the antero-posterior axis of the neurula and spreads simultaneously along this axis towards the rostral and caudal parts of the embryo. The tube is totally closed at stage 20. Embryos which developed in microgravity showed an incomplete closure of their neural tube at cephalic and trunk levels (Fig. 1, right). The disturbance was seen in 13 out of 16 microgravity-exposed embryos (81%) while only 1 out of 22 1***g***-centrifugated embryos (4.5%) exhibited abnormal neurulation. Despite this morphological difference, the epidermal ciliated cells functioned normally (stage 16) and each microgravity embryo rotated randomly clockwise or counter-clockwise around its anteroposterior axis as 1***g***-controls did. The closure of the neural tube was completed at stage 31. On the cephalic level, the five brain subdivisions were morphologically normal; however, microcephaly developed more frequently in microgravity-exposed embryos (12 out of 30, i.e., 40%, 3 of them acephalic) than in 1*g*-control embryos (10 out of 36, i.e., 28%). Sense organs such as eye and ear developed normally (Gualandris-Parisot *et al.*, 2001).

The cytological differentiation of neuronal and glial structures was investigated in neural precursor cells from *Pleurodeles*, isolated in culture immediately after neuronal induction at the early neurula stage. During microgravity exposure on a 16-day FOTON flight, they differentiated without significant abnormalities, and they developed long neurites and normal networks. Some modifications were related to a faster differentiation of cells and to the formation of varicosities along neurites (Duprat *et al.*, 1998; Husson *et al.*, 1998).

Neurotransmitter appearance was insensitive to microgravity. Expression and activity of cholinergic neurons, an early marker of motor system differentiation, and the differentiation of the GABAergic system were similar to ground controls. In particular, embryos that developed in microgravity displayed identical patterns of choline transferase (ChAT) activity at stage 32 and 33 as 1*g*-ground control embryos. Immunostaining for GABA on ground is positive for the first time at stage 32b/33. Accordingly, staining in microgravity was negative at stage 26 but positive at stages 33–34 and 39–40 as in the ground controls (Gualandris-Parisot *et al.*, 2001).

Muscle development

Muscle development of *Pleurodeles* is rather insensitive to microgravity. Typical markers in somites differentiation such as their position and the appearance of striated structures (organised myofibrilles) did not differ from normal development (Husson *et al.*, 1998; Gualandris-Parisot *et al.*, 2001). In non-microcephalic embryos the first spontaneous contractions of the trunk muscles occurred at the same stages in both microgravity-exposed and 1*g*-ground embryos despite an acceleration in appearance of the 1st ciliary beating of the epidermal ciliated cells and, therefore, rotation of the microgravity-embryo around its antero-posterior axis within the vitelline membrane (Table 2). A few hours after landing, young microgravity-exposed larvae displayed normal swimming behaviour (Gualandris-Parisot *et al.*, 2001; Dournon, 2003).

In contrast to the salamander, axial muscles of microgravity exposed tadpoles of *Xenopus laevis* exhibited a variety of abnormalities associated with muscle degeneration. These weight-bearing muscles were abnormally infolded and widely spaced. Fibres were less numerous (48%) in "flight animals" as in 1*g*-controls. Their mean numbers counted in a standardised sampling quadrant of 1,634.53 μm^2 were 10.8 (range 7.67–13.67) for microgravity-exposed tadpoles and 22.43 (range 19.33–27.00) for the 1*g*-tadpoles. In contrast, non-postural muscles of tadpoles such as the *M. orbitohyoideus* which is the primary muscle for depressing the buccal floor during respiration and feeding showed no sign of degeneration (Snetkova *et al.*, 1995). Muscle loss in aquatic animals might be caused by an increased metabolism due to general stress. The higher level of

Table 2

Kinetics of ciliary beating and movement of embryos during development (from Gualandris-Parisot *et al.*, 2001).

	microgravity flight	1*g* flight	1*g* ground
Appearence of 1st beating movements	64 ± 2 h	67 ± 2 h	74 ± 3 h
	stage 16	stage 16	stage 19
End of ciliary beating and onset of	156 ± 3 h	157 ± 3 h	155 ± 3 h
spontaneous contractions	stage 30	stage 30	stage 30
Hatching	180 ± 3 h	230 ± 3 h	260 ± 4 h
	stage 32	stage 33b	stage 34

HSP72 in adult goldfish muscle and spleen after microgravity exposure was considered as a stress response (Mori *et al.*, 1994; Ohnishi *et al.*, 1998).

Heart development is also sensitive to microgravity. In zebrafish (*Danio rerio*), the green flourescent protein (GFP) reporter gene can be fused with specific zebrafish promotors and enhancers inserted into the fish embryos. The resulting transgenic fish expresses GFP at the same times and places as the actual proteins controlled by these regulatory sequences. The amazing thing is that one can observe the reporter protein in living embryos (see Gilbert, 2003). Using this method it was revealed that heart development was significantly affected by placement in ground-based models that recreate some aspects of the microgravity environment, which induced a 23% increase in GFP-associated fluorescence in the heart of transgenic zebrafish (Gillette-Gerguson *et al.*, 2003).

Synapse formation at the developing neuromuscular synapse in cell cultures

Synapse formation in the nervous system is a prerequisite for normal brain function. The process of nerve-induced receptor accumulation is essential for synaptic function. Nerve-associated accumulation of acethylcholine receptors at the developing neuromuscular synapse was used as a model to explore the sensitivity of developing synaptic contacts to ground-based models which recreate some aspects of the microgravity environment. Co-cultures of spinal neurons and myotomal myocytes isolated from *Xenopus laevis* embryos were mounted in the clinostat at different times after the introduction of neurons to the myocyte cultures. Times were chosen so that the formation of nerve-to-myocyte contacts took place long before, immediately before and after the clinostat rotation was started. Acetylcholine receptor patches were identified by rhodamine α-bungarotoxine labelling. The main observation was that nerve-associated ACh receptor patches (NARP) from cultures in which nerve-muscle contact was established before the onset of rotation were unaffected. In contrast, incidence and area of NARPs showed a marked inhibition in cultures in which nerve contact took place during or shortly before placement in a

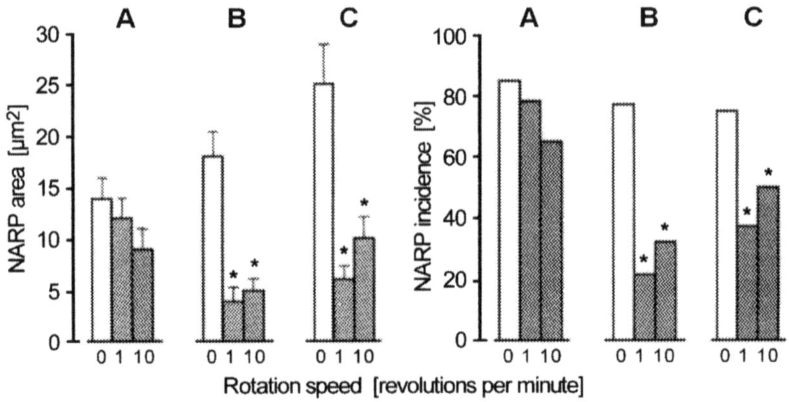

Fig. 2. Effects of clino-rotation on the incidence and area of nerve-induced acetylcholine receptor patches (NARP) in myocytes in mature (A: maturity before clinostat rotation onset), immature (B: synaptic contacts developed just before onset of rotation) and de-novo formed synapses (C: synapses formed during clinostat rotation). Clinostat rotation was performed at 1 or 10 revolutions per minute; 0 indicates no rotation. Note the significant effects in sets B and C and absence if maturation occurred before onset of clinostat rotation (modified from Gruener and Hoeger, 1990).

clinostatt (Gruener and Hoeger, 1990). Thus, the process of synapse formation is sensitive to the gravitational vector with a clear time window of sensitivity (Fig. 2).

These observations were confirmed and extended in space-flown cell cultures. Besides the reduced incidence of ACh receptor aggregates at the site of contact with polystyrene beads, they revealed marked changes in the distribution and organisation of actin filaments (Gruener *et al.*, 1994). Surprisingly, the changes in the receptor's cellular organisation by clinostat rotation did not alter the ACh receptor single channel properties. The mean open-time and conductance of the AChR channel were statistically not different from control values but showed a rotation-dependent trend that suggests a process of cellular adaptation to clinorotation (Reitstetter and Gruener, 1994).

Mineralisation and bone development

In contrast to higher vertebrates, currently there is no evidence for microgravity effects on bone formation in fish and amphibians. Morphometric examinations of the head skeleton of medaka fish (*Oryzias latipes*) revealed no defects after treatment in a 3-D-clinostat (Ijiri, 2003). Despite of these negative observations, there are efforts to establish *Oryzias* as a model to study molecular mechanisms underlying gravity dependent bone loss. Osteoprotegerin (OPG) seems to control the balance between osteoblasts (bone forming cells) and osteoclasts (bone resorbing cells) and, therefore, bone mass. The sequence and expression domains of the OPG genes and the entire genetic network for bone formation are highly conserved between medaka fish and higher vertebrates (Wagner *et al.*, 2003).

Respiratory organ development

The development of the lung reveals, at a first glance, a susceptibility to gravity deprivation; but the reduced lung sizes observed after a space flight are not caused by microgravity. *Xenopus* tadpoles reared on ground, usually come up to the water surface to fill their lungs within 2–3 days after hatching. A negative geotactic behaviour supports the finding of the water surface. In space, geotactical behaviour cannot be performed and the filling of lungs with air is prevented as aquatic animals usually live in closed survival systems where no water surface exists. Thus, as expected, animals reared at microgravity returned from space with smaller lungs compared to 1*g*-tadpoles (Pronych and Wassersug, 1994; Souza *et al.*, 1995).

Developmental characteristics of physiological features

The ultimate step in development is maturity, the formation of a complex organism. It is characterised by a complete harmony between morphological, physiological, molecular and genetic features that cause functions such as beating of the heart, circulation of the blood or behaviour. But also each younger stage possesses this functional stability. To explain micro- or hypergravity induced modifications including features such as behaviour, physiology, anatomy or biochemistry the developmental characteristic for each specific function must be known. It allows to distinguish between developmental acceleration or retardation on one hand and activation of neuroplastic or adaptive processes on the other. From data obtained from aquatic animals which were used for experiments in altered gravity, some characteristics exist which describe the changes for anatomy, physiology, biochemistry and behaviour during the development and maturation from embryonic stage to the adult. Due to the low importance of the aquatic animals for studies on microgravity-effects on muscles, cardiovascular and bone systems, these characteristics are mainly linked to the vestibular system and its underlying peripheral sensory and central neuronal structures.

Developmental characteristics exist in a rather detailed manner for the mouth-breeding cichlid fish *Oreochromis mossambicus*, the Japanese red-bellied newt *Cynops pyrrhogaster* and the clawed toad, *Xenopus laevis*. In *Oreochromis*, the development was described from hatching up to the adult age for the brain creatine kinase activity, a marker for neuronal activation (Slenzka *et al.*, 1993) and for the vestibuloocular reflex (Sebastian and Horn, 1999). In *Xenopus*, one of the most successful animal models in space flight studies, detailed developmental characteristics exist for the vestibular system and its function including the size of the macula utriculi, the number of neurones in the vestibular nuclei, the static vestibuloocular reflex (Horn *et al.*, 1986a,b), the efficiency of vestibular compensation (Rayer *et al.*, 1983) and the brain creatine kinase activity (Slenzka *et al.*, 1993). A developmental characteristic exists also for the fictive swimming of *Xenopus* embryos and young tadpoles, a motor activity which can be recorded from the ventral spinal roots for the first two weeks after hatching (Böser, 2003). In *Cynops*, main emphasis was given to the development of the otolith organs (Koike *et al.*, 1995; Wiederhold *et al.*, 1995).

Within the peripheral structures of the gravity sensing organs, the size and calcification of the otoliths and the size of the sensory epithelium increase continuously (for *Cynops*: Koike *et al.*, 1995; Wiederhold *et al.*, 1995; for *Xenopus*: Horn, 1985; Horn *et al.*, 1986a). In contrast, development of some vestibular nuclei is characterised by a loss of neurones after a period of extensive neuronal proliferation (Horn *et al.*, 1986a). The development of the static vestibuloocular reflex (rVOR) revealed more complex features. In *Oreochromis* and *Xenopus* it first increases upto a certain stage and decreases thereafter. In *Xenopus* it finally maintains a steady state level (Horn *et al.*, 1986a,b), while it increases again in the fish (Sebastian and Horn, 1999). The mechanisms of vestibular compensation that are activated by lesions of the vestibular sense organ and cause a complete or, at least, partial disappearance of lesion-induced movement defects continuously loose their efficiency the older the tadpoles were when the lesion was done (Rayer *et al.*, 1983). The developmental characteristics of the fish brain creatine kinase activity revealed an overall decrease with increasing age but regular occurring extreme fluctuations are superimposed. In *Xenopus laevis*, creatine kinase activity within the brain of adults is at the same level as in very young tadpoles except during the period of extensive neuronal proliferation and brain differentiation when a transient increase can be observed (Slenzka *et al.*, 1993).

Regeneration

Morphological regeneration is a reactivation of development in postembryonic life to restore missing tissue. Its most spectacular aspects are the demonstration of multipotent properties of specific tissue and that the correct positional information is respecified and normal body structures such as complete extremities or retinae are regenerated. Only a few aquatic vertebrates such as the urodeles posses the potency for regeneration. Most investigations about the influence of space flight upon regeneration were performed in the salamander *Pleurodeles waltl*, the latest ones on Russian satellites (BION 10, 1993; BION 11, 1997). The main observation from all these studies was that space flight stimulates the restoration of lens, forelimb, and tail. Exposure to conditions that model some of the aspects of the microgravity environment on earth by rotation in a clinostat has similar beneficial effects (Mitashov *et al.*, 1987, 1996; Grinfeld *et al.*, 1994; Grigoryan *et al.*, 2002).

Tail regeneration was initiated by removing a third part 15 days before the onset of microgravity. At launch, the tail blastemas had formed as a 1 mm thick translucent, convex layer. During microgravity exposure, the blastema elongation continued with similar morphogenetic features as in the ground controls but at a lower pace (length: 3.67 mm in microgravity and 4.36 mm in 1*g*; height: 8.83 mm for microgravity and 9.73 mm for 1*g*). Neuronal tube, cartilage of future vertebrate, muscles and connective tissue formed in flight and ground blastema in the same manner. Molecular markers of the central nervous system such as GFAP (glial fibrillary acidic protein), NF150 (specific intermediate filaments) and TOH (tyrosine hydrolase) were found in both space and ground groups in similar amounts. GFAP and NF150 were detected in the neural tube, TOH in catecholaminergic cells. In contrast, the connective tissue of the blastema of microgravity-exposed salamanders developed more GABA-positive cells than ground controls. Furthermore, the epidermis was absent at the distal end of the blastema and formed waves and thickenings in the other sites whereas the epidermal layer of the ground controls was regular (Grinfeld *et al.*, 1994).

Retina regeneration can be induced by several lesion techniques such as removal of neural retina by microsurgery or optic nerve transsection. An experiment performed on the 2-week Bion 11 flight revealed the intensification of regenerative processes. In particular, the proliferative activity as shown by the number of [^{3}H]-thymidine-labelled cells in the retinal pigmented epithelium, eye growth zone and other retinal areas was 1.2 to 1.5 times higher in microgravity-newts compared to ground controls. The differences were more pronounced in animals lesioned two weeks than four weeks before micro-gravity-onset (Fig. 3). Microgravity experience persists in intact animals for some time because newts flown intact and operated after the flight regenerated faster than 1*g*-ground controls (Grigoryan *et al.*, 2002).

Genetics: the interplay between genes and gravitational environment

Development is the result of the interplay between genes and environment. Most properties of an organism are established during embryonic and neonatal development. In nervous systems important modifications are the result of experience, but the bulk of the responsibility for functional wiring of the brain belongs to genetically pre-programmed aspects. The relationship between genes and behaviour is circular. On the one side, genes control neuronal functions and the development of the brain. On the other side, behaviour, the ultimate and most complex expression of brain activity controls, gene expression and, at the evolutionary time scale, influences the organisation of the genome by selection of the fittest in a changing environment.

Nowadays, neurogenetics has become an important tool which helps to identify those genes responsible for gravity related sensory, neuronal and motor functions, and behaviour. Research in *Oryzias latipes* and *Danio rerio* used this strategy to study the effects of genome variations on expression of behaviour and, vice versa, to study in animals with identical genotype the effect of gravity-dependent stimulation on the expression of the phenotype. The studies were based on the analysis of swimming behaviour or the vestibuloocular reflex.

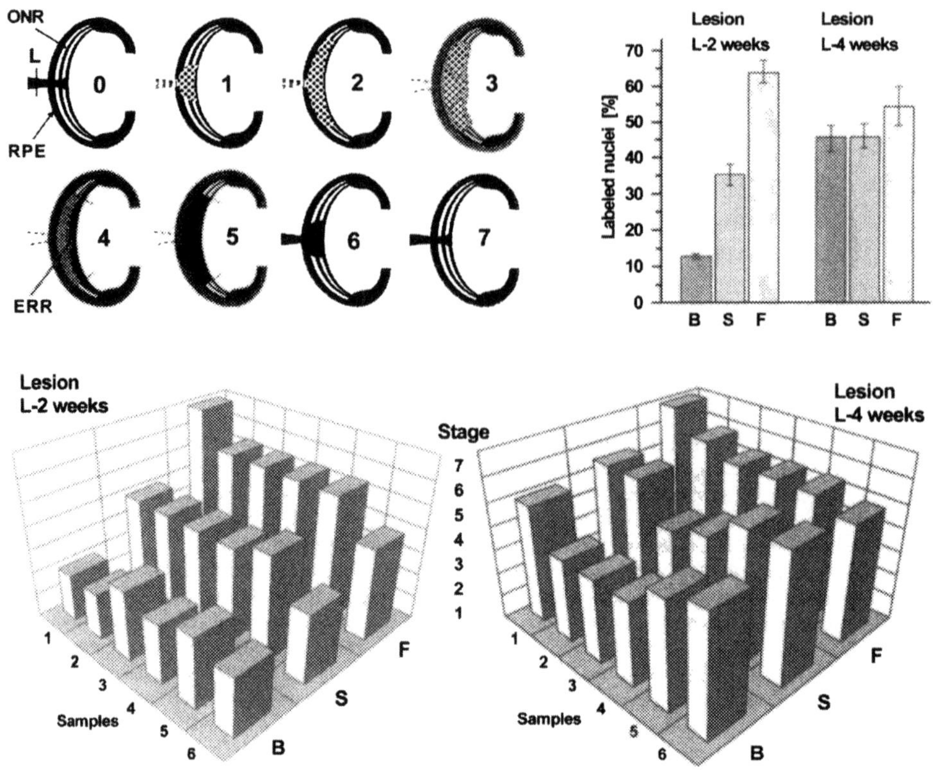

Fig. 3. Retina regeneration under space flight conditions in the newt *Pleurodeles waltl*—Upper left: Successive stages of neuronal regeneration after lesioning of the optic nerve. 0: Operation; stages 1-2: degeneration of original neural retina (ONR); stages 3-5: formation of early retinal regenerate (ERR) by transdifferentiating cells of the retinal pigmented epithelium (RPE) and cells of eye growth zone; stages 6-7: morphogenesis of newly formed retina and regeneration of optic nerve—Upper right: Percentage of [^{3}T]-Tdr-labeled nuclei in the central part of the neural retina—Lower: Extent of retina regeneration in the space flown newts (F) compared to the basal (B) and synchronous controls (S). Newts were operated either 2 or 4 weeks before launch (L-2 weeks and L-4 weeks, respectively) of the satellite (modified from Grigoryan *et al.*, 2002).

In fish, postural control during swimming is under the influence of both gravity and visual cues. Gravity is mediated via the vestibular system. Visual cues force the fish to roll its back towards the centre of light (dorsal light response; von Holst, 1950) or their ventral body side towards the darkest area of the visual environment (ventral substrate response; Meyer *et al.*, 1976). In microgravity, adult and young fish revealed disturbed swimming; they loop and roll around their longitudinal axis even in the presence of directed light.

Some strains of *Oryzias latipes* such as HNI-II rely more on the visual than on the gravitational input (heavily eye-dependent mutants). They possess an effective dorsal light and optokinetic response and do not loop in microgravity; they are more tolerant to microgravity as demonstrated by the absence of

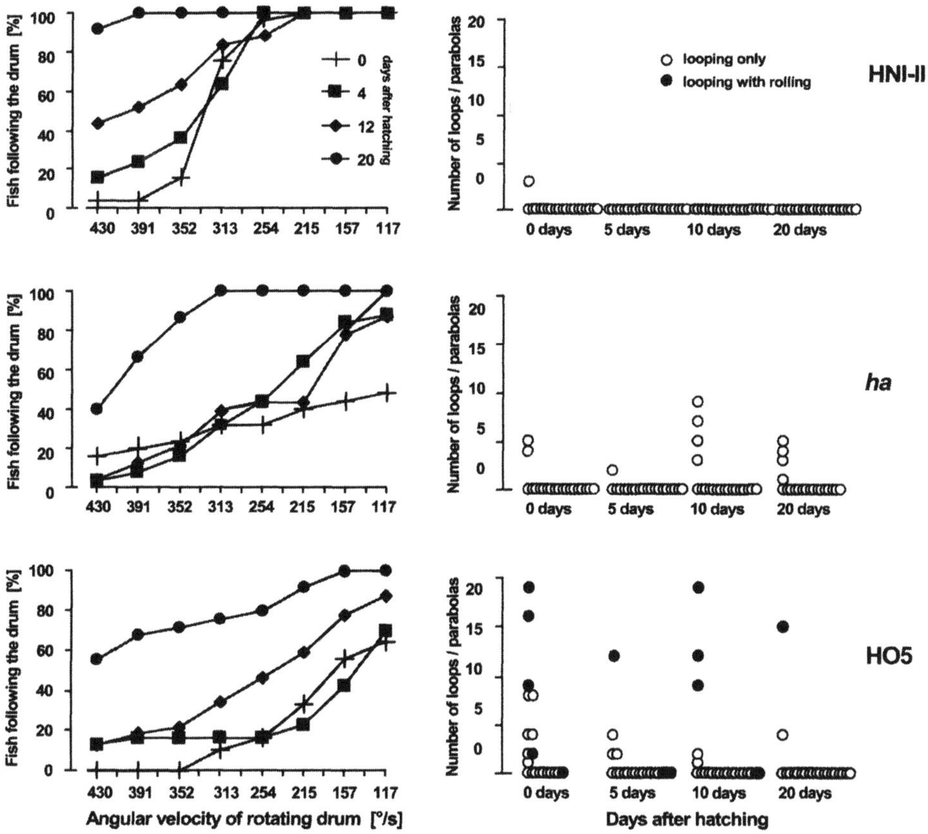

Fig. 4. Strain specific behaviour in medaka fish (*Oryzias latipes*) during exposure to microgravity or during visual stimulation by a moving striped pattern—Left: Relation between the velocity of the moving striped pattern in a rotating drum and the percentage of fish swimming in the direction of pattern movement (optokinetic behaviour)—Right: Characteristics of swimming behaviour during periods of microgravity during parabolic flights—Larval stages were tested at different times after hatching. Note the inverse correlation between expression of optokinetic behaviour and swimming anomalies in microgravity. The strain HNI-II is characterised by good eyesight and an ordinary sense of gravity. The *ha* mutant is not able to develop utricular otoliths and form their saccular otoliths with delay. It reveals a weak sensitivity to microgravity exposure. The HO5 strain is less eye-orientated but more sensitive to gravity. It loops and rolls more frequently during swimming than the strain HNI-II and the *ha* mutant (modified from Furukawa and Ijiri, 2002).

swimming disturbances during microgravity exposure (Fig. 4). The *ha* mutant medaka fish is unable to develop utricular otoliths; its saccular otoliths are formed with delay. It is highly light-dependent in their position control and shows a clear DLR and good optokinetic sensitivity which are, however, less effective than those of the HNI-II strain. This *ha* mutant can, therefore, serve as a substitute model for fish that have spent a life cycle in microgravity. Crossing both strains (HNI-II with good eyesight and the mutant strain *ha* without utricular otoliths) created fish with good eyesight and less gravitational sensitivity (F2 *ha*/*ha*). These fish yielded less looping and no differences in the degree of looping concerning light and dark environment during parabolic flight (Ijiri, 2003).

The ability for adaptation of swimming to microgravity is higher in larval fish than in adults. The strain HO5 is genetically determined to loop in microgravity. Its visual orientation is less developed compared to the strain HNI–II and the mutant *ha* (Fig. 4). However, the swimming behaviour of fry hatched in space (IML-2 mission, 1994) indicated that they use the dorsal light response for postural control despite of the genetic programme for looping (Ijiri, 2000).

Similar to all other vertebrate phyla, fish perform vestibulo-ocular reflexes which are induced by stimulation of the vestibular sense organs. Gravity-related responses are induced by stimulation of the utricular organ (see, Horn *et al.*, 1986b). The *monolith* (*mnl*) mutation in zebrafish inhibits specifically the formation of utricular otoliths. One peculiarity of this mutation is that experimentally six phenotype classes can be formed by immobilising embryos for brief periods in different postures. These phenotypes are characterised by uni- or bilateral absence of otoliths either in the utricule or the saccule. Larvae that did not develop both utricular otoliths (S-S-type) were unable to maintain a dorsal-up posture; they responded with "zigzagging", looping and rolling behaviour on tactile stimulation and did not survive until adulthood. All other phenotypes become adult as the wild-type. Presence of at least one utricular otolith is sufficient for balance and motor coordination and for survival. The S-S larvae showed little or no vestibuloocular reflex, while larvae with one utricular otolith exhibited a slightly reduced response compared to phenotypes with both utricular otoliths (Riley and Moorman, 2000). Homozygous *monolith* (*mnl*) mutants (1 otolith on each side) of zebrafish can be "rescued" by using non-invasive and non-molecular methods, either an immobilisation in a head down position or by incubation in ground based models that duplicate some of the conditions of microgravity during early development (Moorman, 2001).

These studies revealed that the genetic equipment determines the extent of sensitivity to altered gravity. But they also made clear that the gravitational environment affects the development of vestibular morphology and, therefore, the phenotype in gravity orientation. In the worst case, lack of gravity (or even hypogravity) or inability to experience gravity form phenotypes which are unable to survive (see, Riley and Moorman, 2000). Thus related future studies with different strains and mutant fish can help to gain further information about sensory conflicts in weightlessness and related phenomena like space motion sickness (SMS) and kinetosis. They also will help to find out, on the sensory and neuronal level, which mechanisms might cause a less pronounced sensitivity against the microgravity environment in orbit or the hypogravity environment on moon and other planets.

Neurobiology and Behaviour

Aquatic animals were extensively used to study the effects of gravity deprivation on sensory, neuronal and motor systems. Locomotion, vestibular reflexes including their peripheral and central structures, and brain chemistry were analysed. Reports on other sensory systems are rare and revealed no significant effects. In the retina of medaka fish, exposure to some of the conditions of the microgravity environment on earth in the 3-D-clinostat did not affect cell distribution in the photoreceptor layers and gene expression for opsin (Nishiwaki *et al.*, 1999). In contrast, the vestibular system and brain structures involved in vestibular guided physiological and behavioural responses were significantly affected by altered gravity in developing as well as in adult animals.

Otolith development and sensitivity to microgravity

The utricular, saccular and lagenar maculae are the gravity-sensing regions of the inner ear in vertebrates. Macular end organs consist of a sensory neuroepithelium overlaid by a mass of many minute crystallites (otoconia) or/ and by a single massive structure (otolith). Bony fish possess species-specific solid otoliths of constant shape. Their otoliths grow by adding layer by layer of an inorganic phase (mineral) and an organic phase (protein) (Degens *et al.*, 1969; Gauldie, 1993; Pote and Ross, 1993; Fermin *et al.*, 1995); from the resulting "rings" the age of a fish can be determined. The mineral phase is usually calcium carbonate; the crystallographic structures are vaterite, aragonite and/or calcite. Jawless vertebrates, the cyclostomes, have otoconia or otoliths composed of calcium phosphate in the crystallographic form of apatite (Carlström, 1963). Anurans (*Rana esculenta*, *Xenopus laevis*) and urodeles (*Pleurodeles waltl*) possess aragonitic otoconia in their saccule, lagena and endolymphatic sac and calcitic otoconia in their utricle (Marmo *et al.*, 1983a,b; Pote and Ross, 1993; Kido and Takahashi, 1997; Lewis and Pawley, 1981; Pote and Ross, 1993; Wiederhold *et al.*, 1995). An energy dispersive X-ray analysis indicated that both otoconia types are composed of about 95% calcium with trace quantities of sodium, magnesium, phosphorus, sulphur, chloride and potassium.

At the beginning of biological research in space, microgravity effects on the vestibular apparatus and, in particular, on otoliths were studied, preferably in fish. Their age-related growth allowed a clear-cut quantification of microgravity effects. Early studies by Russian scientists revealed that the development of the vestibular apparatus of *Brachyodanio rerio* larvae was not affected by weightlessness (9-days aboard the orbiting complex Salut-5-Soyuz-21). The fine structure of the receptor epithelium and the otolithic apparatus did not show noticeable differences. The ion distribution (potassium, sodium, calcium, phosphorus and sulphur) within the vestibular system also remained unchanged in comparison to control animals (*Brachyodanio rerio* in the spacecraft

Soyuz-22). Similar results were obtained from studies in *Fundulus heteroclitus*. Dimension, shape and relief of the otoliths and the ultrastructure of macular cells remained unaffected after early development in orbit on the Skylab station and the satellite Cosmos-782, even in larvae stages without presence of the vestibular anlage at launch (Vinnikov *et al.*, 1983).

More recent studies have revealed significant differences between microgravity-exposed and 1***g***-ground control fish. In *Danio rerio*, the otolith development was either delayed or slowed down by ground based exposure to models of some aspects of the microgravity environment during early development. Furthermore, the saccular otoliths were significantly smaller and in some animals, there were one or more otoliths absent (Moorman *et al.*, 1999). In swordtail fish (*Xiphophorus helleri*), juveniles and embryos were investigated after two space flights (STS-89 and STS-90). No significant effect was found in the juveniles while in the embryos the size of the utricular otolith was strongly affected by the microgravity exposure. For STS-90, the microgravity-exposed embryos had larger otoliths, for STS-89 the 0***g***-embryos had smaller otoliths compared to their respective ground controls. This contradictory observation correlates with the fact, that embryos from STS-89 were smaller than those from the Neurolab mission STS-90 (Wiederhold *et al.*, 2003).

Hypergravity effects on otolith growth were studied in *Oreochromis mossambicus* in relation to the duration of 3***g***-exposure upto 21 days. Both utricular and saccular otoliths continued growing in a linear way at 3***g***-hypergravity but at a lower rate compared to the 1***g***-controls (Anken *et al.*, 2002). This observation was taken as indicative of the existence of a feedback control of otolith growth in which sensory activity is involved. A clear-cut experiment in the swordtail fish supported this hypothesis. After transsection of its vestibular nerve, calcium incorporation decreased on the operated side. This indicated a feedback mechanism for otolith calcium uptake (Anken *et al.*, 2002). With this animal model, the controversy discussion about feedback mechanisms in otoconia mass regulation in higher vertebrates such as birds and mammals in response to altered gravity (see, Ballarino and Howland, 1984; Hara *et al.*, 1995; Lim *et al.*, 1974; Ross *et al.*, 1985) could be clarified.

In aquatic amphibians, the sensitivity of otoconia to microgravity is more difficult to analyse. Number of otoconia, size and chemical composition have to be determined to get an over-all result. In addition, the quality of the crystals and their number changes. So different aspects were considered in the analyses. Otoconia from microgravity-exposed *Xenopus* tadpoles were reported as 30% larger than those from 1***g***-controls (Lychakov, 1991). The shapes of otoconia were determined in *Xenopus* larvae for two ages after a 9.5-day microgravity-exposure on ISS (Andromède mission, 2001). At launch, embryos and tadpoles were at stages 25/28 and 45; observations revealed no microgravity-effect on the otoconia shape (Horn *et al.*, 2003).

In the newt *Cynops*, otoliths and the area of associated sensory epithelia increase during development. A single clump of otoconia could be first seen at stage 33; no skeleton is formed at this stage. Stage-36 embryos first have distinguishable otoliths, with the utricle in front and the saccule behind. A steady growth of the size of the otoliths is noted in the utricular and saccular otoliths up to stage 55. After separation, the saccular otolith contains more total calcified material than does the utricle. From stage 55 onwards, calcification increases rapidly. This period correlate with the transition in the use of water habitat to land habitats, a time when the body supporting bones grow intensively (Koike *et al.*, 1995). Effects of microgravity on the newt's otoliths and otoconia were studied on IML-2 (1994). Embryos reach orbit before any stones were formed. After the space flight, the mean volume of the otoliths of the utricle and saccule increased with developmental stage nearly at the same rate for both flight and ground animals (Fig. 5). However, the volume of the otoconia made of aragonite and produced in the endolymphatic sac were larger in the flight than in the ground animals, in particular, for stages 50 to 52 (Wiederhold *et al.*, 1997).

In *Pleurodeles*, ground experiments were performed to determine the stages of the first appearence of otoconia in different regions of the inner ear. The crystallographic structures of otoconia and otolithes were defined during development and at adulthood. The saccular otoconia are first in calcite, and then in aragonite; in the utricle and endolymphatic sacs they are permanently in calcite and aragonite. During two space flights, one lasting two weeks, the other five months, *Pleurodeles* larvae were reared aboard Mir and the International Space Station to compare the crystallographic structure of the otoconia with those of ground animals. In adults, after a 5-months flight, biological crystals were altered like those of elderly people (Dournon, 2003). The otoconia were very cavernous with an irregular surface, probably caused by a loss of calcium (Dournon, pers. communication).

The vestibulo-ocular reflex after space flights or hypergravity

In all vertebrate species, postural changes by a lateral roll or nose-down or nose-up tilt elicit eye movements in the opposite direction (roll- or tilt-induced static vestibuloocular reflex, rVOR) (Fig. 6). In eyes with a main direction of vision to the lateral as in young fish and tadpoles, a lateral roll mainly causes a response of the *M. rectus superior* and *inferior* while a torsional response of the eyes is induced in case of the nose-up or nose-down stimulation. The response characteristic which describes the relation between the roll angle and the response of the eye is sine-like. Typical parameters for the description of the rVOR are its amplitude and gain. The rVOR amplitude is the maximal angular movement range of the eye during a complete 360° lateral roll; the rVOR gain is the ratio between a response angle and the roll angle.

During each postural change, a response overshoot occurred. It is probably caused by the simultaneous stimulation of semicircular canal and otolith receptors due to the angular acceleration and postural change, respectively. In *Xenopus* tadpoles and young fish *Oreochromis*, the response overshoot lasts

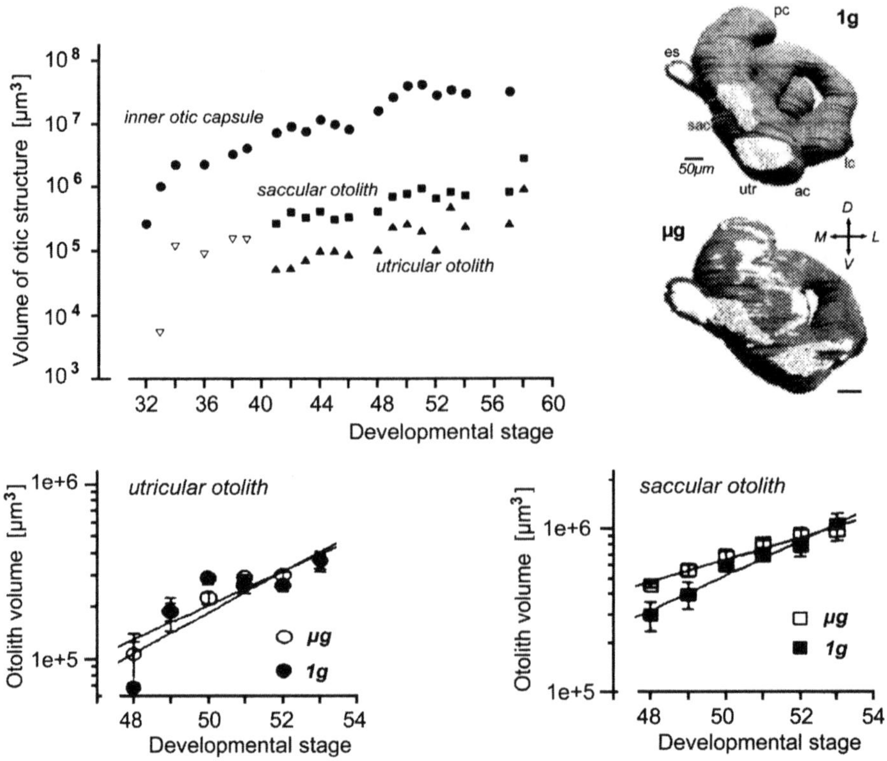

Fig. 5. Otolith development and its sensitivity to microgravity in the newt *Cynops pyrrhogaster*—Upper left: Standard development of the volume of the saccular and uticular otolith (black triangle and square, respectively) and the otic capsule (black circles). Saccular and utricular otolith are first visible in stage 40. In younger stages, only an individual statolith can be seen (white triangle)—Lower: Effects of a 15-day spaceflight (IML-2, 1994) on the development of the saccular and utricular otoliths (open symbols) compared to the ground reared animals (closed symbols). All measurements are from specimen fixed within 5 days of shuttle landing. Vertical bars indicate standard error of means—Upper right: Reconstruction of serial sections through the developing otic vesicle of a ground-reared (1g) and flight-reared (microgravity, µg) stage 52 larvae. Abbreviations: ac, lc, pc, anterior, lateral and posterior canal; es, endolymphatic sac; sac, saccule; utr, utricle; D, dorsal; L, lateral; M, medial; V, ventral; bar indicates 50 µm (modified from Koike *et al.*, 1995; Wiederhold *et al.*, 1997).

2.0 s; thereafter, the eyes maintained a steady state position for more than 10 minutes without any statistically significant change. In *Xenopus* tadpoles and juveniles, the response is usually performed by each animal. In contrast, some early developmental stages of *Oreochromis* are, to a certain percentage, either responders or non-responders to a lateral roll (Sebastian and Horn, 1999).

Both rVOR amplitude and rVOR gain undergo characteristic changes during development of both aquatic species. Knowledge of these developmental characteristics allows to distinguish between developmental retardation or acceleration and sensory adaptation in response to the exposure to altered

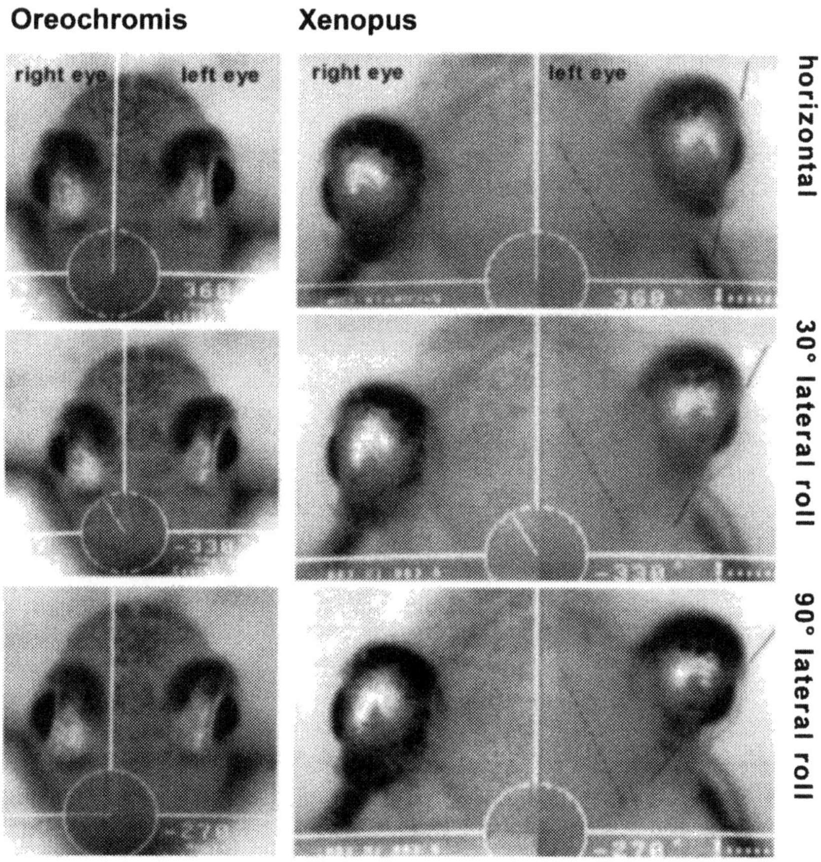

Fig. 6. The roll-induced static vestibulocular reflex in tadpoles (*Xenopus laevis*) (right) and young fish (*Oreochromis mossambicus*) (left)—The pictures show the eye posture for the horizontal posture and roll angles of 30° and 90° to the right side (from top to bottom). The camera was rolled along with the animal so that the back of the animals is always directed upwards in the frames. The response of the eyes was determined using the directions of the eye cup margins (cf. lines in the left eye of the tadpole) and either the vertical or a reference line between two melenophores on the head of the animals (cf. line on the left side of the head). Note that both eye moved to the left with respect to the reference direction during a passive roll to the right side (modified from Sebastian *et al.*, 1996; Horn and Sebastian, 2002).

gravity. Effects of real microgravity during space flights were studied in *Xenopus laevis* and *Oreochromis mossambicus*; these studies were supplemented by exposures to hypergravity. In zebrafish *Danio rerio*, effects of clinostat rotation on the rVOR were studied.

The rVOR in young fish

Young fish (*Oreochromis mossambicus*) were exposed to microgravity for 9 to 10 days during the space missions STS-55 (1993) and STS-84 (1997), or to hypergravity for 9 days. Young animals (stages 11–12) that had not yet developed the rVOR at microgravity or hypergravity onset and older ones (stages 14–16) that had already developed the rVOR were used (for definition of developmental stages of *Oreochromis*, see Anken *et al.*, 1993). After termination of microgravity or hypergravity, the rVOR was recorded for several weeks. In the stage 11/12-fish, the rVOR gain for the roll angles 15°, 30° and 45° was not affected by microgravity if animals were rolled from the horizontal to the inclined posture but was increased significantly if the young fish were rolled in the opposite manner. The rVOR amplitude of microgravity-exposed animals increased significantly by 25% compared to 1***g***-controls during the first post-flight week but decreased to the control level during the second post-flight week. Microgravity had no effect on rVOR gain and rVOR amplitude in stage 14/16 fish. After 3***g***-exposure, both rVOR gain and amplitude were significantly reduced for both stage-11/12 and stage-15 fish. 1***g***-readaptation was completed during the second post-3***g*** week (Fig. 7). A 9-day hypergravity exposure to 2 or 2.5***g*** had no effect on the rVOR. Exposure to all levels of hypergravity tested (2***g***, 2.5***g***, and 3***g***) accelerated the morphological development as assessed by external morphological markers (cf. Anken *et al.*, 1993) and the standard developmental rVOR characteristic. Thus, rVOR modifications include sensitisation by microgravity and desensitisation by hypergravity in responding fish (Sebastian *et al.*, 2001; Horn and Sebastian, 2002).

Zebrafish (*Dario rerio*) exhibited a less clear VOR after exposure to a rotating bioreactor for the nose-up- and nose-down tilt (Moorman *et al.*, 1999, 2002). The sensitivity was restricted to a period which ranged from 24 to 72 h after fertilisation. These studies revealed the existence of a critical period in vestibular development and its duration. This postulation was confirmed by an experiment with zebrafish eggs which were rotated in the bioreactor from 3 to 96 h after the fertilisation except for the period between 24 to 72 h. These larvae developed a normal rVOR (Fig. 8).

The rVOR in Xenopus tadpoles

In *Xenopus laevis* tadpoles, the rVOR was studied after the space shuttle flights STS-55 (1993) and STS-84 (1997) and the Soyuz taxi flight Andromède to the International Space Station (2001). Studies included young animals (stages 25–36) which had not yet developed the rVOR at microgravity-onset and older ones (stages 45) which had already developed the rVOR (for definition of developmental stages of *Xenopus laevis*, see Nieuwkoop and Faber, 1967). The main observations were that the rVOR was modified (Sebastian *et al.*, 1996; Sebastian and Horn, 1998) and that the susceptibility to microgravity covers a period of life during which the efficiency of vestibular compensation is very high (Fig. 9) (see Rayer *et al.*, 1983).

The modification of the rVOR is related to changes of the body shape which developed during the space flight. In particular, exposure to microgravity

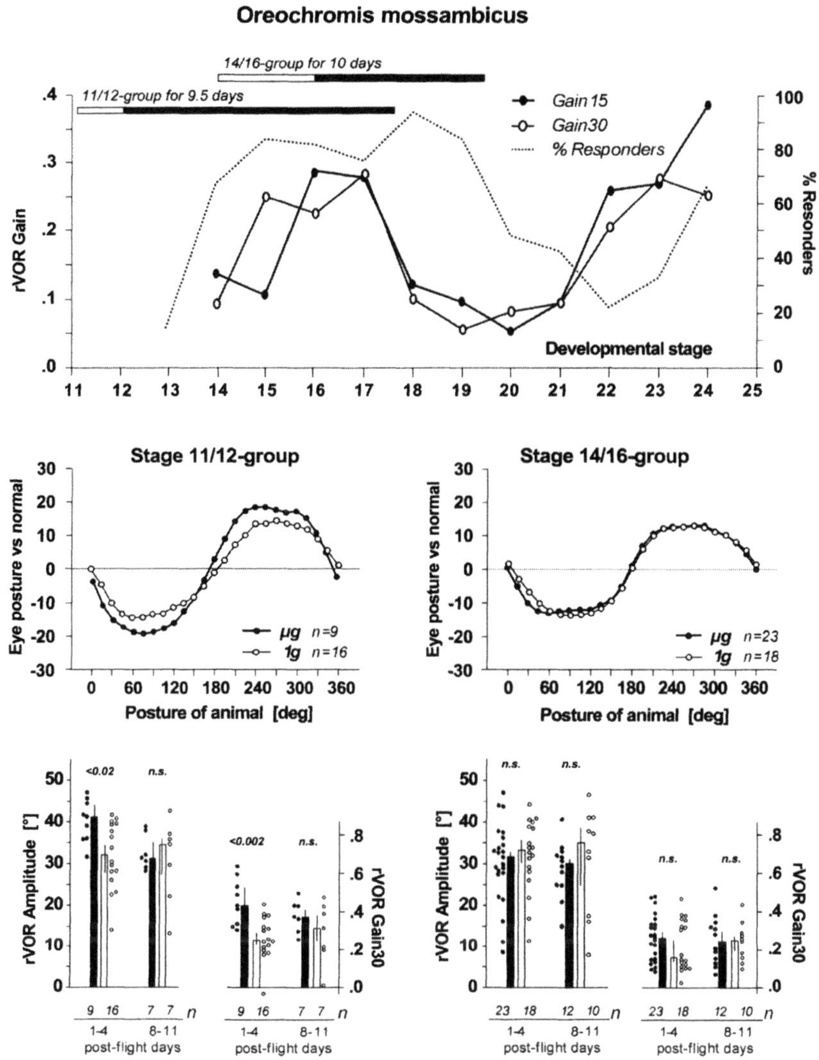

Fig. 7. Development of the roll-induced vestibuloocular reflex in fish (*Oreochromis mossambicus*) and effects of microgravity in young fish—Top: Developmental characteristic for the rVOR gain using lateral roll of 15° and 30°, and for the percentage of responding animals—Middle: Effects of microgravity exposures on the mean reflex characteristics for young fish that had not yet developed the rVOR (stage 11/12) or had developed the rVOR (stage 14/16) at onset of microgravity. Periods of exposures are presented in the graph showing the developmental characteristics. Note the augmentation of the rVOR in the young group by microgravity—Lower: Presentation of individual and mean rVOR gain and rVOR amplitude values obtained from the same experiments with both fish groups (modified from Sebastian and Horn, 1999; Sebastian *et al.*, 2001).

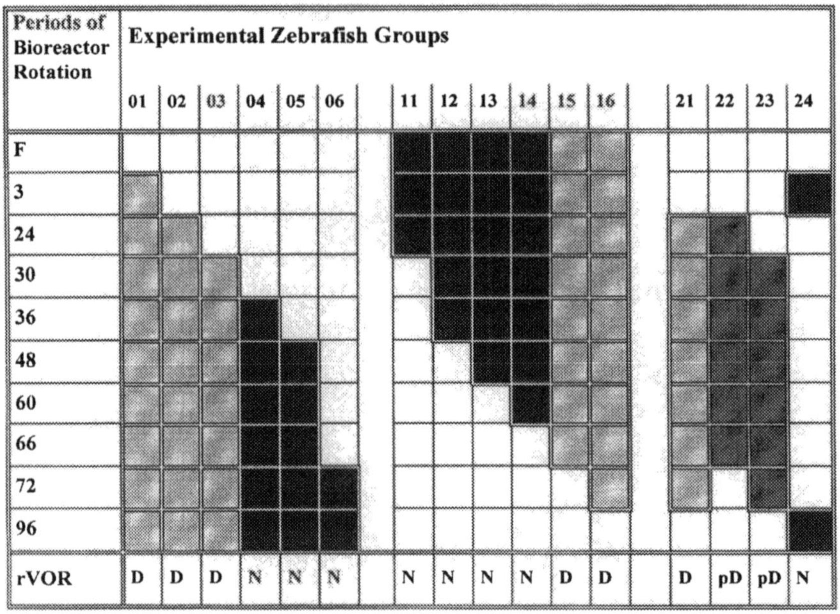

Fig. 8. The critical period in the development of the vestibuloocular reflex of zebrafish (*Danio rerio*)—Embroys were exposed to simulated microgravity by means of a rotating bioreactor; periods of rotation are given on the left in hours after fertilisation (F). Rotation started either at different ages (3, 24, 30, 36, 48, or 72 h after fertilisation) and all animals were tested at the same age (96 h after fertilisation) or rotation was started immediately after fertilisation and measurements were done at different ages (24, 36, 48, 60, 66, 72, or 96 h after fertilisation). Black columns indicate normal (N) rVOR development; light grey columns indicate depressed (D) rVOR development during a period of 5 days after the 96th post-fertilisation hour, and dark grey columns indicate a weak rVOR modification (pD) of short persistance (modified from Moorman *et al.*, 1999).

induced in some tadpoles a hyperextension of the tail (dorsalisation). This malformation occurred if fertilisation of eggs was performed pre-flight but not after in-flight fertilisation (Snetkova *et al.*, 1995; Souza *et al.*, 1995; Sebastian and Horn, 1998). The hyperextension of the tail disappeared after re-entry to 1***g***-conditions (Sebastian and Horn, 1998). All studies revealed a significant depression of the rVOR in tadpoles with tail dorsalisation (Sebastian and Horn, 2001). In contrast, tadpoles with normal tails which had not developed their rVOR at onset of microgravity behaved as their ground reared controls (Fig. 9) while tadpoles with normal tails which had developed their rVOR at onset of microgravity revealed an augmented rVOR after their flight which persisted for 3 days after return to Earth 1***g***-conditions (Horn *et al.*, 2003).

The rVOR was also affected by 3***g***-hypergravity. So far, 6 developmental stages were tested; at onset of hypergravity, they had reached stages 6/9, 11/17, 17/22, 25/28, 33/36 and 45. None of 3***g***-exposed tadpoles developed lordotic tails. The most obvious result was a suppression of the rVOR for many days to weeks. Only tadpoles from the 6/9-group which were exposed for 12 days to 3***g*** showed the same response intensity during the first two days as the 1***g***-control, but thereafter, they were significantly retarded in their further rVOR development (Fig. 9) (Horn and Sebastian, 1996; Horn, 2004).

Thus, similar modifications of the rVOR by altered gravity occurred in fish and frogs during development. The most obvious result is that in both species, periods of life exist during which microgravity induced a sensitisation of the vestibular system while hypergravity always desensitises the vestibular system.

Neurophysiological studies with and without otolith stimulation

Vestibular central neurons can be irreversibly deprived from their otolith input by bilateral labyrinthectomy. Neurophysiological studies in alert guinea pigs (Ris and Godaux, 1998), anaesthetised cats (Ryu and McCabe, 1976) and awake monkeys (Waespe *et al.*, 1992) had revealed that despite absence of any labyrinthine input, vestibular nuclei were able to restore the initially depressed neuronal activity (vestibular compensation, see Schaefer and Meyer, 1974). In contrast to bilateral labyrinthectomy, the otolithic system remains intact in microgravity, however the otolithic mass is unweighted so that the otolith cannot be displaced by body roll and tilt. Absence of weight-stimulation of utricular sensory hair cells makes it likely that otolithic afferents and central vestibular neurons decrease their resting discharges after microgravity-onset. During further space flight, normal activity might be restored due to adaptive, neuroplastic properties of the vestibular system (Precht and Dieringer, 1985; Smith and Curthoys, 1989; Dieringer, 1995).

On 9 November, 1970 two bull frogs were sent into orbit by means of a four-stage Scout rocket assembly launched from the NASA launching site on Wallops Island, Virginia, USA. They were immobilised by cutting the motor leg nerves and had implanted electrodes to record the activity of axons in the VIIIth nerve. The recordings revealed a rapid decrease of the resting discharge of primary neurons after microgravity-onset which was followed by an increase during continuation of flight beyond 1***g***-preflight discharge levels. Until the end of the 7-day space flight, the activity returned back to the 1***g***-control level (Fig. 10, left) (Bracchi *et al.*, 1975).

This transient increase of base activity in the VIIIth nerve points to an up-regulation of sensitivity, i.e., to a sensitisation of the vestibular system during the absence of gravity. A similar sensitisation was observed in the toadfish *Opsanus tau* when the activity of its VIIIth nerve was recorded after two 16-day space flights (STS-90 in 1998; STS-95 in 1999). In these experiments, the otolith system was stimulated by linear acceleration caused by translatory movements. During the first post-flight day, the mean magnitude of the response to this

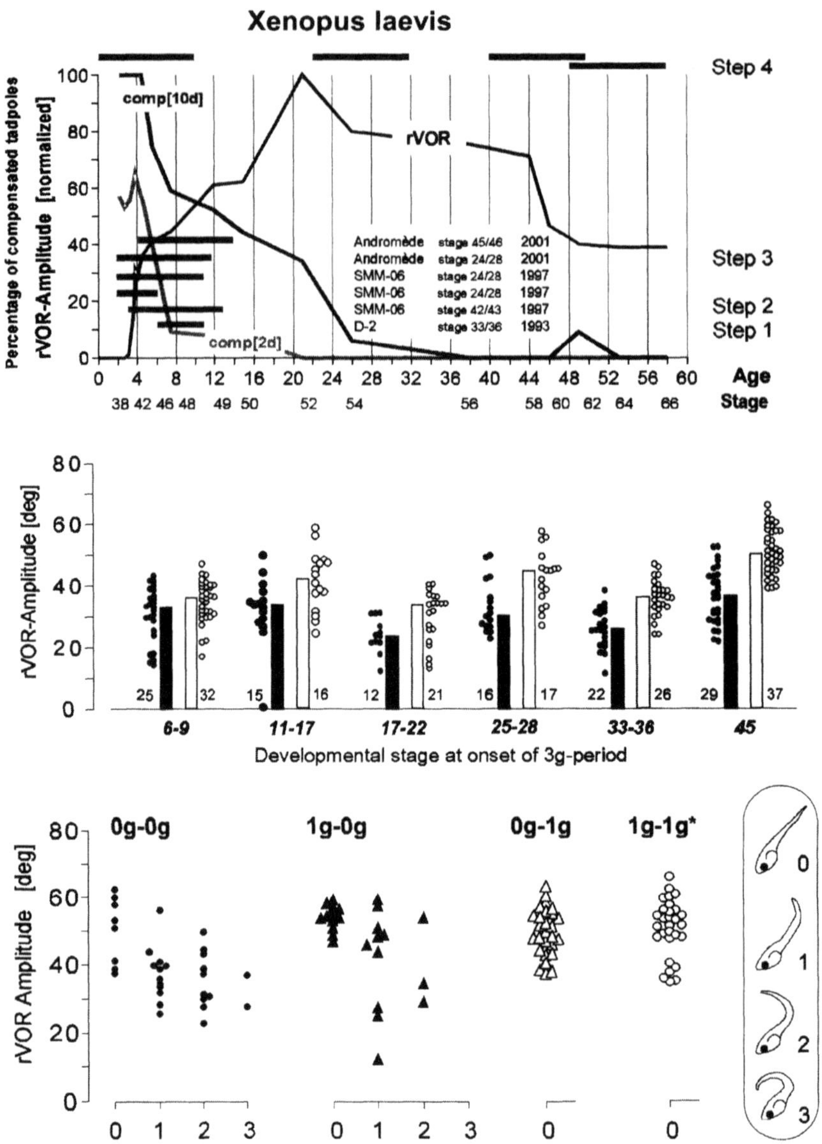

Fig. 9. Development of the roll-induced vestibuloocular reflex (rVOR) in *Xenopus laevis* and effects of altered gravity on its development—Upper: Developmental characteristic of the rVOR amplitude (from Horn *et al.*, 1986) and the ability to compensate for movement defects induced by hemilabyrinthectomy (vestibular compensation) (from Rayer *et al.*, 1983). For rVOR development, the abszissa indicates the age (days after fertilisation) and stage (cf. Nieuwkoop and Faber, 1967) when the rVOR tests were performed; the ordinate shows the normalised rVOR amplitude (= maximal angular movement of the eye during a complete 360° lateral roll). For vestibular compensation, the abszissa indicate the age and stage when one labyrinth was destroyed, and the ordinate the percentage of animals that showed normal swimming behaviour 2 days (comp[2d]) and 10 days (comp[10d]) later. Horizontal bars presents periods during which

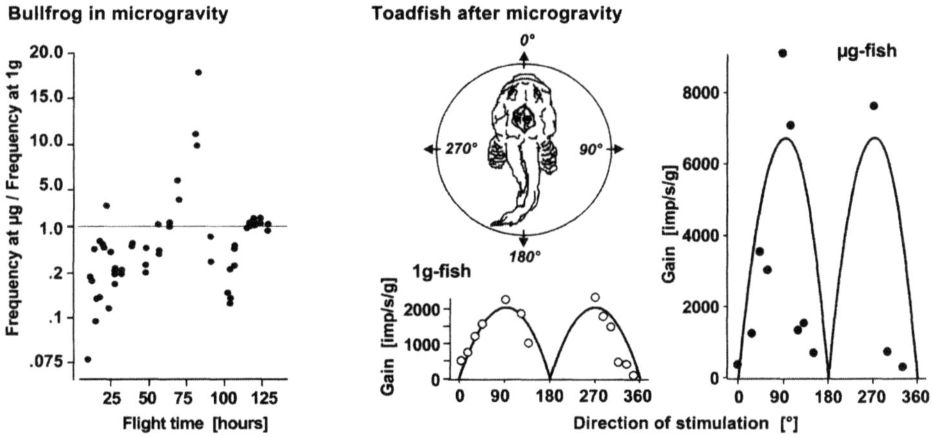

Fig. 10. Effects of microgravity on the activity of vestibular units of the VIIIth nerve—Left: Spontaneous vestibular activity during a spaceflight in the bull frog (*Rana catesbeiana*). The frequency of nerve activity recorded from the flight animal is related to that recorded from the ground animal at the same time. Note the depression of activity during the early periods of flight, the subsequent overcompensation at 75 h of flight, and normalisation shortly before the end of the flight (modified from Bracchi *et al.*, 1975)—Right: Activity in the VIIIth nerve of the toadfish (*Opsanus tau*) after a 16-day spaceflight induced by a translatory movement. Note the higher activity during a side-to-side translation in the fish with microgravity experience compared to the 1G-controls (modified from Boyle *et al.*, 2000).

stimulus was three times greater than for controls (Fig. 10, right). It returned to the level of the ground controls 30 h after the end of the flight (Boyle *et al.*, 2000).

These modifications of both the base activity and the induced activity of units within the vestibular nerve reveal that microgravity exposure increases the sensitivity of the vestibular system. In this respect, the mechanisms of adaptation to microgravity resemble those mechanisms responsible for vestibular compensation in hemilabyrinthectomised animals.

Fig. 9 (*continued*) tadpoles were exposed to microgravity during 3 different missions (step 1, step 2, step 3). Step 4 points to planned spaceflight experiments with old tadpole stages as well as fertilization in microgravity (modified from Rayer *et al.*, 1983; Horn *et al.*, 1986)—Middle: Effects on 3G-hypergravity on the rVOR. The abszissa indicate the stages when 3G exposure starts, and the ordinate the rVOR amplitude. Black and white columns present median values obtained from the 3G- and 1G-tadpoles, respectively; dots and circles individual values from 3G- and 1G-tadpoles, respectively (from Horn, 2004)—Lower: Effects of microgravity on the rVOR amplitude in tadpoles with malformed tail compared to those with normal tails and tadpoles raised under 1G-conditions. Microgravity started when tadpoles had reached stage 24 to 28. Experiment from STS-84. Tadpoles were exposed to microgravity or 1G throughout the mission (0G–0G and 1G–1G, respectively), to microgravity during the first half of the mission and thereafter to 1G (0G–1G), or vice versa (1G–0G). Each symbols represents one tadpole. Extent of tail malformation, cf. right inset (modified from Sebastian and Horn, 2001).

Posture and swimming during and after space flights and hypergravity

Aquatic animals rely on less sensory channels for postural control during locomotion than terrestrial animals. The most important input comes from the gravity sensory systems and vision. Buoyancy supports free swimming so that postural control mechanisms based on weight perception (they are common in terrestrial vertebrates and invertebrates, see Massion 1994; Horn, 2003) can be neglected. For stabilisation of posture also the swim bladder is used. This gravity-dependent control of swimming causes severe swimming disturbances in the absence of gravity in almost all fish and amphibians.

Swimming in fish

There is an overall tendency of fish to show unstable swimming in orbit. Rolling, twisting and looping behaviour is common (von Baumgarten *et al.*, 1975). In some instances, these abnormal movements persist after termination of both microgravity and ground-based models replicating some aspects of the microgravity environment. *Xenopus* tadpoles continued with abnormal swimming for upto 5 days (Neubert *et al.*, 1988; Sebastian and Horn, 1998). Immediately after removal from the rotating bioreactor, zebrafish (*Danio rerio*) swim normally when they were illuminated from above, but disoriented showing loopings and upside down swimming when they were illuminated from below. Their behavioural impairment is probably caused by a delay or depression of otolith development by exposure to ground-based models which recreate some aspects of the microgravity environment (Moorman *et al.*, 1999).

Adult medaka fish revealed a strain-specific sensitivity of swimming in microgravity. Some strains such as HO5, HO4C and HB12A looped and twisted during microgravity even under simultaneous stimulation by light. The strain HNI-II did not loop at all. In contrast, the strain HB32C revealed nearly normal swimming behaviour in microgravity (Ijiri, 1997). Fry of the strains ccT and HO5 which hatched in space (IML-2, 1994) did not loop in microgravity; they turned their back to the light (Ijiri, 2003).

The rolling behaviour of fish in microgravity is clearly related to the asymmetry of the otoliths. Basically, the vestibular system possesses high adaptive capacities (Schaefer and Meyer, 1974). This feature was convincingly demonstrated by Bechterew at the end of the 19th century. He reported experiments in which two labyrinths were successively destroyed at intervals of upto several days. He observed a reversal of the direction of compulsive movements due to the second lesion (Bechterew, 1883). A compensatory process initiated after the first lesion was established with the feature of a left/right asymmetry. This asymmetry became now dominant after the second lesion (Bechterew compensation). Similar mechanisms help to overcome weight asymmetries of left and right otoliths on ground; they are compensated for by physiological mechanisms. In the absence of gravity during orbital or parabolic

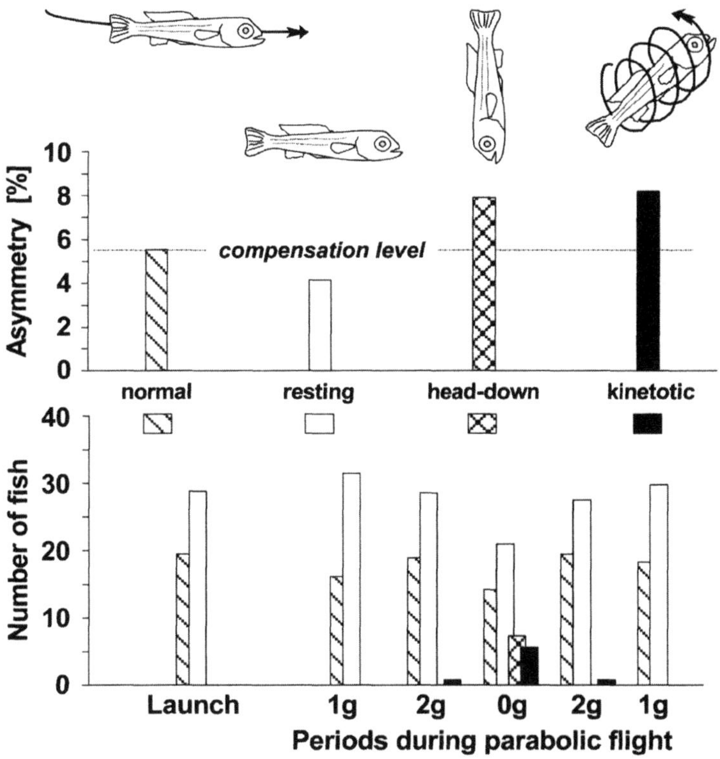

Fig. 11. Effects of microgravity on the swimming behaviour in cichlid fish (*Oreochromis mossambicus*) and its relation to left/right otolith asymmetry—Upper: Types of behaviour—Lower: Occurrence of these behavioural types during the different periods of parabolas. About 10% of fish exhibited rolling (spinning) swimming behaviour exclusively during the microgravity period of parabolas. Left/right otolith asymmetry is largest in these abnormal swimming fish. A left/right otolith asymmetry of up to 5.6% can be compensated by central nervous mechanisms; asymmetries beyond this "compensation level" cause abnormal swimming (modified from Hilbig *et al.*, 2002).

flights, the asymmetry of a left/right compensatory mechanism becomes dominant leading to rotatory swimming behaviour as predicted by the Bechterew approach after bilateral successive labyrinthectomy. In fact, the final proof of this assumption was presented in swordtail fish (*Xiphophorus helleri*) and cichlid fish (*Oreochromis mossambicus*) when they were exposed to microgravity during parabolic flights. Both species revealed rolling behaviour; occurrence and direction were clearly related to the extent of left/right otolith asymmetry (Fig. 11) (Hilbig *et al.*, 2002).

Swimming in amphibians

During and after microgravity exposure, body posture and swimming of adult and larval amphibians is strongly modified. Movements never seen before on

Table 3

Occurrence of dorsalisation in tadpoles (syn. lordotic tadpoles) with microgravity experience. L, tadpoles with lordotic tails; N, tadpoles with normal straight tails—Tadpoles flew on the 9-day STS-84 mission; at launch, they were at stages 25/28. In-flight, tadpoles were exposed to microgravity or simulated 1*g* throughout the mission (microgravity-microgravity and 1*g*–1*g*, respectively), or during the first or second half of the mission to microgravity and during the other half of the mission to 1*g* (microgravity-1*g* and 1*g*-microgravity, respectively). For ground controls, a similar schedule with 1*g*- and 1.4*g*-periods were performed. Experiments were performed 5 h and 14 days post-flight. Each of the 8 samples consisted of 36 embryos at launch, i.e., the total number of embryos at that time was 288 (from Sebastian and Horn, 2002).

Flight	R+5 h	R+14 days	Mort.	Ground	R+5 h	R+14 days	Mort.
microgravity-microgravity	N=14 L=21	N=27 L=1	7	1*g*-1*g*	N=33 L=0	N=29 L=0	4
1*g*-1*g*	N=35 L=0	N=32 L=0	3	1.4*g*-1.4*g*	N=35 L=0	N=28 L=0	7
microgravity-1*g*	N=33 L=0	N=29 L=0	4	1*g*-1.4*g*	N=36 L=0	N=33 L=0	3
1*g*-microgravity	N=15 L=13	N=22 L=1	5	1.4*g*-1*g*	N=28 L=0	N=24 L=0	4

ground in healthy animals occur during microgravity exposure. Terrestrial frogs floating in microgravity stretched four legs out, bent their bodies backwards and expanded their abdomens. Frogs on a surface often bent their neck backwards and walked backwards (Izumi-Kurotani *et al.*, 1994).

Swimming recorded in tadpoles of *Xenopus laevis* in microgravity during space flights revealed loop behaviour, twist (spinning) movements around their longitudinal axis or zigzag-trajectories. Occasionally, these characteristics of abnormal swimming persist for some hours or some days after reentry in 1*g*-conditions. After reentry from microgravity to 1*g*-condition, rotatory swimming disappears rather soon whereas the looping behaviour was maintained in tadpoles for some longer time supported by an abnormal development of their tail. In some percentages of the space flown tadpoles, the tail is bended upwards (lordotic shape) (Neubert *et al.*, 1987; Snetkova *et al.*, 1995). This shape is not a malformation because during further development on ground it disappears (Table 3) (Sebastian and Horn, 1998).

Swimming kinetics of *Xenopus* tadpoles was analyzed after microgravity (STS-47 and clinostat rotation) and during microgravity (Soyuz taxi flight Andromède to the International Space Station) exposure. Tailbeat frequency and swimming velocity were significantly affected by both microgravity and ground-based models that replicate some aspects of the microgravity environment. In particular mean swimming velocity and mean tailbeat frequency were significantly reduced in tadpoles exposed to ground-based models of some microgravity conditions on postflight day 0 compared to 1*g*- and 3*g*-tadpoles. Tadpoles raised in true microgravity exhibited a significant lower tail beat frequency on post-flight day 0 while swimming velocity was not affected throughout the 9-day postflight observation period (Feitek *et al.*, 1998).

Recordings of swimming in microgravity during the Andromède mission to ISS revealed significant differences between flight and ground tadpoles. Young and old tadpoles that were at stages 25/28 and 45, respectively, at onset of microgravity swim longer and faster and have a higher acceleration immediately after onset of swimming than their ground siblings (Dournon *et al.*, 2002).

Neurophysiological studies using the model of fictive swimming

The modifications in swimming behaviour during microgravity exposure including trajectories and duration persist for some days after the reentry to 1***g*** on ground. The neurophysiological consequences were studied in *Xenopus* tadpoles using the model of fictive swimming. In contrast to the free-swimming animal, tadpoles are paralysed during the recording procedure. The motor activity during fictive swimming is measured on the neuronal level extracellularly from the ventral roots (VR). These recordings represent the rhythmical, burst-like activity of the spinal motoneurons that is induced by a tactile stimulation of the tail. Typical parameters of fictive swimming are the duration of the episodes and bursts, the frequency of bursting (which correspond to the time between two bursts), and the rostrocaudal delay which characterises the conduction velocity of activity from rostral to caudal parts of the spinal cord (Fig. 12, upper). Because of the lack of muscular activity in the paralyzed animals no proprioceptive influences are modulating the central oscillator that is producing the motor pattern (Kahn and Roberts, 1982; Kahn et al., 1982; Dale, 1995). Fictive swimming is an excellent model to examine effects of AGF on the locomotor system since the pattern simplicity makes it easy to detect basic changes.

Microgravity exposure manifested in an elongation of the swimming episodes that was significant ($p = 0.05$; mean values for microgravity: 3.0 s and 1***g***: 2.2 s) while the rostrocaudal delay was significantly shorter ($p = 0.05$; mean values for microgravity: 2.8 ms and 1***g***: 4.8 ms). Burst duration was slightly decreased at the rostral recording site 10 myotomes behind the otic vesicle ($p < 0.1$; mean values microgravity: 19 ms and 1***g***: 23 ms) but not in the caudal ventral root located 14 myotomes behind the otic vesicle (not significant; mean values for microgravity: 17 ms and 1***g***: 18 ms). Cycle length at both recording sites was not affected by development under microgravity compared to 1***g*** rearing. The effects could not be demonstrated for the recording days 3 to 6 (Fig. 12, middle left and bottom). Hypergravity at a level of 3***g*** was not as effective as microgravity during space flight (Fig. 12, lower) (Böser, 2003; Böser and Horn, 2002; Böser *et al.*, 2002).

Brain chemistry after space flights and hypergravity

Adaptation to altered gravity includes modifications of cellular energy consumption. Neuronal activity, the formation of synaptic contacts, increased

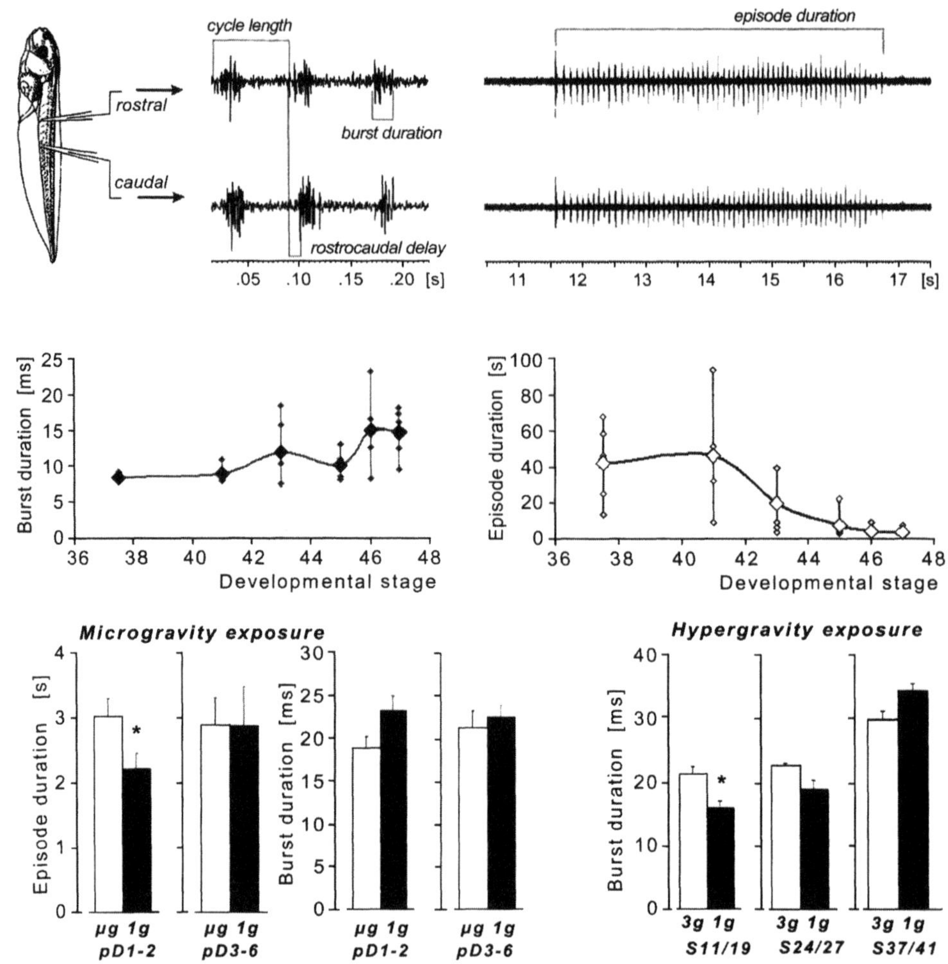

Fig. 12. Fictive swimming in *Xenopus laevis* young tadpoles and its susceptibility to altered gravity—Top: Methods to record fictive swimming from the ventral roots of the spinal cord. An episode of fictive swimming is shown on the right, 3 bursts from this episode with the relevant parameters for analysis on the left—Middle: Developmental characteristics of fictive swimming demonstrated for the parameters "burst duration" and "episode duration". Fictive swimming can only be induced up to developmental stage 47. Note the increase of burst duration (similar characteristics exist for cycle length and rostrocaudal delay) and the decrease of episode duration—Lower: Effects of microgravity on episode and burst duration (left) and 3*g*-hypergravity on burst duration. pD1-2 and pD3-6, recording period between post-flight days 2-3 and 4-7; the recording started 1.5 days after landing of the Soyuz capsule. S11/19, S24/27 and S37/41, stages at onset of the 3*g*-period (modified from Böser *et al.*, 2002; Böser, 2003).

motor activity or circulation of body fluid are energy demanding processes that are detectable by means of markers and stainings for specific enzymes. Thus histochemical methods support the understanding of mechanisms involved in micro- and hypergravity adaptation. Throughout the animal kingdom a lot of enzymes such as glucose-6-phosphate dehydrogenase (G6P-DH), succinate dehydrogenase (SDH), creatine kinase (CK), cytochrome oxidase (CO), NADPH-diaphorase (NADPHD) and Ca^{2+}-ATPase exist which can be used in the study of gravity-related adaptation mechanisms independent on the animal species. G6P-DH and SDH, the limiting enzymes of the Krebs cycle, are important enzymes that maintain energy availability in cells. CK is involved in the mechanism of ATP-regeneration and plays a key role in cellular metabolism, in particular in the muscle and nervous system; its activity is clearly related to development (Slenzka *et al.*, 1993) (Fig. 13). CO characterises basic metabolic activity. NADPHD can be taken as a marker for neuronal plasticity, because it indirectly reflects the activity of the nitric oxide (NO) synthase. Nitric oxide is an intracellular messenger suggested to be involved in the regulation of neuronal plasticity (Krasnow, 1977; Rahmann *et al.*, 1992; Slenzka *et al.*, 1990, 1993; Mori *et al.*, 1994; Anken *et al.*, 1996, 1998; Anken and Rahmann, 1998b).

Most studies about modifications of enzymes after exposure to microgravity or hypergravity were done in the brain of fish (*Oreochromis mossambicus* and *Fundulus heteroclitus*), and the clawed toad *Xenopus laevis*. In *Oreochromis* and

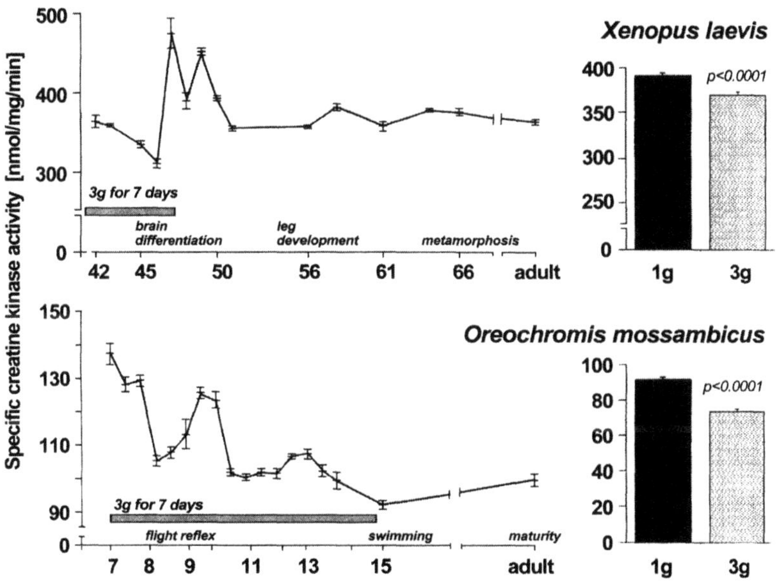

Fig. 13. Creatine kinase activity during development (left) and after a 7-day 3*g* exposure (right) in a fish (*Oreochromis mossambicus*) and an amphibian (*Xenopus laevis*). Abszissae of the developmental characteristics indicate the developmental stages (cf. Nieuwkoop and Faber, 1967; Anken *et al.*, 1993). Note that the 3*g*-exposure covers a period of strong modifications of CK activity during development in both fish and amphibian; this makes the interpretation of the decrease in CK activity after termination of the 3*g*-exposure extremely difficult and demonstrates the necessity to study many developmental stages before any space project is initiated (modified from Slenzka *et al.*, 1993).

Xenopus larvae, CK activity in the whole brain decreased after exposure to hypergravity (2–4***g***). In the fish, the 20% decrease of CK activity was accompanied by a 15% decrease of brain volume. A more detailed analysis revealed that after hypergravity exposure, CK reactivity was decreased for plasma membrane related energy transformation but increased for mitochondrial related energy transformation. In contrast, growth in a clinostat caused a slight increase of plasma membrane related energy transformation. Altered gravity conditions had no effects on adult fish what suggests a higher neuronal plasticity in larval animals or the existence of a sensitive period of gravity-related brain metabolisms (Anken *et al.*, 1998). In killifish (*Fundulus heteroclitus*), CK activity was determined in the cortex of the vestibular cerebellum (*C. eminentia granularis*) after a 19.5-day orbital flight aboard the Cosmos-782 biosatellite. Five days after the end of the microgravity-exposure, individuals hatched during the flight exhibited a significant higher CK activity than the ground fish (Krasnow, 1977).

The reactivity of G6P-DH and SDH in the whole brain of cichlid fish (*Oreochromis mossambicus*) was increased after development in 3***g***. In contrast, G6P-DH reactivity was decreased after development in a clinostat (Anken *et al.*, 1998). A semiquantitative histochemical method (densiometric grey value analysis) revealed that in the young cichlid fish SDH reactivity showed a relation to altered gravity in gravity receptor related brain nuclei (*N. magnocellularis* and *N. oculomotorius superior rectus*) (Fig. 14). In the sensitive nuclei, it was lowest in animals with microgravity-experience and highest in 3***g***-fish (microgravity-orbit < 1***g***-orbit, 1***g***-ground < 1.4***g*** < 3***g***) (Anken *et al.*, 1998).

In the magnocellular nucleus of young *Oreochromis*, CO was also positively correlated with gravity (microgravity-orbit < 1***g***-orbit, 1***g***-ground < 1.4***g*** < 3***g***). Furthermore, energy metabolism after microgravity was decreased in the sensory epithelia of the utricule but not in the saccule. Hypergravity did not influence CO activity in the inner ear, what confirms the hypothesis of compensatory procedures on the CNS level for otolith formation during hypergravity (Anken *et al.*, 1998).

A 9-day hypergravity (2–4***g***) exposure caused a decrease of brain volume (15%) and creatine kinase activity (20%), an increase of cytochrome oxidase but no changes of Ca^{2+}-ATPase in *Oreochromis*. Changes of synaptic ultrastructure in brain nuclei with vestibular afferents (*N. magnocellularis*) were found. These effects can be mediated by conformational changes within the nerve cell membranes. Gangliosides are assumed to modulate membrane-bound functional proteins (e.g., protein kinases, ATPases) for synaptic transmission and long-term neuronal adaptation. Experiments with artificial ganglioside monolayers have shown that their surface potential is easy to influence by alterations of Ca^{2+}-concentration and temperature, in contrast to phospholipides (Slenzka *et al.*, 1990, Rahmann *et al.*, 1992). The NADPH-diaphorase a marker for neuronal plasticity was significantly affected after hypergravity exposure

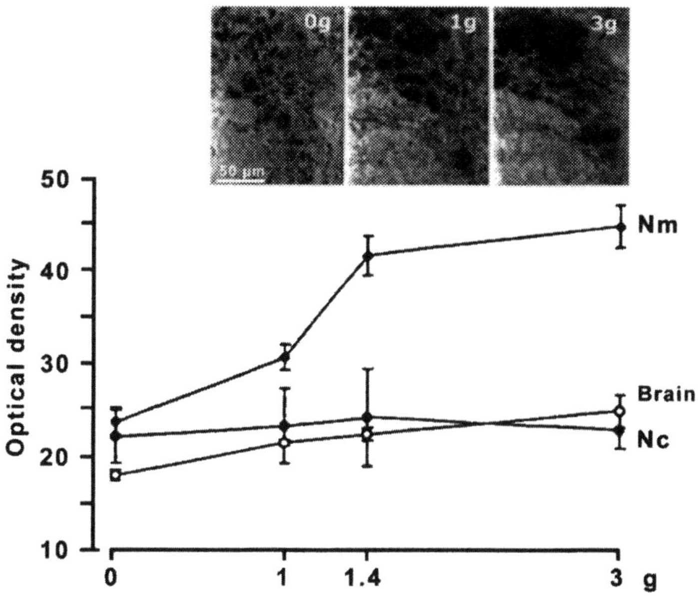

Fig. 14. SDH reactivity in the brain of young cichlid fish after long-term exposure to AGF. Semiquantitative analysis of histochemically demonstrated SDH in whole sections of the total brain (brain) compared to N. magnocellularis (Nm), which receives vestibular (especially utricular) afferents. Other nuclei such as the pretectal N. corticalis (Nc) of the retinohypothalamic system were not affected. IAVG (inverted average grey) values correspond with optical density (cf. inset on top which shows the different staining after 0*g*, 1*g*, and 3*g* exposure in the Nm; scale bar = 50 µm) (modified from Anken *et al.*, 1996, 1998).

for 8 days in brain centres with vestibular relation (*N. magnocellularis*, *N. oculomotorius superior rectus* and cerebellar eurydendroid cells). The results were more distinct in young than in adult fish (*Oreochromis mossambicus*, *Xiphophorus helleri*) indicating a loss of gravity-related plasticity during maturation (Anken and Rahmann, 1998b). Thus, brain metabolism is strongly affected by altered gravity. Modifications occurred mostly in vestibular related nuclei of the brain.

Invertebrate aquatic animals

A number of invertebrate aquatic animals were also used in space research. They focussed on basic developmental processes including fertilisation, embryogenesis, biomineralisation and otoconia formation, and locomotion. Some of these space studies were supplemented by hypergravity observations, mainly concerning otolith and statoconia formation.

Developmental studies were conducted in the classical model organisms, the sea urchins *Paracentrotus lividus* and *Sphaerechinus granularis*. The results

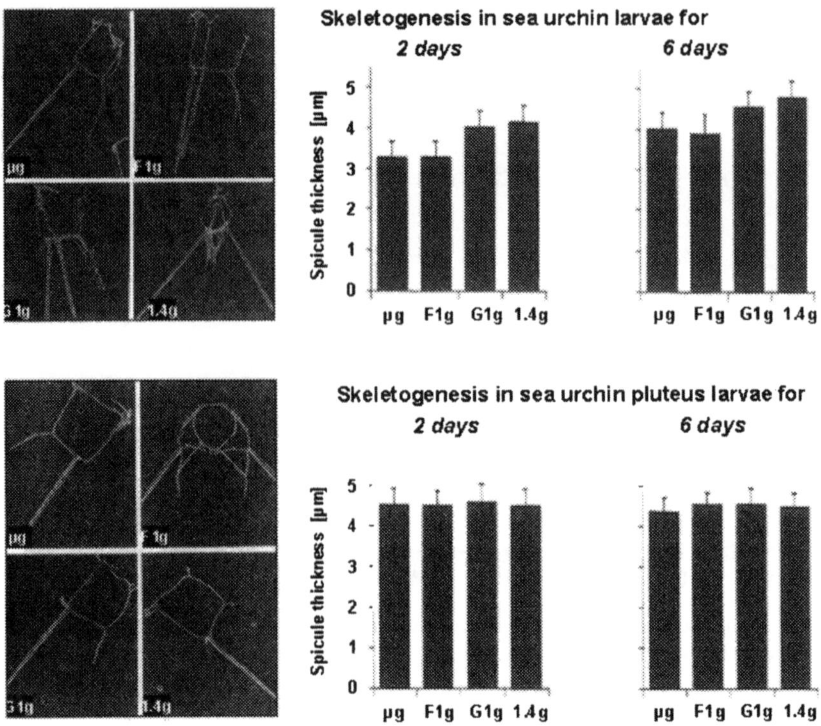

Fig. 15. Skeletogenesis of sea urchins (*Sphaerechinus granularis*) in microgravity. Three days before onset of microgravity, sea urchin embryos were at the blastula stage (upper part) or at the 4-armed pluteus stage (lower part) and were kept at 5° C until launch (STS-76, 1996). Fixation of larvae and plutei 2 days or 6 days after microgravity-onset. Extent of skeletogenesis was determined by means of the thickness of spiculae. Note the difference between flight and ground samples in the young group and lack of effect in the older group. The similarity of results in the young larvae from the inflight groups µg (microgravity) and F1g (1g-centrifugation during flight) point to unspecific effects of flight conditions in addition to microgravity conditions. Spiculae presented on the left are taken from material fixed 6 days after microgravity-onset (modified from Marthy *et al.*, 1999).

indicate that fundamental processes necessary for fertilisation, the subsequent embryogenesis and the bio-mineralisation of spicules occurred normally in the absence of gravity (Fig. 15). That means that the genetic programme dominates or even controls completely the early development in this species. In particular, in an experiment on the MASER 5 sounding rocket flight fertilisation was performed 60 s after the onset of microgravity. Post-flight analysis revealed a 95% fertilisation rate although the elevation of the fertilisation membrane was occasionally weak. During further development under 1*g*-conditions, the cleaving eggs continued embryonic and larval development. Young pluteus larvae were swimming after 4 days, identical to the controls. Larvae lifetimes could be increased to >40 days by feeding with algae. Interestingly, flown virgin eggs could no longer be fertilised. Flown sperm, on the other hand, maintained its fertilisation ability on fresh virgin eggs. A follow-up flight on MASER 6 in 1993 focused on the question whether cleaving eggs (early embryos) are affected by microgravity. Therefore, fertilisation was performed on ground, and early and later cleavage stages were launched. The main result of this MASER 6 flight was that gravity changes are effectively sensed by the individual embryonic cell, but that development of the embryo as a whole is not affected (Marthy, 1997).

Experiments on mineralisation processes in the absence of gravity were performed during orbital flights in the sea urchin (IML-2 mission STS-65 in 1994 and Shuttle-to-Mir mission SMM-03 STS-76 in 1996; Fig. 15) and the pond snail *Biomphalaria glabrata* (STS-89 and STS-90; Marxen *et al.*, 2001), and during parabolic flights in the sea urchin. The observations in sea urchins from the space flight and parabolic flight studies differed which is probably due to the different approaches. For the IML-2 orbital flight, larvae were used. The study revealed that the biomineralisation process, a cascade of developmental events leading from primary mesenchymatic cells (PMC) which are originating from the micromeres at the 16-cells stage to well defined skeletal structures does occur in the absence of gravity during space flight. It also turned out that in this animal species no pronounced de-mineralisation phenomenon occur in the pluteus larvae (Marthy *et al.*, 1996). For the parabolic flight experiments with the regular change between microgravity and hypergravity micromeres were used cultered just before the migration (12 h after fertilisation), before the beginning of spicule formation (24 after fertilisation) or during spicule elongation (36 and 48 h after fertilisation). 24 h post-flight, the length of the spicules were measured. The observation revealed shorter spicules in all microgravity/hypergravity-exposed larvae compared to ground controls; in parallel, the expression of the spicule matrix protein SM30 was also reduced (Izumi-Kurotani and Kiyomoto, 2003). Observations in old embryos of *Biomphalaria* revealed no difference in mineralisation of the shell between microgravity-exposed and 1***g***-ground animals (Marxen *et al.*, 2001).

Studies on the effects of altered gravity on the formation of otoliths and statoconia in aquatic invertebrates revealed some similarities to the observations in the aquatic vertebrates. Microgravity studies were done in *Biomphalaria glabrata* which flew on the 10-day STS-89 and the 16-day STS-90 (Neurolab) missions (Wiederhold *et al.*, 2003). Hypergravity studies were performed in adults and various developmental stages of the marine gastropod *Aplysia californica* (Pedrozo *et al.*, 1996; Pedrozo and Wiederhold, 1994). Both molluscs have in common that their statocysts contain statoconia. The statocyst of *Biomphalaria* is about 150 μm in diameter and its wall contains 13 ciliated mechanoreceptor cells and 3 to 10 statoconia at hatching and upto 400 statoconia when adult. The statocysts of *Aplysia* contains also 13 sensory cells; but statocysts from embryonic *Aplysia* have a single inclusion called the statolith and is retained throughout the life of the animals. But beginning with stage 9 of development, staotconia begin to be produced. After metamorphosis at developmental stage 10, statoconia production reaches the highest level during the life (Pedrozo *et al.*, 1996).

Videotaping of *Biomphalaria* in orbit revealed that the snails were easily dislodged from the aquarium wall. On Earth they spend most of their time attached to the walls. Once separated from the wall they float through the water which gave them in orbit the chance to contact other snails. As these snails are hermaphrodites, mating pairs were often seen floating attached to one another. Five days after eggs are laid, the larvae hatch. Thus, a number of young snails were recovered after landing of the space crafts. The post-flight observations revealed that numbers and volumes of statoconia within the statocysts of young ground animals of the comparable size were smaller than those obtained from the flight animals. In particular, the mean total volume of statoconia in the flight animals was 50% larger than that of the ground animals, the average number of statoconia in flight animals was 37% larger.

Embryonic *Aplysia* were exposed to 2***g***, 3***g*** and 5.7***g*** hypergravity while early metamorphosed animals were treated for 3 weeks with 2***g*** hypergravity. The experiments revealed that statoconia production was inhibited by hypergravity and that also volume was decreased, probably by a down-regulation of urease activity (Pedrozo *et al.*, 1996). In embryonic animals, 2***g***-exposure can cause a reduction of the mean statolith but not in all cases while statocyst size was never affected (Pedrozo and Wiederhold, 1994).

A completely different type of gravity sense organ has developed during evolution in scyphomedusae such as *Aurelia aurita*. Equilibrium control is mainly mediated by the marginal bodies (rhopalia). Rhopaliae are transformed tentacles hanging from the margin of the bell and are protected by surrounding tissue. Endodermal cell in the distal part of the rhopaliae produce statoliths which contain calcium sulphate, carbonate, and phosphate. Non-motile cilia project from sensory cells at the base of each tentacle towards the surrounding wall; they are bent according to the spatial position of the rhopaliae. These organs were considered as a gravity sensing system that help the medusae to maintain each position in space and move on straight courses in this position (Fraenkel, 1925; Bozler, 1926). A neurite plexus mediates the coordination between all rhopaliae. From the comparative point of view, this structural pecularity of sensory cells with cilia located outside the statocyst made *Aurelia* ephyra larvae attractive to study its microgravity ***g***-susceptibility. In fact, some morphological modifications were caused by microgravity including a reduction in the number of lipid droplets in the large spaces near their bases. On the other hand, the neurite plexus and the network of cytoplasmic strands extending to the statocysts were not affected by microgravity. Differences in swimming and orienting in microgravity-ephyrae were not explained by morphological differences in the hair cells or the statocysts (Spangenberg *et al.*, 1996).

Applications of research in aquatic animals to mammals and human

The studies in aquatic animals gave clear evidence for strong effects of altered gravity on many systems with high similarities to observations done in mammals and man such as the sensitisation of the vestibular system by gravity deprivation. A few outstanding observations such as relation between developmental pace and susceptibility to altered gravity might be important for higher vertebrates. The following chapters discuss some aspects which might be of relevance for higher vertebrates and man. In particular, questions about (1) age-related plasticity versus physiological adaptation, (2) kinestosis and space sickness research, (3) rate of development and susceptibility to altered gravity, and (4) auto-regulative principles during development and adulthood are considered.

Age-related plasticity versus physiological adaptation

Sensitisation of physiological and behavioural responses was demonstrated in experiments with toadfish (Boyle *et al.*, 2000), cichlid fish (Sebastian *et al.*, 2001) and clawed toad (Horn *et al.*, 2003). From the physiological point of view, adaptation is the most simple mechanism to explain vestibular sensitisation under microgravity. In general, most receptor cells and neurons decrease their physiological activity after a step-like decrease of the stimulus level, but adapt to a more or less higher activity level during maintained stimulation (*phasic-tonic response pattern*). Vestibular neurons also decrease their resting activity transiently after they were deprived from labyrinthine input, but they recover it to higher levels thereafter (Precht, 1985). Macula activity originated by translatory accelerations during swimming or other movements (Clément and Reske, 1996) might initiate the increase of spontaneous neuronal activity, similar to the initiation of vestibular compensation following hemilabyrinthectomy (Dieringer, 1995). In fact, observations during the STS-84 mission in 1997 and the Andromède flight to the International Space Station in 2001 revealed that tadpoles increased the frequency of swimming but not its total duration during the space flights (Dournon *et al.*, 2002). If neuroplastic capabilities are also used for the adaptation to hypergravity and if adaptive processes are generally guided by genetically pre-programmed set points within the nervous network to which neuronal activity adjusts during development—similar to those for the adaptation to microgravity (Bracchi *et al.*, 1975)—animals with hypergravity experience have to show transiently lower response levels if they are tested in 1***g*** after 3***g***-termination. This prediction was confirmed (Horn and Sebastian, 1996; Sebastian *et al.*, 2001).

Besides the observations about the activity in the VIIIth nerve of *Opsanus tau* and *Rana catesbeiana* (Fig. 10) and the rVOR in *Oreochromis mossambicus* (Sebastian *et al.*, 2001) and *Xenopus laevis* (Horn *et al.*, 2003), several other observations in lower and higher vertebrates support the hypothesis of an adaptive sensitisation or desensitisation of the vestibular system. (1) Creatine kinase activity—a biochemical marker for neuronal activity—at primary vestibular projection sites of the cerebellar cortex in fish (*Fundulus heteroclitus*) was elevated after a space flight (Krasnow, 1977). In neonate swordtail fish (*Xiphophorus helleri*), the otolioth became significantly larger in microgravity during a 16-day space flight (STS-90) compared to their ground-reared siblings (Wiederhold *et al.*, 2003). In the same species, the number of synapses within the *N. descendens*, a vestibular integration centre, became larger during microgravity while the *N. magnocellularis* which receives inputs from the lateral line and the visual *N. corticalis* was not affected (Ibsch *et al.*, 2000). If synapse numbers are directly correlated with the sensitivity of the vestibular system, changes such as these might explain sensitivity modifications. The overcompensation of the activity in the VIIIth nerve of frogs during a space flight (Bracchi *et al.*, 1975) points in the same direction. Also the studies in snails (Wiederhold *et al.*, 2003) are consistent with a sensitisation of the sense of gravity during space flight.

Observations about adaptive sensitisation and desensitisation in higher vertebrates include the 57% increase of the synapse number of Type II hair cells in the rat after microgravity-exposure (Daunton *et al.*, 1991), and the 30% decrease after 2**g**-exposure (Ross, 1992). The sensitisation seems to be a general consequence of gravity deprivation because it was also observed in rat and man. Rats reveal an up-regulation of the immediate early gene *c-fos* within the afferent vestibular nuclei of rats after exposure to microgravity during a space flight (Pompeiano *et al.*, 2002) and a down-regulation of *c-fos* within the efferent parts of the vestibular nuclei (Balaban *et al.*, 2002) while astronauts overestimated their tilt sensation after a 16-day space flight (Clément *et al.*, 2001).

A feature of sensory and motor systems is their susceptibility to modifications of their adequate physical and/or chemical stimuli during a defined period of life. These so-called critical (sensitive) periods were described for sensory and motor systems including vision, hearing, feeling, olfaction or walking (Dews and Wiesel, 1970; Hubel and Wiesel, 1970; van der Loos and Woolsey, 1973; Wiesel, 1982; Knudsen *et al.*, 1984; Oakley, 1993; Walton, 1998). They are characterised (1) by a sensitivity of the developing system to modifications of the adequate environment, preferably to stimulus deprivation, (2) by a clearly defined time window of this sensitivity and (3) by the irreversibility of anatomical, behavioural or physiological modifications induced by these altered environmental conditions.

The studies done so far in the vestibular system of aquatic animals are indicative for the existence of a critical period. They have shown the existence of milestones in vestibular development (Table 4) such as the formation of the ear vesicle and labyrinth (geotactic behaviour of *Fundulus*: see Hoffman *et al.*, 1977, 1978), the appearance of the vestibulo-ocular reflexes (*Xenopus* and *Oreochromis*: Horn and Sebastian, 1996; Sebastian *et al.*, 2001), or the existence of a time window for a sensitivity of the tilt-induced vestibuloocular reflex to

Table 4

Microgravity effects on the development of vestibular function in lower vertebrates based on studies after exposure to real and simulated microgravity

	Response	Sensitivity to microgravity	Insensitivity to microgravity	1*g*-Readaptation	Microgravity	Ref.
Danio rerio (Zebrafish)	rVOR	30–66 hpf	0–24 hpf 72–96 hpf	complete within 5 days	NASA Bioreactor recreating some aspects of microgravity	1, 2, 3
Fundulus heteroclictus	swimming	if no labyrinth was formed preflight	if labyrinth was formed pre-flight	incomplete: re-tests during parabolic flights	*space flight:* Skylab 3 μ*g*-duration 59 days	4, 5
Xenopus laevis	rVOR	*Lordotic tadpoles:*at least up to stage 45 *normal tadpoles:* if preflight rVOR	*normal tadpoles:* if no rVOR pre-flight	complete within 1 to 5 weeks	*space flight:* STS-55, STS-84, Soyuz taxi Andro-mède to ISS μ*g*-duration 9 to 10 days	6, 7, 8
Oreochromis mossambicus	rVOR	if no rVOR preflight	if rVOR pre-flight	complete within 1 week	*space flight:* STS-55, STS-84 μ*g*-duration 9 to 10 days	9, 10

For zebrafish *Danio rerio*, time intervals (hours after fertilisation, hpf) define periods of exposure in the rotating bioreactor which caused a rVOR depression (sensitivity) or not (insensitivity); its otic vesicle is closed for the first time at 18 hpf and first sensory cells appear at 24 hpf (see Haddon and Lewis, 1996). In the fish *Fundulus heteroclitus*, irregular swimming was tested. In the amphibian *Xenopus laevis*, periods of μ*g*-exposures started between stages 24 to 45; its otic vesicle is closed for the first time at stage 27; its rVOR appears for the first time at stage 42 (definition of stages, see Nieuwkoop and Faber, 1967). In the cichlid fish *Oreochromis mossambicus*, microgravity-periods started between stages 11 and 16; its rVOR appears for the first time at stage 13; its ear vesicle containing for the first time a few otoconia can be seen firstly at stage 8 (definition of stages, see Anken *et al.*, 1993). 1*g*-Readaptation starts after microgravity-termination. rVOR, roll-induced vestibuloocular reflex; μg, microgravity, microgravity; lordotic, tail bended dorsally. References: 1 Moorman *et al.*, 1999; 2 Moorman *et al.*, 2002; 3 Haddon and Lewis, 1996; 4 Hoffman *et al.*, 1977; 5 von Baumgarten *et al.*, 1975; 6 Sebastian *et al.*, 1996; 7 Sebastian and Horn, 1998; 8 Sebastian and Horn, 2001; 9 Sebastian *et al.*, 2001; 10 Sebastian and Horn, 1999.

simulated microgravity (zebrafish *Dario rerio*: Moorman *et al.*, 1999, 2002; Haddon and Lewis, 1996; Fig. 8).

However, these observations are not sufficient because two conditions for the demonstration of a critical period are not yet verified: (1) the period with the high susceptibility to environmental modifications covers a limited period during postembryonic development and (2) anatomical, behavioural or physiological modifications persist for long periods or are even irreversible because mostly microgravity- or 3***g***-induced modifications return to normal after re-exposure to 1***g***. Only a few observations indicate the existence of long-lasting effects. For example, slowly developing tadpoles did not readapt to normal development during the observation period of two weeks while fast growing tadpoles of the same age did (Sebastian *et al.*, 1996). Other observations revealed a delayed 1***g***-readaptation in tadpoles which hatched during hypergravity, or when their rVOR is tested during stimulation with small lateral roll angles than with large ones after 3***g***-exposure (Sebastian and Horn, 2001).

In general, research in aquatic animals will give important insight into the existence of critical periods, i.e., age-dependent adaptation and plasticity of the nervous system. Species such as *Xenopus laevis* are important because they share common features of age-related sensitivities in other sensory systems such as vision (to the discovery of a critical period in the formation of intertectal connections lasting up to 2 weeks before metamorphic climax and its dependency of vision [Grant *et al.*, 1992; Keating and Grant, 1992]) with human (Wiesel, 1982). Urodele species such as *Pleurodeles* are important because they offer the possibility for natural fertilisation in orbit so that the complete development of nervous function can be studied in space born animals. Furthermore, fish are excellent models to study vestibular-visual interactions in relation to gravity deprivation because their otolith controlled equilibrium during swimming is supported by the dorsal light response.

Motion sickness and kinetosis research

Motion sickness is a debilitating condition that limits the capacity for work in environments where changes in acceleration are unavoidable, notably at sea (sailors), in the air (aviators), and in microgravity (astronauts). Nausea and kinetosis are typical features and occur in about 65% of astronauts during space flight (Reschke, 1990). Susceptibility to motion sickness is also common in many vertebrate species including aquatic ones and typically varies greatly among individuals and among species. Although many factors have been identified, it remains largely unknown why certain organisms become nauseated by changes in acceleration while others do not (Wassersug *et al.*, 1993). Theories about the origin of motion sickness, space adaptation syndrome and kinetosis and the differences in individual sensitivity and expression suggest that motion sickness is caused by a mismatch between expected and sensed gravity directions (sensory conflict, see Mori *et al.*, 1996), or by a difference between the right and left otolith mass, in particular those of the utricles. A difference in mass results in a different sensitivity to acceleration.

Aquatic vertebrates were selected as model systems to study the basis of these syndroms. Besides the similarity of their vestibular organ with that of mammals and man, they often exhibit kinetotic behaviour characterised by spinning movements and looping response and reveal emetic behaviour. Amphibians are less useful for this type of kinetosis studies although they behave kinetoticly in microgravity (Sebastian and Horn, 1998). Their otoliths are composed of many crystals, which make the analysis of otolith masses difficult. Thus, studies in motion sickness with amphibian concentrated more to the occurrence of emetic behaviour. Frogs as well as salamanders were tested for vomiting at the end of parabolic flights. In sensitive species such as *Rana rugosa*, *R. nigromaculata*, *Hyla japonica* and *Rhacophorus schlegelii*, vomiting did not happen during flights but rather with a delay of 0.5 to 42 h after flight. The emetic behaviour of *Rana* did not change when it was transferred from a terrestrial to an aquatic environment. Vomiting was completely absent in the toad *Xenopus laevis*, the salamander *Cynops pyrrhogaster* and anuran premetamorphic stages (Wassersug *et al.*, 1993; Naitoh *et al.*, 1989).

Fish are the most suitable animal model to study otolith effects on space sickness and kinetosis. They possess compact otoliths which makes a comparison between left and right otolith easy. In fact, the differences in fish otolith sizes are remarkable and goes up to 17% as shown in trout and salmon (Helling *et al.*, 1997) or in cichlid fish (*Oreochromis mossambicus*) (Anken *et al.*, 1998) and cell density but not cell number in kinetotic fish is lower than in normal swimming ones (Bäuerle *et al.*, 2004). Based on such observations, the reasonable hypothesis is that a misbalanced sensitivity of the statolith organs occurs but is totally compensated for by the vestibular system (cf. vestibular compensation: Schaefer and Meyer, 1974) as long as physiological motion pattern takes place. Decompensation leads to kinetosis under non-physiological motion pattern or in the absence or reduction of the gravitational forces. The validity of this asymmetry theory became obvious in experiments with *Oreochromis mosambicus* using parabolic flights. 10% of the fish became kinetotic during the microgravity-periods. The histological analysis of all fish revealed that these 10% of fish had the largest right/left otolith asymmetries (Fig. 11) (Hilbig *et al.*, 2002). This otolith asymmetry hypothesis was also supported by the observation that swordtail fish (*Xiphophorus helleri*) with otolith asymmetry revealed a weaker tolerance to Coriolis stimulation. During constant vertical axis rotation combined with horizontal oscillation, all experimental animals ($n = 22$) showed active compensatory swimming behaviour, while individuals with otoconial imbalance ($n = 3$) entered a passive uncoordinated state at higher stimulation intensity (Helling *et al.*, 2003).

Rate of development and susceptibility to altered gravity

During early periods of life, aquatic animals such as *Xenopus laevis* tend to mature at different pace even if they were taken from the same batch. This fact has some impact on the readaptation capabilities of their rVOR after hypergravity exposure. Tadpoles from one batch started a 10-day 3***g***-exposure when they had reached the stage level between 33 to 36. Tadpoles which developed to stage 46 during hypergravity were unable to readapt to normal rVOR development during the 10-day post-3***g*** recording period, while tadpoles which developed faster and reached the stage 47 during the post-3***g*** recording period did so (Sebastian *et al.*, 1996). This observation is not outstanding. In young swordtail fish (*Xiphophorus helleri*), otolith growth is sensitive to microgravity. The induced modifications were related to the animal's size. Young fish from the STS-89 flight were smaller than those from the Neurolab mission STS-90. The modification of otoliths was opposite for the two flight. For STS-90, the microgravity-embryos had larger otoliths, for STS-89 the 0***g***-embryos had smaller otoliths compared to their respective ground controls (Wiederhold *et al.*, 2003).

These observations revealed the higher risk of developmental retardation for the occurrence of malfunction or malformation of the vestibular system. They might become clinically important due to a general impact of the vestibular system on the development of motor function. Therefore, efforts have to be increased to consider this specific aspect in research using aquatic model animals. They have a large amount of off-springs in one batch and a very high variablity of developmental progress. These studies can deliver fundamental information about plasticity management of the developing organism.

Self-Regulatory effects of genetic programs on the organisms—Is development insensitive to long-term microgravity exposure?

There is no doubt that altered gravity activates adaptive mechanisms which caused normalisation of structure and function during 1***g***-readaptation. A few exceptions exist such as osteoporosis. Another important conclusion from research in aquatic animals is that several developmental stages are strongly affected by microgravity but that they possess the capability to regulate these abnormalities back to normality not only after but even during exposure in microgravity. Thickening of the blastocoel roof at the gastrula stage (Souza *et al.*, 1995; Ubbels *et al.*, 1995; Ubbels, 1997) and disturbances during the neurulation (Gualandris-Parisot, 2001; Dournon, 2003) did not prevent normal development so that larvae with normal morphology and with normal swimming behaviour hatched in microgravity. Normalisation of initially microgravity-induced modifications exists also in the physiological properties of the vestibular system of adults during microgravity conditions (frog: Bracchi *et al.*, 1975; man: Clément, 1998). These observations favour the hypothesis that

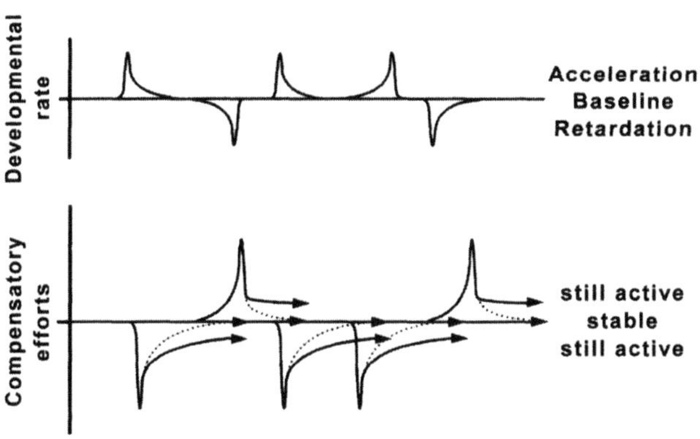

Fig. 16. Normalisation of transient disturbances during development in microgravity (cf. Chapter "Fertilization and early development"). The hypothesis considers the activation of compensatory mechanisms to reach normalisation; they might be based on genetic programs or on a physiological counter-regulation such as sensitisation or desensitisation (modified from Sebastian and Horn, 2001).

despite of AGF-induced morphological and physiological abnormalities, organisms are capable of regulating their development to normality if they are exposed to long-lasting AGF-conditions. One possibility is given by an increase of regulatory efforts in case of developmental retardation, and vice versa, by decreasing their regulatory efforts in case of an increased developmental rate (Fig. 16).

Perspectives

There is always the question about the contribution of studies on both invertebrates and lower vertebrates such as fish and frogs to human benefit. High costs in the preparation and performance of space experiment cause sometimes postulations to cancel research in these species. Recently voices become more and more prominent that claim to use only a few model species. They argue that in model animal species interactions between different organs of the body, i.e., the systemic interactions can better be understood because many specialised researchers investigate "their" specific organs and, thereafter, discuss their results with the other "organ" specialists to find existing interrelations of the microgravity-effects within one body. Aquatic animals have to be among these model species. They have contributed to almost all fields of basic biological research using space. In particular, the knowledge about the effects of microgravity on development would never have been obtained from land-living organisms.

Despite numerous examples about normalisation of morphology and physiology, the manifestation of irreversible structural or physiological changes during long-term exposures in weightlessness is as likely as normal development from an egg to an adult. It depends on the power of genetically and physiologically determined regulatory mechanisms within the developing system that becomes a sustaining and stable self-regulative principle. Alternatively, microgravity-induced modifications occur only transiently, so that after reaching stability, the normalising mechanisms are switched off (Fig. 16). In the worst case, a complete break-down of life stability is caused. Therefore, research in aquatic animal models, preferably on the International Space Station, has to be continued; it will contribute to the understanding of basic adaptive mechanisms to microgravity which might be valid for men. This research will give answers to questions about stable life in space and about the physiological risk of a long-term life in space.

Acknowledgements

Projects of the author mentioned in this article were supported by grants from the Germany Space Agency (DLR) and the German Science Foundation (DFG). All his experiments comply with the "Principles of Animal Care", publication No. 86-23, revised 1985 of the National Institutes of Health, and with the "Deutsches Tierschutzgesetz", BGBl from 17 February, 1993. Permissions for the experiments were given by the Regierungspräsidium of Tübingen (Germany), numbers 399/Ulm, 506/Ulm and 657/Ulm, and by the Animal Care and Use Committee (ACUC) at Kennedy Space Center/Florida.

References

Anken, R. and Bourrat, F. (1998) *Brain atlas of the medaka fish*, INRA editions, Paris.

Anken, R.H., Edelmann, E. and Rahmann, H. (2002) Neuronal feedback between brain and inner ear for growth of otoliths in fish. Adv. Space Res. 30(4), 829–833.

Anken, R.H., Ibsch, M. and Rahmann, H (1998) Neurobiology of fish under altered gravity conditions. Brain Res. Rev. 28(1–2), 9–18.

Anken, R.H., Kappel, T., Slenzka, K. and Rahmann, H. (1993) The early morphogenetic development of the cichlid fish *Oreochromis mossambicus* (Perciformes, Teleostei). Zool. Anz. 231, 1–10.

Anken, R.H. and Rahmann, H. (1998) Influence of long-term hyper-gravity on the reactivity of succinic acid dehydrogenase and NADPH-diaphorase in the central nervous system of fish: a histochemical study. Adv. Space Res. 22(2), 281–285.

Anken, R.H., Slenzka, K., Neubert, J. and Rahmann, H. (1996) Histochemical investigations on the influence of long-term altered gravity on the CNS of developing cichlid fish: results from the 2nd German Spacelab Mission D-2. Adv. Space Res. 17(6–7), 281–284.

Anken, R., Werner, K., Ibsch, M. and Rahmann, H. (1998) Fish inner ear otolith size and bilateral asymmetry during development. Hear. Res. 121, 77–81.

Balaban, E., Centini, C. and Pompeiano, P. (2002) Tonic gravity changes alter gene expression in the efferent vestibular nucleus. NeuroReport 13, 187–190.

Ballarino, J. and Howland, H. (1984) Centrifugation of 2*g* does not affect otoconial mineralisation in chick embryos. Amer. J. Physiol. 246, 305–318.

Bäuerle, A., Anken, R.H., Hilbig, R. and Rahmann, H. (2004) Histology of the utricle in kinetotically swimming fish: a parabolic aircraft flight study. Acta Otolaryngol. 124, 19–22.

Bechterew, W. (1883) Ergebnisse der Durchschneidung des N. acusticus, nebst Erörterung der Bedeutung der semicirculären Kanäle für das Körpergleichgewicht. Pflügers Arch. Ges. Physiol. 30, 312–347.

Böser, S. (2003) *Die Entwicklung motorischer Systeme und ihre Anpassung an veränderte Gravitationsbedingungen beim Krallenfrosch (Xenopus laevis)*, Dissertation, University of Karlsruhe, Germany.

Böser, S. and Horn, E. (2002) Hypergravity exposure affects ventral root activity in tadpoles (*Xenopus laevis*). J. Grav. Physiol. 9, 209–210.

Böser, S., Dournon, C., Gualandris-Parisot, L. and Horn, E. (2002) Ventral root recordings during fictive swimming in tadpoles (*Xenopus laevis*) with microgravity or hypergravity experience. Gravit. Space Biol. Bull. 16, 31.

Boyle, R., Mensinger, A.F., Yoshida, K., Usui, S., Intravaia, A., Tricas, T. and Highstein, S.M. (2000) Neural readaptation to Earth's gravity following return from space. J. Neurophysiol. 86, 2118–2122.

Bozler, E. (1926) Sinnes- und neurophysiologische Untersuchungen an Scyphomedusen. Z. vergl. Physiol. 4, 37–80.

Bracchi, F., Gualierotti, T., Morabito, A. and Rocca, E. (1975) Multiday recordings from the primary neurons of the statoreceptors of the labyrinth of the bullfrog. Acta Otolaryngol. (Stockh) Suppl. 334, 3–27.

Cherdantseva, EM (1987) Embryonic development of guppies in weightlessness [Russian]. Kosmicheskaia Biologiia I Aviakosmicheskaia Meditsina 21(5), 22–25.

Clément, G., Moore, S.T., Raphan, T. and Cohen, B. (2001) Perception of tilt (somatogravic illusion) in response to sustained linear acceleration during space flight. Exp. Brain Res. 138, 410–418.

Dale, N. (1995) An experimentally derived model for the locomotor pattern generator in the *Xenopus* embryo. J. Physiol. 489, 489–510.

Daunton, N.G., Ross, M.D., Fox, R.A., Corcoran, M.L., Cutler, L.K. and Wu, L.C. (1991) Effects of chronic hypergravity on the righting reflex and vestibular endorgan morphology in the rat. Soc. Neurosci. Abstr. 17, 316.

Dews, P.B. and Wiesel, T.N. (1970) Consequences of monocular deprivation on visual behaviour in kittens. J. Physiol. 206, 437–455.

Degens, E.T., Deuser, W.G. and Haedrich, R.L. (1969) Molecular structure and composition of fish otoliths. Mar. Biol. 2, 105–113.

Dieringer, N. (1995) 'Vestibular compensation': Neural plasticity and its relations to functional recovery after labyrinthine lesions in frogs and other vertebrates. Prog. Neurobiol. 46, 97–129.

Dournon, C., Membre, H., Böser, S. and Horn, E. (2002) An European pupil project linked to the scientific aims of the experiment Aquarius-Xenopus on the taxi soyuz flight Andromède to ISS. J. Gravitat. Physiol. 9(1), P375–P376.

Dournon, C. (2003) Developmental biology of urodele amphibians in microgravity conditions. Adv. Space Res. 9, 101–131.

Duprat, A.M., Husson, D. and Gualandris-Parisot, L. (1998) Does gravity influence the early stages of the development of the nervous system in an amphibian? Brain Res. Rev. 28, 19–24.

Feitek, M., Souza, K., Neff, A. and Wassersug, R. (1998) Swimming kinematics and respiratory behaviour of *Xenopus laevis* larvae raised in altered gravity. J. Exp. Biol. 201, 1917–1926.

Fermin, C.D., Martin, D.S., Li, Y.T. and Li, S.C. (1995) The glycan ketatan sulfate in inner ear crystals. Cell. Mol. Biol. 41, 577–591.

Fluck, R.A., Krok, K.L., Bast, B.A., Michaud, S.E. and Kim, C.E. (1998) Gravity influences the position of the dorsoventral axis in medaka fish (*Oryzias latipes*). Dev. Growth Differ. 40(5), 509–518.

Fraenkel, G. (1925) Der statische Sinn der Medusen. Z. vergl. Physiol. 2, 658–690.

Furukawa, R. and Ijiri, K. (2002) Swimming behaviour of larval Medaka fish under microgravity. Adv. Space Res. 30(4), 733–738.

Gallien, L. and Durocher, M. (1957) Table chronologique du développement chez *Pleurodeles waltl* (Michah). Bull. Biol. Fr. Belg. 91, 97–114.

Gauldie, R.W. (1993) Continuous and discontinuous growth in the otolith of *Macruronus novaezelandiae* (Merlucciidae: Teleostei). J. Morphol. 216, 271–294.

Gilbert, S.F. (2003) *Developmental Biology*, Sinauer Associates, Inc., Publishers, Sunderland, Massachusetts.

Gillette-Gerguson, I., Ferguson, D.G., Poss, K.D. and Moorman, S.J. (2003) Changes in gravitational force induce alterations in gene expression that can be monitored in the live, developing zebrafish heart. Adv. Space Res. 32(8), 1641–1646.

Goto, F., Straka, H. and Dieringer, N. (2001) Postlesional vestibular reorganization in frogs: evidence for a basic reaction pattern after nerve injury. J. Neurophysiol. 85, 2643–2646.

Grant, S., Dawes, E.A. and Keating, M.J. (1992) The critical period for experience-dependent plasticity in a system of binocular visual connections in *Xenopus laevis*: its extension by dark-rearing. Eur. J. Neurosci. 4, 37–45.

Grigoryan, E.N., Mitashov, V.I. and Anton, H.J. (2002) Urodelean amphibian in studies on microgravity: effects upon organ and tissue regeneration. Adv. Space Res. 30, 757–764.

Grinfeld, S., Foulquiere, F., Mitashov, V., Bruchlilnskaia, N., Duprat, A.M. (1994) Tissue regeneration in space (spinal cord, muscle and bone) in the amphibian *Pleurodeles waltl.* ESA SP-366, 181–184.

Gruener, R. and Hoeger, G. (1990) Vector-free gravity disrupts synapse formation in cell culture. Amer. J. Physiol. 258, 489–494.

Gruener, R., Roberts, R. and Reitstetter, R. (1994) Reduced receptor aggregation and altered cytoskeleton in cultured myocytes after space-flight. Uchu Seibtutsu Kagaku. 8, 79–93.

Gualandris-Parisot, L., Husson, D., Foulquier, F., Kan, P., Davet, J., Aimar, C.H., Dournon, C. and Duprat, A.M. (2001) *Pleurodeles waltl*, amphibian, urodele, is a suitable biological model for embryological and physiological space experiments on a vertebrate. Adv. Space Res. 28(4), 569–578.

Gualandris-Parisot, L., Husson, D., Bautz, A., Durand, D., Aimar, C., Membre, H., Duprat, A.M. and Dournon, C. (2002) Effects of space environment on the embryonic development up to hatching of salamander eggs fertilised and developed during orbital flights. Biol. Sci. Space 16, 3–11.

Haddon, C. and Lewis, J. (1996) Early ear development in the embryo of the zebrafish, *Danio rerio*. J. comp. Neurol. 365, 113–128.

Hara, H., Sekitani, T., Kido, T., Endo, S., Ikeda, T. and Takahashi, M. (1995) Fine structure of utricle of developing chick embryo exposed to 2***g*** gravity. Acta Otolaryng. (Stockh.) 519, 257–263.

Helling, K., Hausmann, S., Flottmann, T. and Scherer, H. (1997) Untersuchungen zur interindividuellen unterschiedlichen Kinetoseempfindlichkeit. HNO 45, 210–215.

Helling, K., Hausmann, S., Clarke, A. and Scherer, H. (2003) Experimentally induced motion sickness in fish: possible role of the otolith organ. Acta Oto-Laryngol. 123(4), 488–492.

Hilbig, R., Anken, R.H., Sonntag, G., Höhne, S., Henneberg, J., Kretschmer, N. and Rahmann, H. (2002) Effects of altered gravity on the swimming behaviour of fish. Adv. Space Res. 30(4), 835–841.

Hoffman, R.B., Boyd, S.J., von Baumgarten, R.J. and Baky, A.A. (1977) Behavioural analyses of killifish exposed to weightlessness in the Apollo-Soyuz test project. Aviat. Space Environ. Med. 48, 712–717.

Hoffman, R.B., Salinas, G.A., Boyd, S.J., von Baumgarten, R.J. and Baky, A.A. (1978) Effect of prehatching weightlessness on adult fish behaviour in dynamic environments. Aviat. Space Environ. Med. 49, 576–581.

Horn, E. (1985) Funktionelle Entwicklung der Schweresinnesorgane bei Insekten und Vertebraten. DFVLR-Mitt. 85(16), 121–128.

Horn, E. (2003) The development of gravity sensory systems during periods of altered gravity dependent sensory input. Adv. Space Biol. Med. 9, 133–171.

Horn E. (2004) "Critical periods" in vestibular development or adaptation of gravity sensory systems to altered gravitational conditions? Arch. Ital. Biol. 142, 155–174.

Horn, E.R., Dournon, C., Gualandris-Parisot, L., Böser, S. (2003) The development of vestibular structures and functions of tadpoles *(Xenopus laevis)* in the absence of gravity. Soc. Neurosci., Abstract, Program No. 40.6.

Horn, E. and Sebastian, C. (1996) A hypergravity related sensitive period during the development of the roll induced vestibuloocular reflex in an amphibian (*Xenopus laevis*). Neurosci. Lett. 216, 25–28.

Horn, E. and Sebastian, C. (2002) Adaptation of the macular vestibuloocular reflex to altered gravitational conditions in a fish (*Oreochromis* mossambicus). Adv. Space Res. 30(4), 711–720.

Horn, E., Lang, H.G. and Rayer, B. (1986a) The development of the static vestibulo-ocular reflex in the Southern Clawed Toad, *Xenopus laevis* Daudin: I. Intact animals. J. Comp. Physiol. 159A, 869–878.

Horn, E., Mack, R. and Lang, H.G. (1986b) The development of the static vestibulo-ocular reflex in the Southern Clawed Toad, *Xenopus laevis* Daudin: II. Animals with acute vestibular lesions. J. Comp. Physiol. 159A, 879–885.

Hubel, D.H. and Wiesel, T.N. (1970) The period of susceptibility to the physiological effects of unilateral eye closure in kittens. J. Physiol. 206, 419–436.

Husson, D., Gualandris-Parisot, L., Foulquier, F., Grinfeld, S., Kann, P. and Duprat, A.M. (1998) Differentiation in microgravity of neural and muscle cells of a vertebrate (amphibian). Adv. Space Res. 22, 303–308.

Ibsch, M., Anken, R.H. and Rahmann, H. (2000) Weightlessness during spaceflight results in enhanced synapse formation in a fish brain vestibular nucleus. Neurosci. Lett. 296, 13–16.

Ijiri, K. (1997) Explanations for a video version of the first vertebrate mating in space – A fish story. Biol. Sci. Space 11, 153–167.

Ijiri, K. (1998) Development of space-fertilised eggs and formation of primordial germ cells in the embryos of Medaka fish. Adv. Space Res. 21(8–9), 1155–1158.

Ijiri, K. (2000) Vestibular and visual contribution to fish behaviour under microgravity. Adv. Space Res. 25, 1997–2006.

Ijiri, K. (2003) Life-cycle experiments of Medaka fish aboard the International Space Station. Adv. Space Biol. Med. 9, 201–216.

Izumi-Kurotani, A., Yamashita, M., Kawasaki, Y., Kurotani, T., Mogami, Y., Okuno, M., Oketa, A., Shiraishi, A., Ueda, K., Wassersug, R.J. and Naitoh, T. (1994) Behaviour of Japanese tree frogs under microgravity on Mir and in parabolic flight. Adv. Space Res. 14(8), 419–422.

Izumi-Kurotani, A. and Kiyomoto, M. (2003) Morphogenesis and gravity in a whole amphibian embryo and in isolated blastomeres of sea urchins. Dev. Biol. Res. Space 9, 83–99.

Kahn, J.A., Roberts, A. and Kashin, S.M. (1982) The neuromuscular basis of swimming movements in embryos of the amphibian *Xenopus laevis*. J. Exp. Biol. 99, 175–184.

Kahn, J.A. and Roberts, A. (1982) The central nervous origin of the swimming motor pattern in embryos of *Xenopus laevis*. J. Exp. Biol. 99, 185–196.

Keating, M.J. and Grant, S. (1992) The critical period for experience-dependent plasticity in a system of binocular visual connections in *Xenopus laevis*: its temporal profile and relation to normal developmental requirements. Eur. J. Neurosci. 4, 27–36.

Kido, T. and Takahashi, M. (1997) Scanning electron microscopic study of amphibians otoconia. Auris Nasus Larynx 24, 125–130.

Kimmel, C.B., Ballard, W.W., Kimmel, S.R., Ullmann, B. and Schilling, T.F. (1995) Stages of embryonic development of the zebrafish. Dev. Dyn. 203, 253–310.

Knudsen, E.I., Knudsen, P.F. and Esterley, S.D. (1984) A critical period for the recovery of sound localization accuracy following monoaural occlusion in the barn owl. J. Neurosci. 4, 1012–1020.

Koike, H., Nakamura, K., Nishimura, K., Kashima, I., Wiederhold, M.L. and Asashima, M. (1995) Non-invasive assessment of otolith formation during development of the Japanese red-bellied newt, *Cynops pyrrhogaster*. Hear. Res. 88, 206–214.

Komazaki, S. (2004) Gravitational effects on apoptosis of presumptive ectodermal cells of amphibian embryo. J. Exp. Zool. A: Comp. Exp. Biol. 301, 204–211.

Krasnow, I.B. (1977) Quantitative histochemistry of the vestibular cerebellum of the fish *Fundulus heteroclitus* flown aboard the biosatellite COSMOS-782. Aviat. Space Environ. Med. 48, 808–811.

Lewis, E.R. and Pawley, J.B. (1981) Direct SEM study of frozen inner ear. Scanning 4, 131–140.

Lim, D., Stith, J. and Stockwell, C. (1974) Observations on sacculus of rats exposed to longterm hypergravity. Aerospace Med. 45, 705–710.

Llinás, R. and Precht, W. (1976) *Frog Neurobiology*, Springer Verlag, Berlin, Heidelberg, New York.

Lychakov, D.V. (1991) Functional and adaptive changes in the vestibular apparatus in space flight. The Physiologist 34, S205–S206.

Marmo, F., Balsamo, G. and Franco, E. (1983a) The formation of otoliths in the frog *Rana esculenta*. Scanning electron microscopic and X-ray diffraction studies. Acta. Zool. 67, 533–561.

Marmo, F., Balsamo, G. and Franco, E. (1983b) Calcite in the statoconia of amphibians: A detailed analysis in the frog *Rana esculenta*. Cell Tiss. Res. 233, 35–43.

Marthy, H.-J. (1997) Sea urchin eggs under microgravity conditions. ESA SP-1206 81–91.

Marthy, H.-J., Gasset, G., Tixador, R., Schatt, P., Eche, B., Dessommes, A., Giacommi, T., Tap, G. and Gorand, D. (1996) The sea urchin larvae, a suitable model for biomineralisation studies in space (IML-2 ESA Biorack experiment "24-F urchin"). J. Biotechnol. 47, 167–177.

Marthy, H.J., Bacchieri, R., Gasset, G., Eche, B. and Gorand, D. (1999) Sea urchin development in space as revealed by skeletogenesis. ESA SP-1222 139–148.

Massion, J. (1994) Postural control system. Curr. Opinion Neurobiol. 4, 877–887.

Marxen, J.C., Reelsen, O. and Becker, W. (2001) Embryonic development of the freshwater snail *Biomphalaria glabrata* under microgravity conditions (STS-89 mission). J. Grav. Physiol. 8, 29–36.

Meyer, D.L., Heiligenberg, W. and Bullock, T.H. (1976) The ventral substrate response. A new postural control mechanism in fishes. J. Comp. Physiol. 109, 59–68.

Mitashov, V.I., Grigoryan, E.N., Brushlinskaya, N.V., Tuchova, S.Y. and Anton, H.J. (1996) Regeneration of organs and tissues in lower vertebrates during and after space flight. Adv. Space Res. 17(6/7), 241–255.

Mitashov, V.I., Grigoryan, E.N., Tuchova, S.Y. and Cherdanzeva (1987) Organs and tissue regeneration of in amphibian under the space flight conditions. ESA SP-271 299–303.

Moorman, S.J., Burress, C., Cordova, R. and Slater, J. (1999) Stimulus dependence of the development of the zebrafish (*Danio rerio*) vestibular system. J. Neurobiol. 38, 247–258.

Moorman, S.J., Cordova, R. and Davies, S.A. (2002) A critical period for functional vestibular development in zebrafish. Dev. Dyn. 223, 285–291.

Moorman, S.J. (2001) There are two different critical periods for zebrafish vestibular development. FASEB Journal 15(5), A743.

Mori, S., Mitarai, G., Takabayashi, A., Usui, S., Sakakibara, M., Nagatomo, M. and von Baumgartner, R.J. (1996) Evidence for sensory conflict and recovery in carp exposed to prolonged weightlessness. Aviat. Space Environ. Med. 67(3), 256–261.

Mori, S., Mitarai, G., Takagi, S., Takabayashi, A., Usui, S., Nakamura, T., Sakakibara, M., Nagatomo, M. and von Baumgartner, R.J. (1994) Space experiment using large-sized fish: in case of carp in Spacelab-J mission. Acta. Astronautica 33, 41–47.

Naitoh, T., Wassersug, R.J. and Leslie, R.A. (1989) The physiology, morphology, and ontogeny of emetic behaviour in anuran amphibians. Physiol. Zool. 62, 819–843.

Neubert, J., Briegleb, W., Schatz, A., Hertwig, I. and Kruse, B. (1988) The response of structure and function of the gravireceptor in a vertebrate to near weightlessness. Acta Astronautica 17, 257–262.

Neubert, J., Briegleb, W., Schatz, A., Kruse, B., Hertwig, J. and Horn, E. (1987) Synopsis of the spacelab D-1 statolith experiment STATEX. ESA SP-271 241–248.

Nieuwkoop, P.D. and Faber, J. (1967) *Normal Table of Xenopus laevis (Daudin).* North Holland, Amsterdam.

Nishiwaki, Y., Ijiri, K., Satoh, T., Tokunaga, F. and Morita, T. (1999) Retinal Photoreceptors and related gene expression is normal in clinostat-treated fish embryos. Adv. Space Res. 23(12), 2045–2048.

Oakley, B. (1993) The gustatory competence of the lingual epithelium requires neonatal innervation. Dev. Brain Res. 72, 259–264.

Ohnishi, T., Tsuji, K., Ohmura, T., Matsumoto, H., Wang, X., Takahashi, A., Nagaoka, S., Takabayashi, A. and Takahahsi, A. (1998) Accumulation of stress protein 72 (HSP72) in muscle and spleen of goldfish taken into space. Adv. Space Res. 21(8–9), 1077–1080.

Pedrozo, H.A. and Wiederhold, M.L. (1994) Effects of hypergravity on statocyst development in embryonic *Aplysia californica*. Hear. Res. 79, 137–146.

Pedrozo, H.A., Schwartz, Z., Luther, M., Dean, D.D., Boyan, B.D. and Wiederhold, M.L. (1996) A mechanism of adaptation to hypergravity in the statocyst of *Aplysia californica*. Hear. Res. 102, 51–62.

Pompeiano, O., d'Ascanio, P., Centini, C., Pompeiano, M. and Balaban, E. (2002) Gene expression in rat vestibular and reticular structures during and after space flight. Neuroscience 114, 135–155.

Pote, K.G. and Ross, M.D. (1993) Utricular otoconia of some amphibians have calcitic morphology. Hear. Res. 67, 189–197.

Precht, W. and Dieringer, N. (1985) Neuronal events paralleling functional recovery (compensation) following peripheral vestibular lesions. *In* Berthoz, A. and Melvill Jones, G. (eds.). *Adaptive mechanisms in gaze control: Facts and theories*, pp. 251–268. Elsevier Science Publ BV, Amsterdam, New York, Oxford.

Pronych, S. and Wassersug, R.J. (1994) Lung use and development in *Xenopus laevis* tadpoles. Can. J. Zool. 72, 738–743.

Rahmann, H., Slenzka, K.H., Körtje, K.H. and Hilbig, R. (1992) Synaptic plasticity and gravity: Ultrastructural, biochemical and physiochemical fundamentals. Adv. Space Res. 12(1), 163–172.

Rayer, B., Cagol, E. and Horn, E. (1983) Compensation of vestibular-induced deficits in relation to the development of the Southern Clawed Toad, *Xenopus laevis* DAUDIN. J. Comp. Physiol. 151, 487–498.

Reitstetter, R. and Gruener, R. (1994) Vector-averaged gravity does not alter acetylcholine receptor single channel properties. Uchu Seibtutsu Kagaku. 8, 71–78.

Reschke, M.F. (1990) Statistical prediction of space motion sickness. *In* Crampton, G.H. (ed.). *Motion and space sickness*, pp. 263–316. CRC Press, Boca Raton, Fl.

Riley, B.B. and Moorman, S.J. (2000) Development of utricular otoliths, but not saccular otoliths, is necessary for vestibular function and survival in zebrafish. J. Neurobiol. 43, 329–337.

Ris, L. and Godaux, E. (1998) Neuronal activity in the vestibular nuclei after contralateral or bilateral labyrinthectomy in the alert guinea pig. J. Neurophysiol. 80, 2352–2367.

Ross, M.D., Donovan, K. and Chee, O. (1985) Otoconial morphology in space-flown rats. The Physiologist 28(Suppl.), 219–226.

Ross, M.D. (1992) Synaptic plasticity in utricular maculas of rats exposed to microgravity. ASGSB Bull. 6, 100.

Ryu, J.H. and McCabe, B.F. (1976) Central vestibular compensation. Effects of the bilateral labyrinthectomy on neural activity in the medial vestibular nucleus. Arch. Otolaryngol. 102, 71–76.

Schaefer, K.P. and Meyer, D.L. (1974) Compensation of vestibular lesions. *In* Kornhuber, H.H. (ed.). *Vestibular system: Part 1: Basic mechanisms (Handbook of Sensory Physiology Vol. IV/1)*, pp. 463–490. Springer Verlag, Berlin, Heidelberg, New York.

Sebastian, C. and Horn, E. (1998) The minimum duration of microgravity experience during space flight which affects the development of the roll induced vestibuloocular reflex in an amphibian. Neurosci. Lett. 253, 171–174.
Sebastian, C. and Horn, E. (1999) Light-dependent supression of the vestibuloocular reflex during development. NeuroReport 10, 171–176.
Sebastian, C. and Horn, E. (2001) Features of vestibuloocular reflex modulations induced by altered gravitational forces in tadpoles (*Xenopus laevis*). Adv. Space Res. 28, 579–588.
Sebastian, C., Eßeling, K. and Horn, E. (1996) Altered gravitational experience during early periods of life affects the static vestibulo-ocular reflex of tadpoles of the Southern Clawed Toad, *Xenopus laevis* Daudin. Exp. Brain Res. 112, 213–222.
Sebastian, C., Eßeling, K. and Horn, E. (2001) Altered gravitational forces affect the development of the static vestibuloocular reflex in fish (*Oreochromis mossambicus*). J. Neurobiol. 46, 59–72.
Shi, D.-L. and Boucout, J.C. (1995) The chronological development of the urodele amphibian *Pleurodeles waltl* (Michah). Int. J. Dev. Biol. 39, 427–441.
Slenzka, K., Anken, R., Bäuerle, A., Körtje, K.H., Paulus, U. and Rahmann, H. (1990) Morphological, electronmicroscopical and biochemical aspects of hypergravity conditions during early ontogenetic development of cichlid fish. ESA SP-307, 321–328.
Slenzka, K., Appel, R. and Rahmann, H. (1993) Brain creatine kinase activity during ontogeny of the cichlid fish *Oreochromis mossambicus* and the clawed toad *Xenopus laevis*, influence of gravity? Neurochem. Int. 22, 405–411.
Smith, P.F. and Curthoys, I.S. (1989) Mechanisms of recovery following unilateral labyrinthectomy: a review. Brain Res. Rev. 14, 155–180.
Snetkova, E., Chelnaya, N., Serova, L., Saveliev, S., Cherdanzova, E., Pronych, S. and Wassersug, R.J. (1995) The effects of space flight on *Xenopus laevis* larval development. J. Exp. Zool. 273, 21–32.
Souza, K.A., Black, S.D. and Wassersug, R.J. (1995) Amphibian development in the virtual absence of gravity. Proc. Natl. Acad. Sci. USA 92, 1975–1978.
Spangenberg, D.B., Coccaro, E., Schwarte, R. and Lowe, B. (1996) Touch-plate and statolith formation in graviceptors of ephyrae which developed while weightless in space. Scanning Microscopy 10, 875–888.
Takabayashi, A., Ohara, K., Ohmura, T., Watanabe, S., Mori, S., Tanaka, M. and Sakuragi, S. (1997) Mechanisms of vestibular adaptation of fish under microgravity. Uchu Seibutsu Kagaku 11(4), 351–354.
Ubbels, G.A. (1997) Fertilization and development of *Xenopus* eggs on sounding rockets and in a clinostat. ESA SP-1206 25–36.
Ubbels, G.A., Reijnen, M., Meijerink, J. and Narraway, J. (1995) The role of gravity in the establishment of the dorso-ventral axis in the amphibian embryo. ESA SP-1162, 175–185.
van der Loos, H. and Woolsey, T.A. (1973) Somatosensory cortex: structural alterations following early injury to sense organs. Science 179, 395–398.

Vinnikov, Y.A.A., Gazenko, O.G., Lychakov, D.V. and Palmbach, L.R. (1983) Formation of the vestibular apparatus in weightlessness. *In* Romand, R. (ed.). *Development of auditory and vestibular systems*, pp. 537–560. Academic Press, New York.

von Baumgarten, R.J., Simmonds, R.C., Boyd, J.F. and Garriott, O.K. (1975) Effects of prolonged weightlessness on the swimming pattern of fish aboard Skylab 3. Aviat. Space Environ. Med. 46, 902–906.

von Holst, E. (1950) Die Arbeitsweise des Statolithenapparates bei Fischen. Z. vergl. Physiol. 32, 60–120.

Waespe, W., Schwarz, U. and Wolfensberger, M. (1992) Firing characteristics of vestibular nuclei neurons in the alert monkey after bilateral vestibular neurectomy. Exp. Brain Res. 89, 311–322.

Wagner, T.U., Renn, J., Riemensperger, T, Volff, J.N, Koster, R.W., Goerlich, R., Schartel, M. and Winkler, C. (2003) The teleost fish medaka (*Oryzias latipes*) as a genetic model to study gravity dependent bone homeostasis *in vivo*. Adv. Space Res. 32(8), 1459–1465.

Walton, K. (1998) Postnatal development under conditions of simulated weightlessness and space flight. Brain Res. Rev. 28, 25–34.

Wassersug, R.J., Izumi-Kurotani, A., Yamashita, M. and Naitoh, T. (1993) Motion sickness in amphibians. Behav. Neur. Biol. 60, 42–51.

Wiederhold, M.L., Yamashita, M., Larsen, K.A., Batten, J.S., Koike, H. and Asashima, M. (1995) Development of the otolith organs and semicircular canals in the Japanese red-bellied newt, *Cynops pyrrhogaster*. Hearing Res. 84, 41–51.

Wiederhold, M.L., Gao, W.Y., Harrison, J.L. and Hejl, R. (1997) Development of gravity-sensing organs in altered gravity. Grav. Space Biol. Bull. 10, 91–96.

Wiederhold, M.L., Gao, W., Harrison, J.L. and Parker, K.A. (2003) Early development of gravity-sensing organs in microgravity. *In* Buckey, J.C., Jr. and Homick, J.L. (eds.). *The Neurolab spacelab mission: Neuroscience research in space*, pp. 123–132. NASA SP-535, Houston Texas.

Wiesel, T. (1982) Postnatal development of the visual cortex and influence of environment. Nature 299, 583–591.

Yamashita, M., Izumi-Kurotani, A., Imamizo, M., Koike, H., Okuno, M., Pfeiffer, C.J., Komazaki, S., Sasaki, F., Ohira, Y., Kashima, I., Kikuyama, S., Ohnishi, T., Mogami, Y. and Asashima, M. (2001) Japanese red-bellied newts in Space-AstroNewt experiment on Space Shuttle IML-2 and Space Flyer Unit. Biol. Sci. Space 15, 96–103.

Yokota, H., Neff, A.W. and Malacinski, G.M. (1994) Early development of *Xenopus* embryos is affected by simulated gravity. Adv. Space Res. 14, 249–255.

Experimentation with Animal Models in Space
G. Sonnenfeld (editor)

DOI: 10.1016/S1569-2574(05)10011-2

Primates in Space Flight

Tana M. Hoban-Higgins, Edward L. Robinson and
Charles A. Fuller

Section of Neurobiology, Physiology and Behavior, University of California, Davis, CA, USA

Introduction

It is through the combination of human and animal experimentation that science has come so far in unlocking the mysteries of physiology and medicine. Space science has used this powerful combination of experimental models in its attempts to learn how the body adapts to the absence of gravity. As evident from a cursory examination of this book, many years of work have produced a substantial body of data. To give an example, during orbital space flight the gravitational force is essentially removed, and under these "unloaded" conditions, the musculoskeletal system atrophies (Zerath *et al.*, 1990; 1991; 1996; 2000; Arnaud *et al.*, 2000; 2002; Belozerova *et al.*, 2000; 2002; 2003; Fitts *et al.*, 2000a; 2000b; Mazin *et al.*, 2000; Oganov *et al.*, 2000; Rodionova *et al.*, 2000; 2001; 2002; Roy *et al.*, 2000a; 2000b; 2001). The knowledge gained from space flight and ground experiments has led to the design and use of countermeasures, special techniques or exercises that attempt to preclude, or at least minimize, such adverse effects of space flight.

Historically, the first and most basic question was whether organisms could survive away from the surface of the earth. This is what the Montgolfier brothers were asking when they placed a sheep, a duck and a rooster in a basket and sent them aloft in 1793. After it had been established that the space flight environment was survivable, the safety of each space vehicle and its associated equipment had to be verified prior to human use. In addition, the idea that physiological systems might exhibit subclinical changes that would not be detectable in humans highlighted the need for space flight studies on animals. Although we now have a long history of human space flight, animals remain important for studying the effects of the space environment. During space flight, animal subjects can be maintained in controlled conditions where the only experimental variable is the microgravity space flight environment, whereas studies of humans are confounded by issues of operational demands and mandatory countermeasures (French, 1968; Sulzman, 1986).

The desire to study an animal subject with close phylogenetic ties to humans led researchers to utilize non-human primate models. Initially, the weight and size of the subject were major constraints. These factors were among those that led to the consideration and use of the squirrel monkey as a flight candidate (Brizzee *et al.*, 1977; Daunton *et al.*, 1980; Matsunami, 1997). An adult male representative of this small, diurnal South American primate will attain a mass of about 1 kg as an adult. Another New World primate that served as an early flight subject was the Cebus monkey, also known as the capuchin (Winget, 1977). While slightly larger (1.5 to 3 kg as an adult), this species is very docile and intelligent. Despite the abilities of the squirrel and cebus monkey, they were not the ideal model as they are both New World primates. Humans are Old World primates and are most closely related to the great apes. Chimpanzees were chosen as flight subjects for equipment tests performed prior to the first space flights by American astronauts in the Mercury series of the early 1960s (Henry and Mosely, 1963a). However, despite the close genetic ties between chimpanzees and humans and their intelligence, there were several distinct disadvantages to using chimpanzees. Specifically, their relatively large size and, more importantly, their extreme strength made them unsuitable candidates for use with much of the equipment developed for space flight (Keeling, 1977). An extensively studied species of Old World monkey is the rhesus macaque (Golarz de Bourne *et al.*, 1977; Perachio, 1977). A wealth of background data exists for this species, as it is a commonly used non-human primate model, and the rhesus has proved to be very intelligent as well as highly motivated. The rhesus is an adaptable species, as evidenced by its cohabitation with humans in its natural habitat. Two other species of macaque, the pig-tailed macaque and the toque macaque (cynomolgus monkey), were also studied early during space flight (Pace *et al.*, 1977; Sandler, 1977; see Table 1).

Early rocket flights

As with many other endeavors, non-human primates served as human surrogates in the first space flights (Henry *et al.*, 1952; see Sandler, 1977 for review). Prior to these early primate flights there were conflicting opinions regarding the possible physiological responses to microgravity. For example, it was not known if it would even be possible to sleep, eat or perform essential tasks in the orbital environment (Gauer and Haber, 1950; Walrath J, 1959). Although it was generally assumed that the respiratory and cardiovascular systems would be able to function in microgravity, there was great concern about how the body would respond to the altered vestibular input of weightlessness (Gauer and Haber, 1950; Haber and Gerathewohl, 1951). It was considered likely that the conflict between the presumably normal visual input and the reduced labyrinth and proprioceptive input would result in disorientation and nausea sufficient to preclude individuals from performing

Table 1

Chronological list of primate ballistic and orbital experiments. Mission name and date, monkey type and sex, country

Mission	Date	Species	Name	Sex	Country
V-2 No 37	06/1948	Rhesus	Albert I	m	US
V-2 No 47	06/1949	Rhesus	Albert II	m	US
V-2 No 32	09/1949	Cynomolgus	Albert III	m	US
V-2 No 31	12/1949	Cynomolgus	Albert IV	m	US
Aerobee USAF-12	04/1951	Cebus	Albert V		US
Aerobee USAF-19	09/1951	Rhesus	Albert VI	m	US
Aerobee USAF-26	05/1952	Cebus	Mike	m	US
		Cebus	Patricia	f	
AM-13 Bioflight 1	12/1958	Squirrel	Old Reliable	m	US
AM-18 Bioflight 2	05/1959	Rhesus	Abel	m	US
		Squirrel	Baker	f	
LJ-2	12/1959	Rhesus	Sam	m	US
LJ-1B	01/1960	Rhesus	Miss Sam	f	US
MR-2	01/1961	Chimpanzee	Ham	m	US
SP Pod 13	11/1961	Squirrel	Goliath	m	US
MA-5	11/1961	Chimpanzee	Enos	m	US
SP Pod 6 (36E)	12/1961	Rhesus	Scatback		US
Vesta 4	03/1967	Pig-tailed	Martine	f	FRA
Vesta 5	03/1967	Pig-tailed	Pierrette	f	FRA
Biosatellite 3	06/1969	Pig-tailed	Bonny	m	US
Bion 6 (K1514)	12/1983	Rhesus	Abrek	m	RUS
		Rhesus	Bion	m	
Spacelab 3	04/1985	Squirrel	3165	m	US
		Squirrel	384-80	m	
Bion 7 (K1667)	07/1985	Rhesus	Verny	m	RUS
		Rhesus	Gordy	m	
Bion 8 (K1887)	09/1987	Rhesus	Drema	m	RUS
		Rhesus	Yerosha	m	
Bion 9 (K2044)	09/1989	Rhesus	Zhakonya	m	RUS
		Rhesus	Zabiyaka	m	
Bion10 (K2229)	12/1992	Rhesus	Krosh	m	RUS
		Rhesus	Ivasha	m	
Bion 11	12/1996	Rhesus	Lapik	m	RUS
		Rhesus	Multik	m	

their normal duties. Another area of concern was the effects of the forces generated during launch, however, since these could be calculated, the abilities of an organism to resist acceleration could be tested on the ground using centrifuges or sleds (Gauer and Haber, 1950; Pickering *et al.*, 1959).

The first primates were launched into space on a series of eight US vertical rocket flights (Henry *et al.*, 1952; Pickering *et al.*, 1959; Klein, 1981; Souza *et al.*, 1995). They were lightly sedated using morphine and nylon webbing was used to secure these monkey astronauts to sponge rubber beds located in the nose cone of the rocket. Respiration was measured by a thermocouple located in a rubber facemask. Electrocardiograms were measured using subcutaneous electrodes. For the last three subjects, arterial and venous pressures were measured using cannulae and pressure transducers. All data were transmitted to ground stations (Holloman Air Development Center and White Sands Proving Ground).

Albert I, a rhesus monkey, was launched on June 11, 1948 in a V-2 rocket and reached an apogee of 63 km. During ascent his heart rate dropped from the normal 190 beats per minute to 110 beats per minute. His respiratory rate was 60 breaths per minute. During his brief exposure to microgravity, heart rate increased to the preflight level and respiration remained at 60 to 65 breaths per minute. Similar results were seen in all eight of the monkeys studied in this series of flights.

A second rhesus monkey, Albert II, followed Albert I into space a year later aboard another V-2 rocket. His flight reached an apogee of 134 km. Two additional V-2 rockets carrying primates were launched in 1949, one in September and the second in December. Each housed a single anesthetized cynomolgus macaque (Albert III and IV).

The next three primate flights employed the newly developed Aerobee rocket, a high-altitude free-flight rocket stabilized using fixed fins (see Fig. 1). In April of 1951, a cebus monkey (Albert V) reached an altitude of 61 km during a brief flight. A rhesus monkey (Albert VI) was launched in September of the same year. Arterial and venous pressure measurements were recorded during his ascent to 71 km. Not surprisingly, arterial pressure rose during the

Fig. 1. An Aerobee rocket. This type of rocket was used in three early experiments in which four non-human primates were exposed to microgravity. Credit: NASA Glenn Research Center.

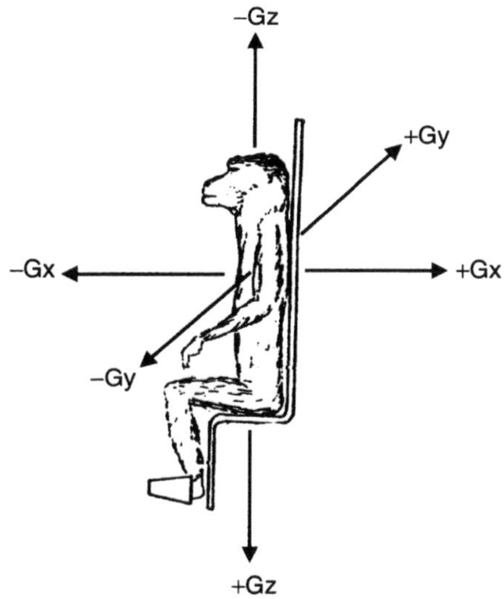

Fig. 2. Diagram of seated monkey with *g* vectors indicated. Credit: After Figure 1 in Pace *et al.*, 1971.

firing of the booster rocket during launch. During subgravity exposure, arterial pressure slowly but continually decreased until the opening of the parachute and the return to 1 ***g***.

The last of the Aerobee primate flights carried two cebus monkeys, Patricia and Mike. The animals were oriented such that one would experience the acceleration vector in the $+G_z$ direction (seated) and the other in the $+G_x$ direction (supine), as illustrated in Fig. 2. The flight launched in May of 1952 and reached an apogee of only 26 km. Both animals survived the flight and recovery and evidenced no significant changes in heart rate during the flight.

The Jupiter rocket was used for the Bioflights of the late 1950s (Graybiel *et al.*, 1959; Sandler, 1977; Souza *et al.*, 1995). In Bioflight 1, an unanesthetized squirrel monkey named Old Reliable was launched in December of 1958. The physiological variables recorded included heart rate, heart sounds and body temperature. Cabin environmental data were also recorded, including ambient temperature, cabin pressure and radiation. An altitude of approximately 480 km was reached during the flight which ended due to mechanical failure of the rocket.

Bioflight 2 carried two primate astronauts, a rhesus monkey named Abel and a squirrel monkey named Baker (Graybiel *et al.*, 1959). Animal position and couch design mimicked those proposed for the Mercury missions. EMG from two muscles was added to the suite of physiological data recorded (muscle not specified). In addition, Abel was trained to press a lever in response to a light signal. Unfortunately the data from this performance task were lost. After

Fig. 3. Liftoff of a Little Joe rocket. This type of rocket was used to power both the Sam and Miss Sam flights of 1959 and 1960. These missions tested equipment used in the Mercury program. Credit: NASA Johnson Space Center.

a 15 min flight into space, Abel and Baker returned to Earth, landing in Antigua. Heart rate and respiration rose during launch in all three of the Bioflight animals, then returned to baseline during microgravity exposure. The monkeys had only slight variations in body temperature, and cardiovascular and respiratory rates over the flight, with the squirrel monkey Baker presenting the most stable readings.

The flights of Sam and Miss Sam, both rhesus monkeys, tested Mercury equipment intended for use with human astronauts and were launched using Little Joe solid fuel rockets (Sandler, 1977; Souza *et al.*, 1995; see Fig. 3). Sam's flight took place during December of 1959 and Miss Sam's in January of 1960. The flights lasted between 11 and 15 min and both reached an altitude of 85 km. Both animals performed the same lever-pressing task as Abel had on an earlier flight, with Sam besting Miss Sam.

The Mercury series

From the flights conducted thus far, it was apparent that primates could survive both the acceleration forces produced during launch and return to Earth as well as the microgravity space flight environment, at least for brief periods. The next primate astronauts would be chimpanzees who would verify the workings of the Mercury capsule (Henry, 1963; Henry and Mosely, 1963a, b; Simmonds, 1977; Souza *et al.*, 1995). Performance tests were also used to establish that astronauts would be able to perform necessary tasks in spite of the physical and psychological demands of space flight (Robles *et al.*, 1963a, b). In addition to providing answers to these important questions, the MR-2 and MA-5 missions would provide an end-to-end operational test of the equipment, personnel and procedures necessary to successfully launch and recover an astronaut in a Mercury capsule (Henry and Mosely, 1963b).

The MR-2 mission was launched on January 31, 1961 (Henry, 1963; Stingely *et al.*, 1963). Its sole occupant, Ham, occupied a couch in a Mercury capsule powered by an Atlas rocket. The peak acceleration reached during launch was 17.0 ***g*** during the firing of the escape rocket. At apogee, Ham experienced 6-½ min of 0 ***g***. Reentry deceleration achieved a maximum of 14.6 ***g***. Throughout the flight, ECG, respiration and body temperature were recorded. Respiration and heart rate increased during launch, then decreased over the period of weightlessness. Both parameters were within the normal range recorded on Earth at 1 ***g***. Both variables increased during reentry and returned to baseline values after landing. An immediate postflight differential white blood cell count revealed a stress response, similar to that seen in young chimpanzees after exposure to high ambient temperature. However, during the flight, capsule ambient temperature ranged from 58 to 66°F (Ward and Briutz, 1963). White blood cell numbers returned to normal values within two weeks postflight and Ham became the primary backup candidate for the MA-5 mission.

Ham performed two different tasks during a ballistic flight lasting under 17 min (Brown and Iwan, 1963; Robles *et al.*, 1963a). In a continuous avoidance task, he was required to press a lever with his right hand at least once every 15 s. He completed this task successfully during both launch and weightlessness, however the number of lever presses per time declined substantially during reentry. It was not clear if this decline was a result of the deceleration forces experienced during reentry (5 s of > 14 ***g*** and 10 s of > 10 ***g***) or were due to prior exposure to the weightless environment. A flight verification experiment was carried out two weeks after the MR-2 flight to determine if the decrement in performance on the continuous avoidance task was a result of the activities surrounding flight rather than the flight itself. Ham was allowed the same number of hours of sleep, was instrumented in the same manner and performed the same tasks as during flight. His performance during the postflight verification test was similar to that observed in prior ground-based trials conducted without simulation of flight activities. Similar findings of a decrease in the number of lever presses had been seen in following exposure to radial acceleration. It was therefore concluded that the acceleration that accompanied reentry likely had been a greater factor in the decrease in performance on the continuous avoidance task than microgravity exposure.

In addition to the continuous avoidance task, Ham was presented with a discrete avoidance task every two minutes to measure reaction time. This task required that he press a second lever within 5 s of the presentation of a blue light. His performance on this task was comparable to preflight control tests.

A second chimpanzee, Enos, was launched in November of 1961 for a 3-h, 20-min orbital flight (Henry and Mosely, 1963b; Mosely and Henry, 1963; Stingely and Mosely, 1963; Ward, 1963). Blood pressure measurement was added to the suite of physiological variables monitored, and urine and feces were collected. Heart and respiration rates were increased inflight, but did not exceed control values recorded during a ground-based centrifuge test. Similarly, blood pressure was also elevated inflight, but was also within the preflight baseline levels established for this subject. Over the course of the flight, the animal experienced premature ventricular contractions; these were not associated with the timing of performance tasks or with the observed physiological changes.

In addition to the avoidance tasks used in the previous Mercury flight, Enos was required to perform three additional tasks (Brown and Iwan, 1963; Robles *et al.*, 1963b). Six-minute rest periods were provided between tasks. The combined avoidance tasks lasted for 15 min while the other three were each of 10 min maximum duration. The tasks were presented in a repeating sequence with the avoidance tasks presented four times and the other three tasks three times. The first of the new tasks required a delayed response to the presentation of a symbol in order to receive a water reward. For this task, the interval between symbol presentation and lever pressing was higher than the established preflight baseline. The second new task was a fixed ratio task in which 50 lever presses were required for the delivery of a food pellet. This task was accomplished with the same relative success as the preflight baseline. The third new task was an oddity task, three symbols were presented and Enos had to press the lever corresponding to the symbol that was different from the other two. The first session of this task was performed with close to baseline accuracy. However, the center lever malfunctioned during the second session, preventing the animal from performing this task successfully. The center lever was not required for the other three tasks and performance on the other tasks was not compromised. Since the animal maintained a continued high level of performance and did not attempt to respond during the six-minute rest periods between task sessions, it was clear that the unexpected stress of the lever failure did not compromise the animal's psychomotor abilities. Thus, the MA-5 flight demonstrated that complex tasks could be performed, not only in the microgravity space flight environment, but also during the acceleration and deceleration of launch and landing. In addition, it was also shown that eating and drinking were possible in microgravity.

French rocket flights

During March of 1967, two female pig-tailed macaques were carried in vertical flight on Vesta rockets as part of a series of launches conducted by CERMA at the Hammaguir site in the Sahara. In addition to heart rate, respiration and body temperature, electroencephalographic (EEG) data were recorded from the hippocampus and cortex, and muscle activity (EMG) from the biceps, triceps and trapezius. The animals were observed by camera and were trained to move a lever in response to a light signal. From launch to recovery each mission lasted 15 min. These flights are notable in that they represent the first telemetered space flight EEG data.

Biosatellite III

The early ballistic and orbital flights had demonstrated that primates (including humans) could survive in the microgravity environment, albeit for relatively brief time frames. Concern therefore potentially shifted to the debilitating effects of living in microgravity for weeks to months. Human bed rest studies suggested that prolonged space flight might result in broad deterioration, including a loss of skeletal muscle and bone, and circulatory adjustments resulting in orthostatic hypotension (Lamb, 1966; Pace, 1963). With a planned duration of 30 days, Biosatellite III (Fig. 4) was intended to be the first long-duration primate space flight (Adey *et al.*, 1969; Adey, 1972; Adey and Hahn, 1971; Jenkins, 1965). The physiological variables measured included cardiovascular responses, respiration, EEG, EMG, electrooculogram (EOG) and body and brain temperatures. In all, 33 channels of physiological information were recorded. Food and water intake were recorded and performance was assessed using a psychomotor test system. Urine samples were

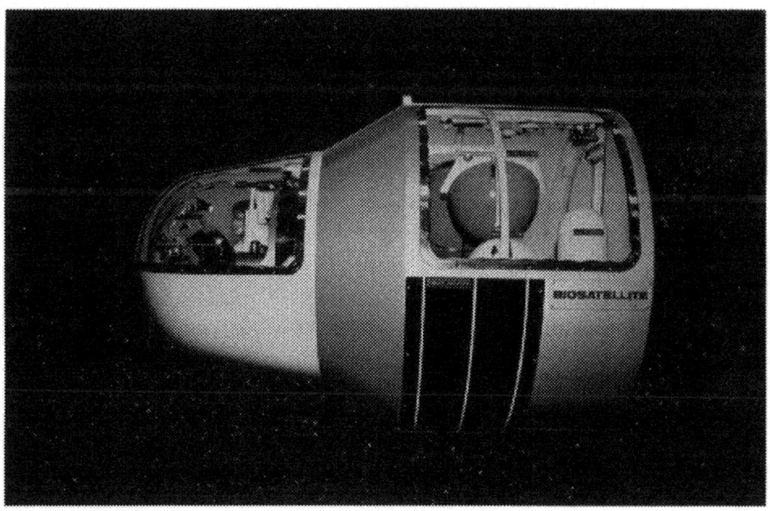

Fig. 4. Artist's rendering of the Biosatellite III vehicle. The monkey occupied the forward part of the capsule and the research equipment was in the rear section. Credit: NASA Ames Research Center.

collected and analyzed in flight (Pace *et al.*, 1971). In addition to studying bone, muscle, cardiovascular, sleep and metabolic changes, these data would be examined for the presence and persistence of daily (circadian) rhythms (Pittendrigh, 1965). The presence of daily rhythms in an environment free of Earth's environmental cycles would constitute additional evidence that they are endogenously generated rather than externally imposed.

Five male pig-tailed macaques were trained as flight subjects. The four backup animals not chosen for flight provided ground control measurements. Biosatellite III was launched on June 28, 1969 with the pig-tailed macaque Bonny on board. Although all data systems operated successfully, the mission was terminated after 8.8 days in response to the physical state of the monkey who evidenced both declining body temperature and heart rate (Meehan and Rader, 1973).

Continuous recording of EEG, EOG and EMG allowed the first determination of sleep states in space flight (Hanley and Adey, 1971; Hoshizaki *et al.*, 1971; Walter *et al.*, 1971; Hanley, 1973). The Biosatellite III mission provided the first confirmation of the presence of rapid eye movement (REM) sleep in microgravity. As the majority of dreaming occurs in REM, this was thought to be important for the continued well-being of humans during long-term space flight. However, the flight monkey had an abnormal and fragmented sleep–wake cycle. Further, the timing of sleep was delayed in relationship to the light–dark cycle. This alteration in the timing of sleep suggested a possible loss of synchrony between the animal's internal clock and his environment, particularly the capsule light–dark cycle. Analysis of respiration, heart rate, temperature and urinary calcium excretion rhythms revealed periodicities of 25.5 to 30 h despite the presence of a 24.0 h light–dark and task presentation schedule (Hahn *et al.*, 1971; Hoshizaki *et al.*, 1973). This observation of altered circadian rhythmicity in the space flight environment would be confirmed in later flights.

The radiographic method was used to examine bone density at 17 skeletal sites (Mack, 1971). As is commonly seen following exposure to microgravity, the Biosatellite III flight animal had a decrease in bone density. The control animals, who were studied in flight couches, also showed decreased bone density, but to a lesser degree than the flight animal. Urinary calcium excretion was decreased in the flight animal, as well as in one of the controls (Pace *et al.*, 1971). This hypocalciuria was also present preflight, and was attributed to inhibition of urinary calcium excretion due to the decreased urine pH resulting from the casein-based diet (Durham *et al.*, 1970).

Nonhuman primates on the Space Shuttle

Squirrel monkeys next flew in space on board the Space Shuttle Challenger as part of the Spacelab 3 mission (Fuller, 1985; Souza *et al.*, 1995). This was a verification flight for the Research Animal Holding Facility (RAHF) and a test of procedures for providing animal care and collecting physiological data in flight in anticipation of future Shuttle missions carrying experiments with this primate model (Perry and Reid, 1983; Phillips, 1988). Two uninstrumented squirrel monkeys were flown in a RAHF with a capability to house four of these primates. A pelletized diet was provided on demand when a tap switch was operated and water was dispensed through a lixit system. The animals were less active inflight than had been observed preflight. One animal appeared to adapt well to the microgravity environment, while the other appeared to suffer from space adaptation syndrome and initially consumed no food. On the fifth day of the mission, a crew member presented food to the animal, after which it resumed feeding on its own.

The Russian Bion series

At the same time as Spacelab 3, the Russians expanded their successful Biosatellite series to include primate subjects. In fact, most of the space flight data from non-human primates have been obtained from the Russian Bion flights using rhesus macaques (Ilyin, 1981; 1988; Gazenko and Ilyin, 1984; 1986; 1987; Ilyin *et al.*, 1991; 2000; Souza *et al.*, 1995; Korolkov *et al.*, 1996; Sulzman, 1996). Two male rhesus macaques were flown on each of the six Bion missions. Originally referred to as the Cosmos Biosatellites, the flights with primate experiments (Bion 6–11) ranged in duration from 5 to almost 14 days and were conducted from 1983 through 1997. Each was launched using a modified Vostok rocket. The primate housing unit utilized during the Bion missions was the Bios-primate capsule. This hardware was carefully designed so that it could be used for multiple flights with minimal refurbishment. This not only reduced the monetary costs of the Bion flights, but also reduced the amount of time required to prepare for each new mission. Each Bion capsule contained one monkey and two capsules were placed side by side within the Bion space capsule (Fig. 5). Control or "flight simulation" experiments were performed for each mission using a mockup of the Bion capsule. The main life support and animal health monitoring systems were developed for Bion 6 and, over successive missions, additional hardware was added to allow investigators to answer specific questions about how the physiology and behavior of the non-human primate adapts during space flight and readapts following return to Earth.

Bion 6 (Cosmos 1514), the first of the Cosmos series to carry two rhesus monkeys was a 5-day flight (Gazenko and Ilyin, 1984). The primary research objective of this flight was to test the hardware and examine cardiovascular responses to prolonged microgravity exposure. With the onset of microgravity, mean arterial pressure increased and carotid artery blood flow velocity transiently decreased, however both parameters returned to near baseline levels over the next several hours (Sandler *et al.*, 1987). Adaptation to the space flight environment took place over several days. Both animals appeared to

Fig. 5. Bion capsule. Twelve rhesus monkeys were flown on six missions during the Bion program to study the effects of the microgravity environment on physiology and behavior. The flights ranged from 5 to 14 days in duration. Credit: TsSKB Progress in Samara.

suffer from space adaptation syndrome during the first two days of flight. The expected cephalic fluid shift occurred and was accompanied by a loss of body water from both the interstitial and plasma compartments, resulting in an increased hematocrit (Zhidkov *et al.*, 1987). However, autoregulatory mechanisms were adequate to maintain blood flow to the head (Krotov *et al.*, 1987).

Heart rate was slightly lower than in the preflight control study with the lowest heart rate occurring between the hours of 8 pm and midnight. Axillary temperature was lower during flight as were skin temperatures (Klimovitsky *et al.*, 1987; Sulzman *et al.*, 1992). These observations, combined with the lower heart rate suggest decreased heat loss and a corresponding reduction in metabolic rate in microgravity. There was a loss of synchronization of circadian rhythms, as had been seen in the Biosatellite III monkey (Klimovitsky *et al.*, 1987; Sulzman *et al.*, 1992). The circadian rhythms of axillary temperature and activity did not maintain a 24.0 h period, despite the presence of a 24 h light–dark cycle.

In addition to providing physiological data, the Bion 6 monkeys performed three behavioral tests: a test of eye–head and hand coordination, a foot pedal motor task and a test of autonomic response to vertical oscillation (Kozlovskaya *et al.*, 1985). Adaptation to the microgravity environment was followed by readaptation after landing, showing plasticity in these systems.

Bion 7 (Cosmos 1667) was a 7 day flight conducted in 1985 (Hines and Skidmore, 1994) and the subsequent Bion 8 (Cosmos 1887) flight in 1987 was 12.5 days in duration (Gazenko and Ilyin, 1986). The results from Bion 7 and 8 provided confirmation of the results of the studies on Bion 6. Bion 8 had originally been planned as a 14-day mission, however, one of the feeders malfunctioned and the decision was made to have the capsule land earlier than scheduled. The 14-day Bion 9 (Cosmos 2044) flight was launched in 1989 (Connolly *et al.*, 1994). New sensors were developed for this flight to test the response of semicircular canal and otolith afferents to microgravity, and alterations of circadian and temperature regulation and metabolism, among other systems. The 11.6-day Bion 10 (Cosmos 2229) mission was carried out in 1992 (Ilyin *et al.*, 2000). Again, additional sensor capabilities were developed for this flight. The next Bion mission, 11, was not given a Cosmos number. Bion 11 was launched in December of 1996 and recovered 13.7 days later in January of 1997 (Golov *et al.*, 2000; Ilyin *et al.*, 2000). Studies were conducted by international teams of scientists in several discipline areas: neuromuscular and behavioral, neurosensory, regulatory and metabolism, and skeletal (Kozlovskaya *et al.*, 2000).

The data accumulated over the Bion missions has given researchers greater insight into the physiological and behavioral changes that occur during prolonged exposure to microgravity. Comparison of results between space flight and flight simulation experiments provided the ability to differentiate between the effects of space flight and those of confinement in the Bion capsule. In addition to the changes seen in the cardiovascular system, decreased blood cell formation was seen in bone marrow (Burkovskaya and Korolkov, 2000) as well as changes in immune function (Pochkhua, 1992; Lesnyak *et al.*, 1993; Sonnenfeld *et al.*, 1996). Reduced body water content (Lobachik *et al.*, 2000), as well as alterations in circadian rhythms were observed throughout the Bion series (Sulzman *et al.*, 1992; Fuller *et al.*, 1996; Alpatov *et al.*, 2000). Other results from the Bion missions are summarized below.

Musculoskeletal: Over the Bion missions, the skeletal system responded to microgravity with a decrease in bone formation (Zerath *et al.*, 1990; 1991; 1996; 2000; 2002; Rodionova *et al.*, 2000; 2001; 2002). This was accompanied by a loss of bone mineral density (Oganov *et al.*, 2000; Zerath *et al.*, 2002). The net effect of these responses was to decrease bone volume. Both calcitonin and parathyroid hormone levels fell during space flight while serum cortisol levels increased (Arnaud *et al.*, 2000). The reduction in the calcium regulatory hormones also acted to reduce 1,25-dihyroxyvitamin D levels (Arnaud *et al.*, 2002). These responses were also seen in the flight simulation experiments, although to a much lesser degree. Despite adequate dietary amounts of calcium, there was a decrease in intestinal calcium absorption during space flight. Postflight experiments revealed an increase in the turnover of both mineralized and non-mineralized connective tissue (Martinez *et al.*, 2000).

Unloading affects skeletal muscle as well as bone. In general, slow or type I fibers of the antigravity muscles are most affected by space flight (Roy *et al.*, 2000a; Kischel *et al.*, 2001). Fiber atrophy in the Bion animals was accompanied by a decrease in the force generated per cross-sectional area (Fitts *et al.*, 2000a, 2000b). Structural remodeling was observed at the soleus myotendinous junction (Carnino *et al.*, 2000). Average EMG generated by goal directed movement was reduced inflight (Falempin *et al.*, 2000). The neuromuscular system also showed adaptation during space flight (Roy *et al.*, 1996; Recktenwald *et al.*, 1999; 2000; Edgerton *et al.*, 2000b; Hodgson *et al.*, 2000). Altered coordination between muscles occurred as well as changes in EMG and gait during postflight locomotion tests.

Regulatory: As had been observed during the Biosatellite III mission, alterations in circadian rhythms during space flight were also present during the Bion flights (Alpatov *et al.*, 2000; Fuller *et al.*, 1996). The timing of the brain temperature rhythm was delayed relative to both the light–dark cycle as well as to other rhythms. These changes did not occur during the flight simulation experiments. In addition, there was a general vasoconstriction and a decrease in heart rate in flight, suggesting a decreased metabolic rate (Hoban-Higgins *et al.*, 2000). Overall energy expenditure was lower in space flight than in ground control experiments as shown by turnover of doubly labeled water (Stein *et al.*, 1996). Most of the Bion animals had a decrease in body temperature inflight, however, thermoregulatory ability was not compromised (Klimovitsky *et al.*, 2000).

Neurovestibular: Adaptation of the neurovestibular system inflight was assessed using both an eye–head–hand coordination task and by testing responses to a vertical oscillation stimulus (Sirota *et al.*, 1987; Kozlovskaya *et al.*, 1989; Antsiferova *et al.*, 2000; Shlyk *et al.*, 2000a, b). Postflight tests were also conducted to determine the time course of reacclimation to the 1 ***g*** environment on Earth (Cohen *et al.*, 1992; Dai *et al.*, 1994; 1996; 1998). In microgravity, in order to fix their gaze on a target, the monkeys moved their head less and their eyes more than on Earth. During Bion 11, the effects of otolith stimulation were greater early in the flight (Badakva *et al.*, 2000). This stimulation elicited less activity within the median vestibular nuclei as the flight progressed. As a whole, these data suggest that neural plasticity exists in both the vestibular receptors and the central nervous system in response to altered ***g*** (Correia *et al.*, 1992; Slenzka, 2003).

Conclusion

The first monkey was launched towards space over 50 years ago. Although experiments using non-human primates were first conducted to ensure that living beings could survive in space flight, they have subsequently provided a wealth of data illuminating the effects of exposure to microgravity on physiology and behavior. Space flight studies using non-human primates have both paved the way for the presence of humans in space, and have been important tools for identifying risks and developing countermeasures to the adverse effects of space flight. These studies have also contributed to the advancement of science and produced benefits for those of us who remain on Earth (Edgerton *et al.*, 2000a).

Acknowledgement

This work was supported by NSBRI grant NCC9-58-2333 (CAF).

References

Adey, W.R. (1972) Studies on weightlessness in a primate in the Biosatellite 3 experiment. Life Sci. Space Res. 10, 67–85.

Adey, W.R. and Hahn, P.M. (1971) Introduction–Biosatellite 3 results. Aerosp. Med. 42(3), 273–280.

Adey, W.R., Crocket, A.T., Mack, P.B., Meehan, J.P. and Pace, N. (1969) Bioosatellite 3 preliminary findings. Science 166, 492–493.

Alpatov, A.M., Hoban-Higgins, T.M., Klimovitsky, V.Y., Tumurova, E.G. and Fuller, C.A. (2000) Circadian rhythms in Macaca mulatta monkeys during Bion 11 flight. J. Gravit. Physiol. 7(1), S119–123.

Antsiferova, L.I., Shlyk, G.G. and Ignatenko, A.V. (2000) Effect of microgravity on the implementation of conditioned reflex skills of rhesus monkeys. J. Gravit. Physiol. 7(1), S95–98.

Arnaud, S.B., Navidi, M., Deftos, L., Thierry-Palmer, M., Dotsenko, R., Bigbee, A. and Grindeland, R.E. (2002) The calcium endocrine system of adolescent rhesus monkeys and controls before and after spaceflight. Am. J. Physiol. Endocrinol. Metab. 282(3), E514–21.

Arnaud, S.B., Navidi, M., Deftos, L., Buckendahl, P., Dotsenko, M.A., Bengtson, S., Bigbee, A. and Grindeland, R.E. (2000) Calcium metabolism in Bion 11 monkeys. J. Gravit. Physiol. 7(1), S153.

Badakva, A.M., Zalkind, D.V., Miller, N.V. and Riazansky, S.N. (2000) Head pitch movement and vestibular neuronal activity in response to otolith stimulation of monkeys in space. J. Gravit. Physiol. 7(1), S99–105.

Belozerova, I.N., Nemirovskaya, T.L., Shenkman, B.S. and Kozlovskaya, I.B. (2003) Characteristic of changes in the structure and metabolism of the vastus lateralis muscles in monkeys after space flight. Neurosci. Behav. Physiol. 33(7), 735–740.

Belozerova, I.N., Nemirovskaia, T.L., Shenkman, B.S. and Kozlovskaia, I.B. (2002) Changes in structure and metabolism in the m. vastus lateralis in monkeys after a space flight Ross Fiziol Zh Im I M Sechenova. 88(4), 424–431.

Belozerova, I.N., Nemirovskaya, T.L. and Shenkman, B.S. (2000) Structural and metabolic profile of rhesus monkey m. vastus lateralis after spaceflight. J. Gravit. Physiol. 7(1), S55–58.

Brizzee, K.R., Ordy, J.M. and Kaack, B. (1977) The squirrel monkey as a candidate for spaceflight. *In* Simmonds, R.C. and Bourne, G.H. (eds.). *The Use of Nonhuman Primates in Space*, pp. 281–328. NASA Conference Publication 005. NASA, Washington, DC.

Brown, E.J. and Iwan, R.D. (1963) Behavioral apparatus for the MR-2 and MA-5 flights. *In* Henry, J.P. and Mosely, J.D. (eds.). *Results of the Project*

Mercury Ballistic and Orbital Chimpanzee Flights, pp. 15–20. NASA SP-39, Washington, DC.
Burkovskaya, T.E. and Korolkov, V.I. (2000) Hemopoiesis in bone marrow of monkeys after spaceflight. J. Gravit. Physiol. 7(1), S129–134.
Carnino, A., Rofino, S., Chopard, A. and Marini, J.-F. (2000) Effects of a 14-day spaceflight on soleus myotendinous junction ultrastructure in rhesus monkey. J. Gravit. Physiol. 7(1), S65–68.
Cohen, B., Kozlovskaya, I., Raphan, T., Solomon, D., Helwig, D., Cohen, N., Sirota, M.G. and Yakushin, S. (1992) Vestibuloocular reflex of rhesus monkeys after spaceflight. J. Appl. Physiol. 73(2 Suppl), 121S–131.
Connolly, J.P., Grindeland, R.E. and Ballard, R.W. (eds.). (1994) *Final Reports of the U.S. Experiments Flown on the Soviet Biosatellite Cosmos (2044)*, NASA TM 108802, NASA Ames Research center, Moffett Field, CA.
Correia, M.J., Perachio, A.A., Dickman, J.D., Kozlovskaya, I.B., Sirota, M.G., Yakushin, S.B. and Belozerova, I.N. (1992) Changes in monkey horizontal semicircular canal afferent responses after spaceflight. J. Appl. Physiol. 73(2 Suppl), 112S–120.
Dai, M., Raphan, T., Kozlovskaya, I. and Cohen, B. (1998) Vestibular adaptation to space in monkeys. Otolaryngol Head Neck Surg. 119(1), 65–77.
Dai, M., Raphan, T., Kozlovskaya, I. and Cohen, B. (1996) Modulation of vergence by off-vertical yaw axis rotation in the monkey: normal characteristics and effects of space flight. Exp. Brain Res. 111(1), 21–29.
Dai, M., McGarvie, L., Kozlovskaya, I., Raphan, T. and Cohen, B. (1994) Effects of spaceflight on ocular counterrolling and the spatial orientation of the vestibular system. Exp. Brain Res. 102(1), 45–56.
Daunton, N., Greene, L.O. Jr., Fox, R.A. (1980) Squirrel monkey as a candidate for studies of space sickness and vestibular function in spacelab. *In: Space Environment Workshops for Life Scientists. Presented at Fed Amer Soc Exp Biol and Aerosp Med Assoc.*
Durham, R.M., Tejada, R., Parker, M. and Cockett, A.T. (1970) Reduction of urinary precipitates through manipulation of diet in *Macaca nemestrina*. Aerosp. Med. 41(3), 259–263.
Edgerton, V.R., Roy, R.R., Hodgson, J.A., Day, M.K., Weiss, J., Harkema, S.J., Dobkin, B., Garfinkel, A., Konigsber, E. and Kozlovskaya, I. (2000a) How the science and engineering of spaceflight contribute to understanding the plasticity of spinal cord injury. Acta Astronaut. 47(1), 51–62.
Edgerton, V.R., Roy, R.R., Recktenwald, M.R., Hodgson, J.A., Grindeland, R.E. and Kozlovskaya (2000b) Neural and neuroendocrine adaptations to microgravity and ground-based models of microgravity. J. Gravit. Physiol. 7(3), 45–52.
Falempin, M., Canu, M.H., Langlet, C. and Kozlovskaya, I.B. (2000) Effect of microgravity on the electromyographic activity of two upperlimb muscles during a goal-directed movement and during locomotion. J. Gravit. Physiol. 7(1), S69–70.

Fitts, R.H., Desplanches, D., Romatowski, J.G. and Widrick, J.J. (2000a) Spaceflight effects on single skeletal muscle fiber function in the rhesus monkey. Am. J. Physiol. Regul. Integr. Comp. Physiol. 279(5), R1546–1557.

Fitts, R.H., Romatowski, J.G., De La Cruz, L., Widrick, J.J. and Desplanches, D. (2000b) Effect of spaceflight on the maximal shortening velocity, morphology, and enzyme profile of fast- and slow-twitch skeletal muscle fibers in rhesus monkeys. J. Gravit. Physiol. 7(1), S37–38.

French, J.D. (1968) On the need for basic biomedical reseach in the national space program. Biosci. 8(1), 24–26.

Fuller, C.A. (1985) Early adaptation to altered gravitational environments in the squirrel monkey. Physiol. 28(6 Suppl), S201–202.

Fuller, C.A., Hoban-Higgins, T.M., Klimovitsky, V.Y., Griffin, D.W. and Alpatov, A.M. (1996) Primate circadian rhythms during spaceflight: results from Cosmos 2044 and 2229. J. Appl. Physiol. 81(1), 188–193.

Gauer, O. and Haber, H. (1950) Man under gravity-free conditions. Chapter VI, German Aviation Medicine, World War II, Volume I. US Air Force Publications, Supt. Documents.

Gazenko, O.G. and Ilyin, E.A. (1987) Physiological investigations of primates onboard biosatellites Cosmos-1514 and Cosmos-1667. Physiologist. 30(1 Suppl), S31–35.

Gazenko, O.G. and Ilyin, E.A. (1986) Investigations onboard the biosatellite Cosmos-1667. Adv. Space Res. 6(12), 101–106.

Gazenko, O.G. and Ilyin, E.A. (1984) Investigations on-board the biosatellite Cosmos-83. Adv. Space Res. 4(10), 29–37.

Golarz de Bourne, M.N., Bourne, G.H. and McClure, H.M. (1977) The rhesus monkey (*Macaca mulatta*) as a candidate for spaceflight. *In* Simmonds, R.C. and Bourne, G.H. (eds.). *The Use of Nonhuman Primates in Space*, pp. 245–276. NASA Conference Publication 005, NASA, Washington, DC.

Golov, V.K., Magedov, V.S., Skidmore, M.G., Hines, J.W., Kozlovskaya, I.B. and Korolkov, V.I. (2000) Bion 11 mission hardware. J. Gravit. Physiol. 7(1), S27–35.

Graybiel, A., Holmes, R.H., Beischer, D.E., Champlin, G.E., Pedigo, G.P., Hixson, W.C., Davis, T.R.A., Barr, N.L., Kistler, W.G., Niven, J.I., Wilbarger, E., Stulken, D.E., Augerson, W.S., Clark, R. and Berrian, J.H. (1959) An account of experiments in which two monkeys were recovered unharmed after ballistic space flight. Aerosp. Med. 30(12), 871–931.

Haber, J. and Gerathewohl, S.J. (1951) Physics and psychophysics of weightlessness. J. Aviation Med. 22, 180–189.

Hahn, P.M., Hoshizaki, T. and Adey, W.R. (1971) Circadian rhythms of the *Macaca nemestrina* monkey in Biosatellite 3. Aerosp. Med. 42(3), 295–304.

Hanley, J. (1973) Wake and sleep cycles in the biosatellite III primate in earth orbital flight. Physiologist 16(2), 209–211.

Hanley, J. and Adey, W.R. (1971) Sleep and wake states in the Biosatellite 3 monkey: visual and computer analysis of telemetered electroencephalographic data from earth orbital flight. Aerosp. Med. 42(3), 304–313.

Henry, J.P. (1963) Synopsis of the results of the MR-2 and MA-5 flights. *In* Henry, J.P. and Mosely, J.D. (eds.). *Results of the Project Mercury Ballistic and Orbital Chimpanzee Flights*, pp. 1–2. NASA SP-39, Washington, DC.

Henry, J.P., Ballinger, E.R., Maher, P.J. and Simons, D.G. (1952) Animal studies of the subgravity state during rocket flight. J. Aviat. Med. 23, 421–432.

Henry, J.P. and Mosely, J.D. (eds.). (1963a) *Results of the Project Mercury Ballistic and Orbital Chimpanzee Flights*, NASA SP-39, Washington, DC.

Henry, J.P. and Mosely, J.D. (1963b) Antecedents and planning aspects of the MR-2 flight. *In* Henry, J.P. and Mosely, J.D. (eds.). *Results of the Project Mercury Ballistic and Orbital Chimpanzee Flights*, pp. 3–6. NASA SP-39, Washington, DC.

Hines, J.W. and Skidmore, M.G. (eds.). (1994) *Final Report of the U. S. Experiment Flown on the Soviet Biosatellite Cosmos 1667*, NASA TM 108803. NASA Ames Research Center, Moffett Field, CA.

Hoban-Higgins, T.M., Stein, T.P., Dotsenko, M.A., Korolkov, V.I. and Fuller, C.A. (2000) Energy metabolism of *Macaca mulatta* during spaceflight. J. Gravit. Physiol. 7(1), S145–148.

Hodgson, J.A., Riazansky, S.N., Goulet, C., Badakva, A.M., Kozlovskaya, I.B., Recktenwald, M.R., McCall, G., Roy, R.R., Fanton, J.W. and Edgerton, V.R. (2000) Rhesus leg muscle EMG activity during a foot pedal pressing task on Bion 11. J. Gravit. Physiol. 7(1), S87.

Hoshizaki, T., Hahn, P.M. and Adey, W.R. (1973) Circadian rhythms and sleep-wake activity in the Biosatellite monkey. Physiologist 16(2), 202–208.

Hoshizaki, T., Durham, R. and Adey, W.R. (1971) Sleep-wake activity patterns of a *Macaca nemestrina* monkey during nine days of weightlessness. Aerosp. Med. 42(3), 288–295.

Ilyin, E.A. (1988) The Cosmos biosatellite: achievements and research outlook. Kosm Biol Aviakosm Med. 22(6), 41–51.

Ilyin, E.A. (1981) Future investigations onboard Soviet biosatellites of the Cosmos series. Acta. Astronaut. 9–10, 1149–1157.

Ilyin, E.A., Korolkov, V.I., Skidmore, M.G., Viso, M., Kozlovskaya, I.B., Grindeland, R.E., Lapin, B.A., Gordeev, Y.V., Krotov, V.P., Fanton, J.W., Bielitzki, J.T., Golov, V.K., Magedov, V.S. and Hines, J.W. (2000) Bion 11 mission: primate experiments. J. Gravit. Physiol. 7(1), S9–18.

Ilyin, E.A., Korolkov, V.I., Kozlovskaya, I.B., Lobachik, V.I. and Truzhennikov, A.N. (1991) The monkey in space flight. Physiologist 34(1 Suppl), S49–51.

Jenkins, D.W. (1965) The NASA Biosatellite Program. Life Sci. Space Res. 3, 230–240.

Keeling, K.R. (1977) The chimpanzee as a candidate for spaceflight. *In* Simmonds, R.C. and Bourne, G.H. (eds.). *The Use of Nonhuman Primates in Space*, pp. 277–280. NASA Conference Publication 005. NASA, Washington, DC.

Kischel, P., Stevens, L., Montel, V., Picquet, F. and Mounier, Y. (2001) Plasticity of monkey triceps muscle fibers in microgravity conditions. J. Appl. Physiol. 90(5), 1825–1832.
Klein, H.P. (1981) U. S. biological experiments in space. Acta. Astronaut. 8(9-10), 927–938.
Klimovitsky, Vy., Alpatov, A.M., Hoban-Higgins, T.M., Utekhina, E.S. and Fuller, C.A. (2000) Thermal regulation in *Macaca mulatta* during space flight. J. Gravit. Physiol. 7(1), S149–152.
Klimovitsky Vla, Alpatov, A.M., Sulzman, F.M., Fuller, C.A. and Moore-Ede, M.S. (1987) Circadian rhythms and temperature homeostasis in monkeys during a flight on the Kosmos 1514 biosatellite Kosm Biol Aviakosm Med. 21(5), 14–18.
Korolkov, V., Helwig, D., Visoand, M. and Connolly, J. (1996) Cosmos 2229 mission. Overview. J. Appl. Physiol. 81(1), 186–187.
Kozlovskaya, I.B., Grindeland, R.E., Viso, M. and Korolkov, V.I. (2000) Bion 11 science objectives and results. J. Gravit. Physiol. 7(1), S19–25.
Kozlovskaya, I.B., Ilyin, E.A., Sirota, M.G., Korolkov, V.I., Babayev, B.M., Belozerova, I.N. and Yakushin, S.B. (1989) Studies of space adaptation syndrome in experiments on primates performed on board of soviet biosatellite "Cosmos-1887". Physiologist 32(1 Suppl), S45–48.
Kozlovskaya, I.B., Barmin, V.A., Kreidich, Yu.V. and Repin, A.A. (1985) The effects of real and simulated microgravity on vestibulo-oculomotor interaction. Physiol. 28(6 Suppl), S51–56.
Krotov, V.P., Sandler, G., Badakva, A.M., Khainz, D. and Magedov, V.S. (1987) Features of changes in blood pressure and blood flow in the common carotid artery of monkeys exposed aboard the biosatellite Cosmos-1514. Kosm Biol Aviakosm Med. 21(3), 69–74.
Lamb, L.E. (1966) Status of knowledge of weightlessness. *In Space Research, Directions for the Future, part III*, pp. 191–213. Space Sciences Board NAS-NRC, Washington, DC.
Lesnyak, A.T., Sonnenfeld, G., Rykova, M.P., Meshkov, D.O., Mastro, A. and Konstantinova, I. (1993) Immune changes in test animals during spaceflight. J. Leukoc. Biol. 54(3), 214–226.
Lobachik, V.I., Chupushtanov, S.A., Pischulina, G.N. and Nosovsky, A.M. (2000) The effect of a 14-day flight on body fluids in primates. J. Gravit. Physiol. 7(1), S135–137.
Mack, P.B. (1971) Bone density changes in a *Macaca nemestrina* monkey during the Biosatellite 3 project. Aerosp. Med. 42(8), 828–833.
Martinez, D.A., Buckendahl, P.E., Grindeland, R.E., Dotsenko, M.A., Bigbee, A.J. and Vailas, A.C. (2000) Evaluation of bone and collagen metabolism by assessing urinary biomarkers in non-human primates. J. Gravit. Physiol. 7(1), S169–170.
Matsunami, K. (1997) Squirrel monkey – an ideal primate model of space physiology. Biol. Sci. Space 11(2), 87–111.

Mazin, M.G., Kiselyova, E.V., Nemirovskaya, T.L. and Shenkman, B.S. (2000) Ultrastructure of skeletal muscles of rhesus monkeys after spaceflight. J. Gravit. Physiol. 7(1), S59–62.

Meehan, J.P. and Rader, R.D. (1973) Cardiovascular observations in the *Macaca nemestrina* monkey in Biosatellite III. Physiologist 16(2), 184–193.

Mosely, J.D. and Henry, J.P. (1963) Summary of results of the MR-2 flight. *In* Henry, J.P. and Mosely, J.D. (eds.). *Results of the Project Mercury Ballistic and Orbital Chimpanzee Flights*, pp. 33–34. NASA SP-39, Washington, DC.

Oganov, V.S., Bakulin, A.V., Novikov, V.E. and Murashko, L.M. (2000) Change and recovery of bone mass and acoustic properties of rhesus monkeys after Bion 11 spaceflight. J. Gravit. Physiol. 7(1), S163–168.

Pace, N. (1963) The effects of weightlessness on mammals. *In Space Biology, Proc. 24th Biol. Colloquium*, pp. 65–74. Oregon State University, Corvallis, Oregon.

Pace, N., Rahlman, D.F., Kodama, A.M., Grunbaum, B.W. and Mains, R.C. (1977) The pig-tailed monkey (*Macaca nemestrina*) as a space-flight candidate. *In* Simmonds, R.C. and Bourne, G.H. (eds.). *The Use of Nonhuman Primates in Space*, pp. 343–356. NASA Conference Publication 005. NASA, Washington, DC.

Pace, N., Grunbaum, B.W., Kiepert, D.W., Rahlman, D.F. and Smith, G.D. (1971) Urine excretion rates of calcium, creatine and creatinine in the test monkeys and flight monkey used for NASA Biosatellite III. EPL 71-1. University of California Berkeley, Berkeley, CA.

Perachio, A.A. (1977) Vestibular functions and sleep in space experiments with nonhuman primates. *In* Simmonds, R.C. and Bourne, G.H. (eds.). *The Use of Nonhuman Primates in Space*, pp. 179–196. NASA Conference Publication 005. NASA, Washington, DC.

Perry, T.W. and Reid, D.H. (1983) Spacelab mission 4–the first dedicated life sciences mission. Aviat. Space Environ. Med. 54(12 Pt 1), 1123–1128.

Phillips, R.W. (1988) Animals in biomedical space research. Acta. Astronaut. 17(2), 249–251.

Pickering, J.E., Brown, W.L., Stallings, H.D., Benon, R., Zellmer, R.W., Bitter, H.L., Carr, R.M. and McDowell, A.A. (1959) Primates in space. Report No. 2. *In Bioastronautics – Advances in Research, School of Aviation Med. USAF*, Randolph AFB, Texas.

Pittendrigh, C.S. (1965) On the biological probems to be atacked with a series of U S satellites in 1966. Life Sci. Space Res. 3, 206–214.

Pochkhua, M.A. (1992) The humoral immunity indices of rhesus monkeys under the action of extreme factors in space flight. Aviakosm Ekolog Med. Jul-26(4), 50–54.

Recktenwald, M.R., Hodgson, J.A., Roy, R.R., Riazansky, S.N., McCall, G., Kozlovskaya, I.B., Washburn, D.A., Fanton, J.W. and Edgerton, V.R. (2000) Quadrupedal locomotion in rhesus monkeys after 14 days of spaceflight. J. Gravit. Physiol. 2000 Jan, 7(1), S71.

Recktenwald, M.R., Hodgson, J.A., Roy, R.R., Riazanski, S., McCall, G.E., Kozlovskaya, I., Washburn, D.A., Fanton, J.W. and Edgerton, V.R. (1999) Effects of spaceflight on rhesus quadrupedal locomotion after return to 1G. J. Neurophysiol. 81(5), 2451–2463.

Robles, Frederick, H. Jr., Grunzke, M.E. and Reynolds, H.H. (1963a) Performance aspects of the MA-5 flight. *In* Henry, J.P. and Mosely, J.D. (eds.). *Results of the Project Mercury Ballistic and Orbital Chimpanzee Flights*, pp. 39–52. NASA SP-39, Washington, DC.

Robles, Frederick, H. Jr., Grunzke, M.E. and Belleville, R.E. (1963b) Performance aspects of the MR-2 flight. *In* Henry, J.P. and Mosely, J.D. (eds.). *Results of the Project Mercury Ballistic and Orbital Chimpanzee Flights*, pp. 21–24. NASA SP-39, Washington, DC.

Rodionova, N.V., Oganov, V.S. and Zolotova, N.V. (2002) Ultrastructural changes in osteocytes in microgravity conditions. Adv. Space Res. 30(4), 765–770.

Rodionova, N.V. and Oganov, V.S. (2001) Morpho-functional adaptations in the bone tissue under the space flight conditions. J. Gravit. Physiol. 8(1), P87–88.

Rodionova, N.V., Shevel, I.M., Oganov, V.S., Novikov, V.E. and Kabitskaya, O.E. (2000) Bone ultrastructural changes in Bion 11 rhesus monkeys. J. Gravit. Physiol. 7(1), S157–161.

Roy, R.R., Zhong, H., Talmadge, R.J., Bodine, S.C., Fanton, J.W., Kozlovskaya, I. and Edgerton, V.R. (2001) Size and myonuclear domains in Rhesus soleus muscle fibers: short-term spaceflight. J. Gravit. Physiol. 8(2), 49–56.

Roy, R.R., Zhong, H., Bodine, S.C., Pierotti, D.J., Talmadge, R.J., Barkhoudarian, G., Kim, J., Fanton, J.W., Kozlovskaya, I.B. and Edgerton, V.R. (2000a) Fiber size and myosin phenotypes of selected rhesus lower limb muscles after a 14-day spaceflight. J. Gravit. Physiol. 7(1), S45.

Roy, R.R., Zhong, H., Bodine, S.C., Talmadge, R.J., Fanton, J.W., Kozlovskaya, I.B. and Edgerton, V.R. (2000b) Myonuclear number and size of rhesus soleus fibers after 14 days of actual or simulated flight. J. Gravit. Physiol. 7(1), S63.

Roy, R.R., Bodine, S.C., Pierotti, D.J., Kim, J.A., Talmadge, R.J., Barkhoudarian, G., Fanton, J.W., Kozlovskaya, I. and Edgerton, V.R. (1999) Fiber size and myosin phenotypes of selected Rhesus hindlimb muscles after a 14-day spaceflight. J. Gravit. Physiol. 6(2), 55–62.

Roy, R.R., Hodgson, J.A., Aragon, J., Day, M.K., Kozlovskaya, I. and Edgerton, V.R. (1996) Recruitment of the Rhesus soleus and medial gastrocnemius before, during and after spaceflight. J. Gravit. Physiol. (1), 11–15.

Sandler, H., Krotov, V.P., Hines, J., Magadev, V.S., Benjamin, B.A., Badekeva, A.M., Halpryn, B.M., Stone, H.L. and Krilov, V.S. (1987) Cardiovascular results from a rhesus monkey flown aboard the Cosmos 1514 spaceflight. Aviat. Space Environ. Med. 58(6), 529–536.

Sandler, H. (1977) A review of animal flight experiments. *In* Simmonds, R.C. and Bourne, G.H. (eds.). *The Use of Nonhuman Primates in Space*, pp. 3–22. NASA Conference Publication 005. NASA, Washington, DC.

Shlyk, G.G., Ignatenko, A.V., Efimova, M.Y. and Zavadskaya, O.I. (2000a) Behavior and performance of rhesus monkeys during Bion 11 flight. J. Gravit. Physiol. 7(1), S81–85.

Shlyk, G.G., Shirvinskaya, M.A., Stekolin, I.Y., Efimova, M.Y. and Zavadskaya, O.I. (2000b) General and special training of rhesus monkeys flown on Bion biosatellites. J. Gravit. Physiol. 7(1), S75–79.

Simmonds, R.C. and Bourne, G.H. (eds.). (1977) *The Use of Nonhuman Primates in Space. Proceedings of a symposium held at Ames Research Center, Dec 2–4, 1974*, NASA Conference Publication 005. NASA, Washington, DC.

Sirota, M.G., Babayev, B.M., Belozerova, I.B., Nyrova, A.N., Yakushin, S.B. and Kozlovskaya, I.B. (1987) Characteristics of vestibular reactions to canal and otolith stimulation at an early stage of exposure to microgravity. Physiologist 30(1 Suppl), S82–84.

Slenzka, K. (2003) Neuroplasticity changes during space flight. Adv. Space Res. 31(6), 1595–1604.

Sonnenfeld, G., Davis, S., Taylor, G.R., Mandel, A.D., Konstantinova, I.V., Lesnyak, A., Fuchs, B.B., Peres, C., Tkackzuk, J. and Schmitt, D.A. (1996) Effect of space flight on cytokine production and other immunologic parameters of rhesus monkeys. J. Interferon Cytokine Res. 16(5), 409–415.

Souza, K., Hogan, R. and Ballard, R. (1995) *Life Into Space. Space Life Sciences Experiments NASA Ames Reseach Center. 1965–1990*, NASA Reference Publication 1372. NASA Ames Research Center, Moffett Field, CA.

Stein, T.P., Dotsenko, M.A., Korolkov, V.I., Griffin, D.W. and Fuller, C.A. (1996) Energy expenditure in rhesus monkeys (Macaca mulatta) during spaceflight using doubly labeled water ($^{2}H_{2}{}^{18}O$). J. Appl. Physiol. 81(1), 201–207.

Stingely, N.E. and Mosely, J.D. (1963) MA-5 operations. *In* Henry, J.P. and Mosely, J.D. (eds.). *Results of the Project Mercury Ballistic and Orbital Chimpanzee Flights*, pp. 35–38. NASA SP-39, Washington, DC.

Stingely, N.E., Mosely, J.D. and Wheelwright, C.D. (1963) MR-2 operations. *In* Henry, J.P. and Mosely, J.D. (eds.). *Results of the Project Mercury Ballistic and Orbital Chimpanzee Flights*, pp. 7–14. NASA SP-39, Washington, DC.

Sulzman, F.M. (1996) Overview. J. Appl. Physiol. 81(1), 3–6.

Sulzman, F.M. (1986) *Evaluation of the need for a large primate research facility in space*, NASA-CR-179661. NASA, Washington, DC.

Sulzman, F.M., Ferraro, J.S., Fuller, C.A., Moore-Ede, M.C., Klimovitsky, V., Magedov, V. and Alpatov, A.M. (1992) Thermoregulatory responses of rhesus monkeys during spaceflight. Physiol. Behav. 51(3), 585–591.

Walrath, J. (1959) Weightlesness. Cornell Engin. 25, 14–15.

Walter, D.O., Berkhout, J.I., Buchness, R., Kram, E., Rovner, I. and Adey, W.R. (1971) Digital computer analysis of neurophysiological data from Biosatellite 3. Aerosp. Med. 42(3), 314–321.

Ward, W.E. (1963) Medical and physiological aspects of the MA-5 flight. *In* Henry, J.P. and Mosely, J.D. (eds.). *Results of the Project Mercury Ballistic and Orbital Chimpanzee Flights*, pp. 53–58. NASA SP-39, Washington, DC.

Ward, W.E. and Briutz, W.E. Jr. (1963) Medical and physiological aspects of the MR-2 flight. *In* Henry, J.P. and Mosely, J.D. (eds.). *Results of the Project Mercury Ballistic and Orbital Chimpanzee Flights*, pp. 25–32. NASA SP-39, Washington, DC.

Winget, C.M. (1977) The capuchin monkey as a candidate for spaceflight. *In* Simmonds, R.C. and Bourne, G.H. (eds.). *The Use of Nonhuman Primates in Space*, pp. 329–342. NASA Conference Publication 005. NASA, Washington, DC.

Zerath, E., Grynpas, M., Holy, X., Viso, M., Patterson-Buckendahl, P. and Marie, P.J. (2002) Spaceflight affects bone formation in rhesus monkeys: a histological and cell culture study. J. Appl. Physiol. 93(3), 1047–1056.

Zerath, E., Holy, X., Andre, C., Renault, S., Noel, B., Delannoy, P., Hott, M. and Marie, P.J. (2000) Effects of Bion 11 14-day space flight on monkey iliac bone. J. Gravit. Physiol. 7(1), S155–156.

Zerath, E, Novikov, V., Leblanc, A., Bakulin, A., Oganov, V. and Grynpas, M. (1996) Effects of spaceflight on bone mineralization in the rhesus monkey. J. Appl. Physiol. 81(1), 194–200.

Zerath, E., Holy, X., Malouvier, A., Caissard, J.C. and Nogues, C. (1991) Rat and monkey bone study in the Biocosmos 2044 space experiment. Physiologist 34(1 Suppl), S194–195.

Zerath, E., Nogues, C., Borne, M. and Sourdaine, P. (1990) Bone effects of 13 days of weightlessness on rat and monkey some results of biocosmos 1887 and ground simulations. Physiologist 33(1 Suppl), S94–95.

Zhidkov, V.V., Abrosimov, S.V., Endeka, D.K., Borisov, G.I. and Lobachik, V.I. (1987) Effect of space flight factors on hydration homeostasis in monkeys, Kosm Biol. Aviakosm Med. 21(5), 19–22.

List of Main Authors

Jacopo Aguzzi
Neuroscience Institute,
Morehouse School of Medicine,
Atlanta, GA, USA

Jeffrey R. Alberts
Department of Psychology,
Indiana University,
Bloomington, IN, USA and
Department of Obstetrics and Gynecology,
Wake Forest University School of Medicine,
Winston Salem, NC, USA

Anatole M. Badakva
Institute of Biomedical Problems,
Moscow, Russia

Stephen Keith Chapes
Division of Biology and Department of
Diagnostic Medicine and Pathobiology,
Kansas State University,
Manhattan, KS 66506, USA

Bernard Cohen
Department of Neurology,
Mount Sinai School of Medicine,
New York, USA

Mingjia Dai
Department of Neurology,
Mount Sinai School of Medicine,
New York, USA

Stephen B. Doty
Hospital for Special Surgery,
USA

Galina Durnova
Institute of Biomedical Problems,
123007, Moscow, Russia

Roman Reddy Ganta
Division of Biology and Department of
Diagnostic Medicine and Pathobiology,
Kansas State University,
Manhattan, KS 66506, USA

Ruth K. Globus
NASA Ames Research Center,
Moffett Field, CA 94035-1000,
USA

Richard E. Grindeland
Life Sciences Division,
NASA – Ames Research Center,
Moffett Field, 94035,
USA

Daniel C. Holley
Department of Biological Sciences,
San Jose State University,
1 Washington Square,
San Jose, CA 95192, USA

Gay R. Holstein
Department of Neurology,
Mount Sinai School of Medicine,
New York, USA

Emily Morey-Holton
NASA Ames Research Center,
Moffett Field, CA 94035-1000,
USA

Eberhard R. Horn
University of Ulm,
Germany

Eugene A. Ilyin
Russian State Research Center –
Institute of Biomedical Problems,

Russian Academy of Sciences,
Moscow, Russia

Alexander Kaplansky
Institute of Biomedical Problems,
123007, Moscow, Russia

Inessa Kozlovskaya
Institute of Biomedical Problems,
Moscow, Russia

Emily Morey-Holton
NASA Ames Research Center,
Moffett Field, CA 94035-1000,
USA

April E. Ronca
Department of Psychology,
Indiana University, Bloomington, IN, USA
and Department of Obstetrics and Gynecology,
Wake Forest University School of Medicine,
Winston Salem, NC, USA

Michael G. Skidmore
Life Sciences Division,
NASA – Ames Research Center,
Moffelt Field, 94035, USA

Gerald Sonnenfeld
Department of Biological Sciences and
Vice President for Research,
Binghamton University,
State University of New York,
Binghamton,
New York, USA

David L. Tomko
National Aeronautics and Space Agency,
USA

Gianluca Tosini
Neuroscience Institute,
Morehouse School of Medicine,
Atlanta, GA, USA

Laurence Vico
INSERM E0366,
Paris, France

Charles E. Wade
United States Army Institute of
Surgical Research,
Fort Sam, Houston,
TX 78234, USA

Life Sciences Division,
NASA – Ames Research Center,
Moffett Field, CA 94035,
USA

Thomas Wronski
University Florida,
Gainsville, FL,
USA

Sergei Yakushin
Department of Neurology,
Mount Sinai School of Medicine,
New York, USA

Keyword index

www.ingramcontent.com/pod-product-compliance
Ingram Content Group UK Ltd.
Pitfield, Milton Keynes, MK11 3LW, UK
UKHW020237290926
15470UKWH00002B/583

9 780444 519078